ALL OUR OWN WATER

LANDSCAPE EVOLUTION, CAVES AND HYDROGEOLOGY OF GOWER

Peter Kokelaar

Rain approaching Oxwich Bay

ISBN 978-1-3999-0335-6

© 2021 Peter Kokelaar

All rights reserved. No part of this book may be reproduced, stored in a retrieval system or transmitted in any form or by any means, electronic, mechanical, photocopying, recording, scanning or otherwise, without permission from the author.

E-mail: bpkokelaar@gmail.com
Further information: kokelaargower.com

Design, layout and graphics by the author
Photographs by the author unless otherwise attributed
Printed in Wales by Gomer, Llandysul

Disclaimer: Inclusion in this book must not be taken to imply a right of public access or any measure of safety in exploration.

Cover photographs (attributed with details in the text):
Front
(Top) South Gower coast at The Knave
(Middle left) Cave formations in Ogof Ffynnon Wyntog
(Middle right) Microrills in limestone at Caswell
(Bottom) Bathymetry off the south Gower coast

Back
(Top left) Patella 'beach' deposit, 123,000 years old
(Top right) Barns Cave taking Llethrid flood overflow water
(Bottom) Geology of Gower

Title page photograph:
Rain approaching Oxwich Bay (Helen Kokelaar)

CONTENTS

PREFACE v

1 OVERVIEW AND INTRODUCTION
1.1 Outstanding natural beauty 1
1.2 Places and gaps 2
1.3 Geology and deep time 2

2 LANDSCAPES CUT IN TIME: PENEPLAINS AND PLATFORMS
2.1 Introduction 9
2.2 Absolute and relative levels 9
2.3 Views on views 10
2.4 Imagined ancient landscapes 16
2.5 Beneath the sea 20
2.6 Coastal platforms, modern and ancient 21
2.7 Lundy: a vital insight 31
2.8 Relations with sea level 33
2.9 When and why the sea came up 35
2.10 The headless valley problem 37
2.11 The perforation of Great Tor 40
2.12 Cutting the modern coast 42
2.13 Loads of ice 46
2.14 Rough times on Patella Beach 47
2.15 More storms and cement 57
2.16 Summary 60

3 GLACIATION OF GOWER
3.1 Introduction 69
3.2 Cryptic limits 72
3.3 Black Mountain Rosetta Stone 73
3.4 Rotherslade revisited 76
3.5 New map of Last Glacial Maximum limits on Gower 78
3.6 Pwll-du Lobe: taking the plunge 82
3.7 Ilston Lobe 83
3.8 Ice, ritual and ancient preference 86
3.9 Forces of nature 89
3.10 Paviland Lobe 90
3.11 Little Reynoldston subglacial sink hole 95
3.12 The Bulwark landslide 96
3.13 In recession: kettles and kames 98
3.14 Recessional moraines and drains 100
3.15 Blue Anchor ice margin 104
3.16 Accommodating talus 105
3.17 The big picture 108

4 CAVE DEVELOPMENTS
4.1 Plenty of time 113
4.2 Little did we know 113
4.3 Magical caves at The Knave 115
4.4 Green Cwm and the Last Glacial Maximum underground 124
4.5 Tooth Cave: a stygian realm of curiosities 132
4.6 Debris flow, scratches and snails 140
4.7 Inception of caves in the limestones around Green Cwm 150
4.8 Remarkable record of a desert above 151
4.9 One weird depression 153
4.10 Caves lost in time 154
4.11 Speleothem formation ages 156
4.12 Recovering Llethrid 157

5 SINKS AND SPRINGS
5.1 All our own water: an introduction 169
5.2 Sinks and springs of central Gower 172
5.3 Sorts of springs 173
5.4 Lost accessible depths and misfit springs 175
5.5 Gower water supplies 175
5.6 Wellhead spring water supply 176
5.7 The case for an exotic water supply 181
5.8 Input and output: fingerprints in clay 183
5.9 Enough water? 184
5.10 Storms, turbidity and Wellhead spring discharge behaviour 186
5.11 Enigma variations 189
5.12 Anomalous spring-water temperatures 195
5.13 Holy Well springs on Cefn Bryn 198
5.14 Quick and dirty evaluation of till properties 201
5.15 Rough calculation of Cefn Bryn perched aquifer potential 201
5.16 The great drought of 2018 204

5.17 Air versus spring temperature relations throughout one year 205
5.18 Chemical signatures of spring types 207
5.19 Tufa and other oddities 217
5.20 Sinks and springs: what have we learned? 221

6 STORMS, DUNES AND ROCKSLIDES
6.1 Introduction 229
6.2 Driven out by sand 233
6.3 Comings and goings of dunes 234
6.4 Transient images of missing sand 237
6.5 Beach stripping and revelation by storm 241
6.6 The beautiful sand monster: Helwick 245
6.7 Pwll-du storm beaches 247
6.8 Oxwich rockslides: the full story 252

7 APPENDICES
7.1 ACKNOWLEDGEMENTS 259
7.2 REFERENCES 261
7.3 SPRINGS GAZETTEER 266
7.4 REVIEW OF WELLHEAD DATA BY GEOFF WILLIAMS 274
7.5 PLACENAMES INDEX 282

Loughor Estuary tidal salt marsh of north Gower. Now virtually full of sediment, this wide valley 23,000 years ago was occupied by a glacier that was hundreds of metres thick and flowed down from far to the north in central Wales. The straight channel, Bennett's Pill, was hand cut to convey coal in and limestone out by boat, to and from the quay and hamlet of Landimore (Llandîmôr).

To Peter Sambrook, friend and kindred spirit, whose outlook made discovery worthwhile.

Preface This study started life as an investigation of the local knowledge that springs supplying water on Gower must be partly fed by rainwater that 'goes down' in the limestone hills to the north, in Carmarthenshire and Breconshire. Accordingly, it 'comes up' on Gower having followed the folded layer of limestone beneath the South Wales Coalfield and under the Loughor Estuary. We were taught this in school in the 60s and it has long been widely accepted. The title reveals my finding, but, ranging around on Gower again in my retirement, it soon became clear that there was much more I did not understand and, moreover, could not read about to find explanations. This book presents several new discoveries and explanations, some of significance depending on one's focus. Happily free of academic pressures, I decided to highlight the most interesting new perspectives and put everything in one accessible place.

Focussing on new views, this work is not geographically comprehensive; it is wide ranging and might best be considered an extended essay. It is about the physical nature of Gower, written, I hope, so as to be accessible to any non-specialist prepared to grapple lightly with some science. The content in parts is 'involved' where it provides previously lacking full and up-to-date explanations, especially regarding the 60 million years or so of the landscape evolution. It is heavily illustrated with photographs and diagrams, many with long captions. Gower's physical features are essentially visual, so the large figures and their detailed explanations should complement the text. I tread thin ice in places and there will be mistakes and omissions. This book is to share freely the recent insights, ideas and discoveries; above all it is for lovers of Gower and for their enjoyment.

The introduction has only a brief account of the geology, as there is sufficient known and published, and there are assorted maps and keys to cope with geological-time terminology and place names. I have tried to make the text informal but could not avoid using references, which are collected at the end. I try to explain things 'along the way', not to derail, and I use additional explanatory notes numbered and placed at the ends of their chapters. The many people who have generously provided essential help and support are duly acknowledged, although this does not adequately reflect my sincere gratitude for their various skills, insights and good humour.

Lastly here, but importantly, the reader is respectfully reminded that some explorations and investigations referred to in this book are potentially hazardous. Gower big tides, cliffs and caves especially beg caution and serious planning for safety, in some instances involving specialist equipment and skills. Further, many sites have restricted access on a permanent or seasonal basis and several caves are subject to landowners' specific permissions. Inclusion in this book must not be taken to imply a right of public access or any measure of safety in exploration. That said, go and enjoy...

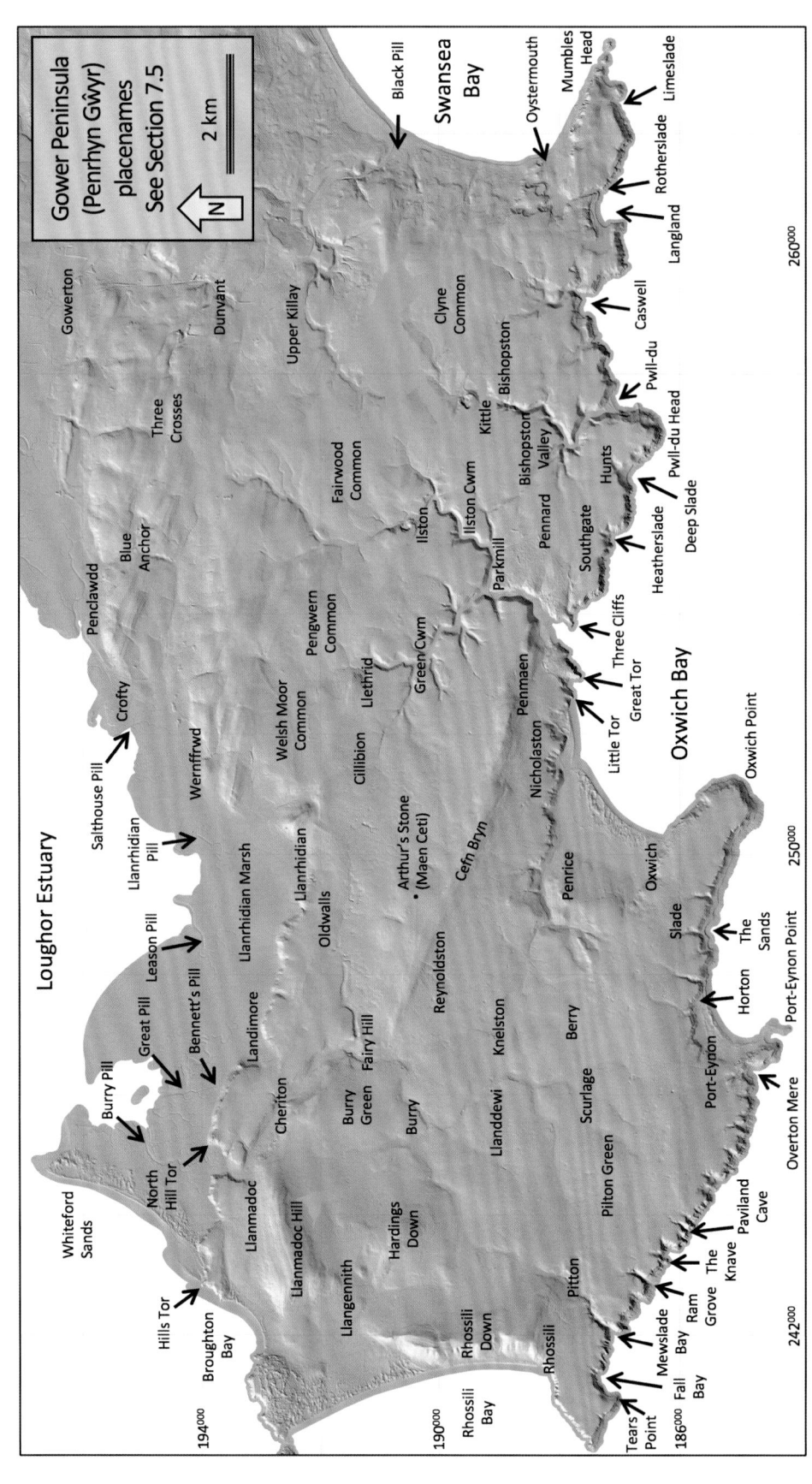

1 OVERVIEW AND INTRODUCTION

1.1 Outstanding natural beauty

Beauty is in the eye of the beholder – the mind's eye – but there are, thank goodness, many like minds. In 1956 passionate and dedicated men and women were able successfully to make the public case that peninsular Gower warranted recognition and care as an Area of Outstanding Natural Beauty; it was the first one. Perhaps just seeing for oneself is enough – beauty in the mind's eye – but some folk might wonder how it came to be like that. Such an appreciation can add fascination to the beauty.

This essay attempts to add some fascination and even awe to the beauty of Gower, on all scales, from the entire peninsula in its coastal setting to the patterns etched on rocks due to the molecular behaviour of rainwater. What exactly are those renowned platforms at 200, 400 and 600 feet above sea level? How did they happen? Where did all that sand come from? Where exactly was that famous limit of the Welsh Ice responsible for dumping exotic boulders all over the place during the last glaciation? And, just how old are those limestone caves that held remains of the woolly mammoth and giant hippopotamus, as well as our distant ancestors?

Droves of locals and visitors enjoy just being on Gower, and they will continue to do so, but to learn and appreciate more about our peninsula seems a worthy pursuit in being potentially rewarding, as I have found. I grew up ranging widely on Gower, essentially oblivious to how little I understood about its features and development through time. With climbing and caving friends we happily explored the cliffs, slept out, swam, saw phosphorescent breaking tide edges come and go at night, and watched dark storms advance to soak us. From school we knew that there were discoveries of 'bone caves' with important evidence of ancestors who had seen and done what we were still seeing and doing, possibly equally hungry then and at times without the sea being there. We were told about grand ideas concerning relations of the land to sea level, and, of course, about the geological structure of the South Wales Coalfield where, at that time, men still toiled underground. For ourselves, we investigated caves and springs, testing connections with lurid green fluorescein dye, and we 'put up' new climbs on the limestone cliffs, getting to appreciate the bedding planes, fractures and fossils we clung to.

One thing that particularly fascinated me as a school-boy caver was the widely accepted scholarly inference that Gower's main water supplies had to be supplemented by rainfall that went down in the limestone hills far to the north, passed beneath the Loughor Estuary, and came up in the springs here. I wanted to understand that then and it was still bugging me in my retirement when, prompted by a haughty dismissal to the effect that 'everyone knows it *has to* come from outside of Gower', I resolved to look into it. This report is the result, rather grown beyond the initial investigation. The title reflects my finding, but it turned out that there was a lot of other fascinating stuff remaining to discover and try to explain.

This work reflects a thoroughly enjoyable, self-indulgent episode of enquiry and discovery for the author in retirement back home on Gower.[1] In my own mind's eye I find considerable enhancement of the beauty in discovering the what and how of the Gower scenery, and I hope to share at least some of the fascination here. I confess to greedily harvesting the fresh fruits of technology in using available LiDAR[2] map imagery, which 'sees through' vegetation and, with computer enhancement, reveals ground features that are impossible to see otherwise. Similarly I have used wonderful sea-floor sonar images, now available at 2-metre resolution, and of course Google Earth too, all freely available with due acknowledgement. The National Library of Scotland presents an easily accessible wide range of ages of maps at various scales, all 'georeferenced' to both Google Earth and high resolution location fixes to the Ordnance Survey National Grid and to Latitude and Longitude. It is all just so easy now to see and register the previously unseen; I have been like a hungry child left alone in a sweet shop.

1.2 Places and gaps

This work focuses primarily on *new* perspectives and findings concerning the natural *physical environment* of peninsular Gower. It is an essay, albeit a very long one, because its content is selected for novelty and potential fascination, as opposed to being a comprehensive or systematic treatise. Some myths are dispelled and the 'arguments' – i.e., the logic and evidence – then are given quite fully. Long threads of reasoning can be tedious, but brief dismissal of the work and beliefs of others would be disrespectful; hopefully, balance is achieved and justice done here to the pioneers of Gower research. Nevertheless, the reader is encouraged to think critically, because this author not infrequently treads thin ice and there inevitably will be errors. The provision of references to previous work is as lean as seems reasonable. Some interpretations simply seem best to fit the available observations, commonly rather few, but new light may be shed through further enquiry and contentious views thus discarded or, better, replaced. Much is hypothesis in need of critical testing; nothing is very serious.

There are places and features on Gower that will be of interest to many people but are not considered here – the gaps. There are adequate descriptions elsewhere of the many archaeological features that are so fine and abundant in recording human activities on peninsular Gower. These are discussed only briefly in so far as they figure in the natural environment, as for example regarding the origin of Maen Ceti, Arthur's Stone, on Cefn Bryn.

Similarly, there is precious little here on the raw-material extractive industries so clearly evident on Gower, as these too are described elsewhere, for example regarding the exceptional extent of quarrying and export of limestone, and the trade in red ochre known as 'reddle'. Quarrying is addressed in a few places, however, in the context of significant or misunderstood impacts on the visual landscape. And there is practically no account made here of the wonderful flora and fauna when it is so well covered elsewhere, especially piecemeal in the journal 'Gower' and bound together in Jonathan Mullard's 'Gower' (2006).

It has been impossible to avoid names of places all over Gower and, for the unfamiliar reader, important locations are referred to with their Ordnance Survey National Grid Reference, mostly six- or eight-figure numbers prefixed SS. However, it is hoped that the included maps and annotated figures will allow the reader to go with the flow on the page and not have to keep picking up a separate map. Probably the most useful map, whether in an armchair or in the field, is the Ordnance Survey 1:25,000 2½ inches to the mile (4 cm to 1 km) Explorer Map 164. A place-names index is provided finally, so readers may see what may be of interest where they go.

Place names used are mostly those on the OS 1:10,560, 6 inches to the mile maps published in 1964. For the record, this author is a staunch supporter and respecter of the Welsh language, having learned a little in school here. He at least remembers the Welsh for "this is book one" and sight-reads Welsh, singing loudly in chapel blissfully unaware of the detailed meaning of the verses. The Welsh placenames here are mostly those from the maps mentioned, and it is no intended disrespect that the comparatively recent 're-Welshing' of some Gower place names has not necessarily been followed.[3]

1.3 Geology and deep time

The geology of peninsular Gower is well known (e.g., George 1940), so here we simply consider its setting, its outline history, and its time framework where we have to introduce formal names (Figs 1 and 2). Gower peninsula lies at the southwest margin of the South Wales Coalfield. It is useful to view Gower geology as closely similar to that of the Vale of Glamorgan, with which it is continuous under Swansea Bay, except that in the eastern area there remains a significant cover of Triassic and Jurassic rocks. Gower had such a cover, but it was eroded away in the interval since about 64 million years ago (Section 2.4).

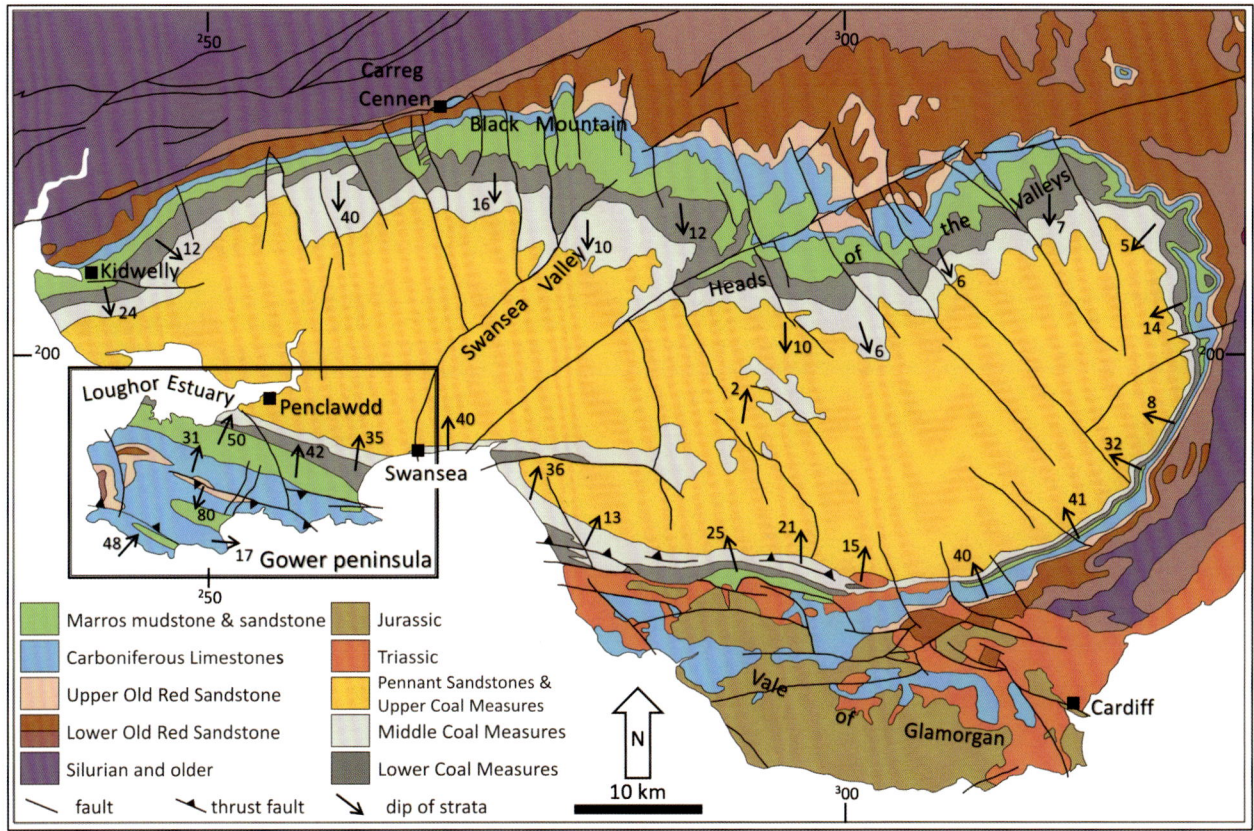

Figure 1. Simplified geological map of the South Wales Coalfield (Fig. 101 shows cross-sections). The key shows oldest rocks at bottom left and youngest at top right. The oval pattern defines the coalfield syncline, or structural 'basin', with strata around the margins dipping inwards (see dip symbols). In contrast to Gower, the Vale of Glamorgan has extensive cover of Triassic and Jurassic strata overlying the older strata. We can reasonably suppose that such a cover once existed on Gower.

The Carboniferous rocks of South Wales are folded into a broad syncline – a structural basin – with the youngest, uppermost layers in the middle (the gold-ornamented Pennant Sandstones) and the oldest and lowest around the edges (the blue-ornamented Limestones). We loosely refer to the Carboniferous rocks on the south side of the coalfield as 'the south crop'[4], and those on the north side 'the north crop'. The folding and faulting that caused uplift and erosion of Gower rocks, as well as forming the coalfield syncline, climaxed in the interval of about 300-290 million years ago (Fig. 2). This crustal compression and shortening here was only one phase of the Variscan Orogeny, which was a protracted episode of mountain building (orogenesis) due to continental collision farther south in Europe. Interestingly, the northwards advance towards South Wales of the Variscan deformation and associated uplift was heralded by the youngest Carboniferous strata, the Pennant Sandstones. These, in contrast to previous units, record sediment coming from the south; growing mountains there had started to shed eroded material northwards before itself becoming folded and uplifted.

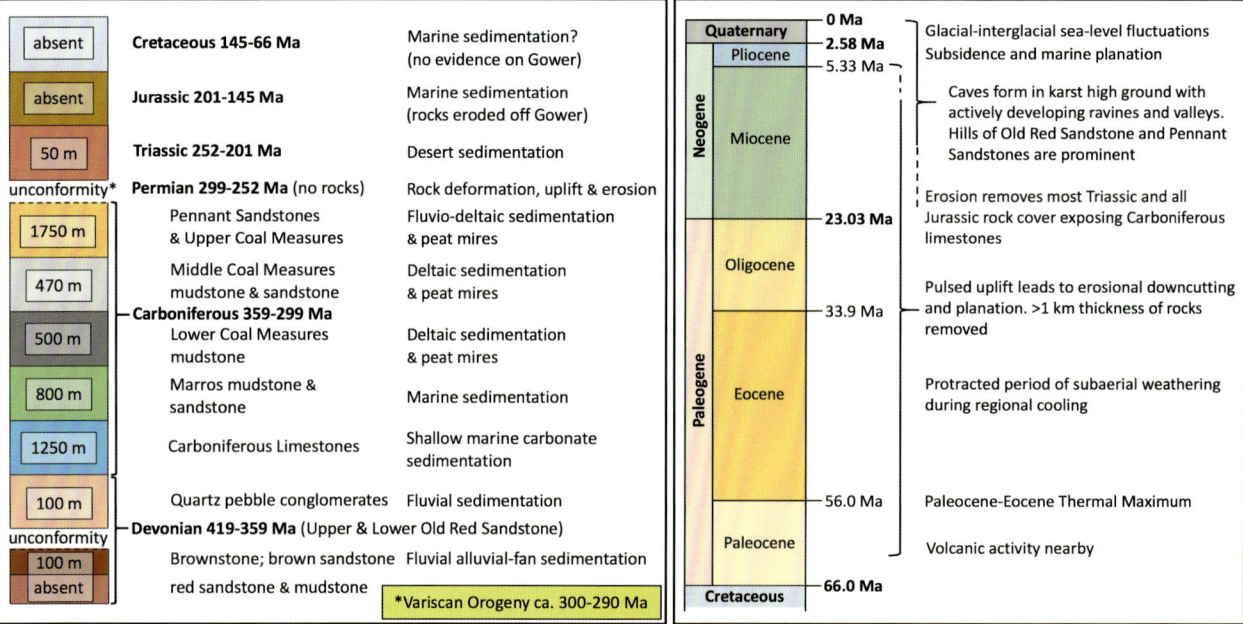

Figure 2. Simplified geological succession and history applicable to Gower in the context of the South Wales Coalfield (Fig. 1). Formal geological period names are in bold as are their time spans; Ma stands for mega annum and means million years, e.g., 66 Ma means 66,000,000 years (conventionally in terms of Earth history, before 'now'). The left-hand panel shows main rock-type divisions of the Devonian and Carboniferous Periods alongside the main environments and processes represented through time. The 'Variscan Orogeny' at about 300-290 Ma is the main phase of folding and faulting that uplifted and caused erosion of the Devonian (Old Red Sandstone) and Carboniferous rocks in this region. Rock unit thicknesses are from British Geological Survey (2011). The right-hand panel shows the succession of names and time divisions since 66 Ma, alongside processes that affected Gower. Apart from thin Quaternary glacial deposits, the interval since 66 Ma was dominated by uplift and erosion, so there are no representative rocks here.

The geological map of peninsular Gower (Fig. 3) presents the main geological rock and age units as they form a pattern that resulted from initially being mostly flat-lying layers and then crumpled and broken – folded and faulted – owing to collision of continents 300-290 million years ago. The trend of the rock layers on the map of Gower, roughly west-northwest to east-southeast, is known as the 'strike'; the slope on those layers, down at a right-angle (90°) to the strike, is the 'dip', generally indicated by direction arrows on the map and shown in the cross-section beneath. Towards the northeast the rock layers mostly dip towards northeast and overlying younger layers appear in succession; e.g., the Lower Coal Measures are overlain by Middle Coal Measures and in turn by Pennant Sandstones. The notorious 'backbone' of Gower is the narrow ridge of Old Red Sandstone rocks forming Cefn Bryn (SS518889), which trends northwest from the middle of Gower towards the similar but more massive Old Red Sandstone prominences of Llanmadoc Hill (SS430924) and Rhossili Down (SS420888) in the west. Around here the pattern of rock layers gets more complicated.

Overlying the Old Red Sandstone conglomerates and sandstones is the most substantial unit of the geology, the Carboniferous Limestones, at least 1250 m thick. These dominate the spectacular cliff scenery and underlie much of Gower where the karstic terrain[5] is mostly obscured by glacial deposits although evident in the numerous sink holes and quarries. There are several distinctive rock divisions representing shallow, warm-water carbonate reef environments, including fossiliferous layers with abundant crinoids, corals and shells[6], with calcareous mudstones and a few sandstones, mudstones and coals. Depositional bedding forms distinctive layering patterns in cliffs and on the beach platforms, while abundant fractures form cross-cutting gullies and narrow valleys.

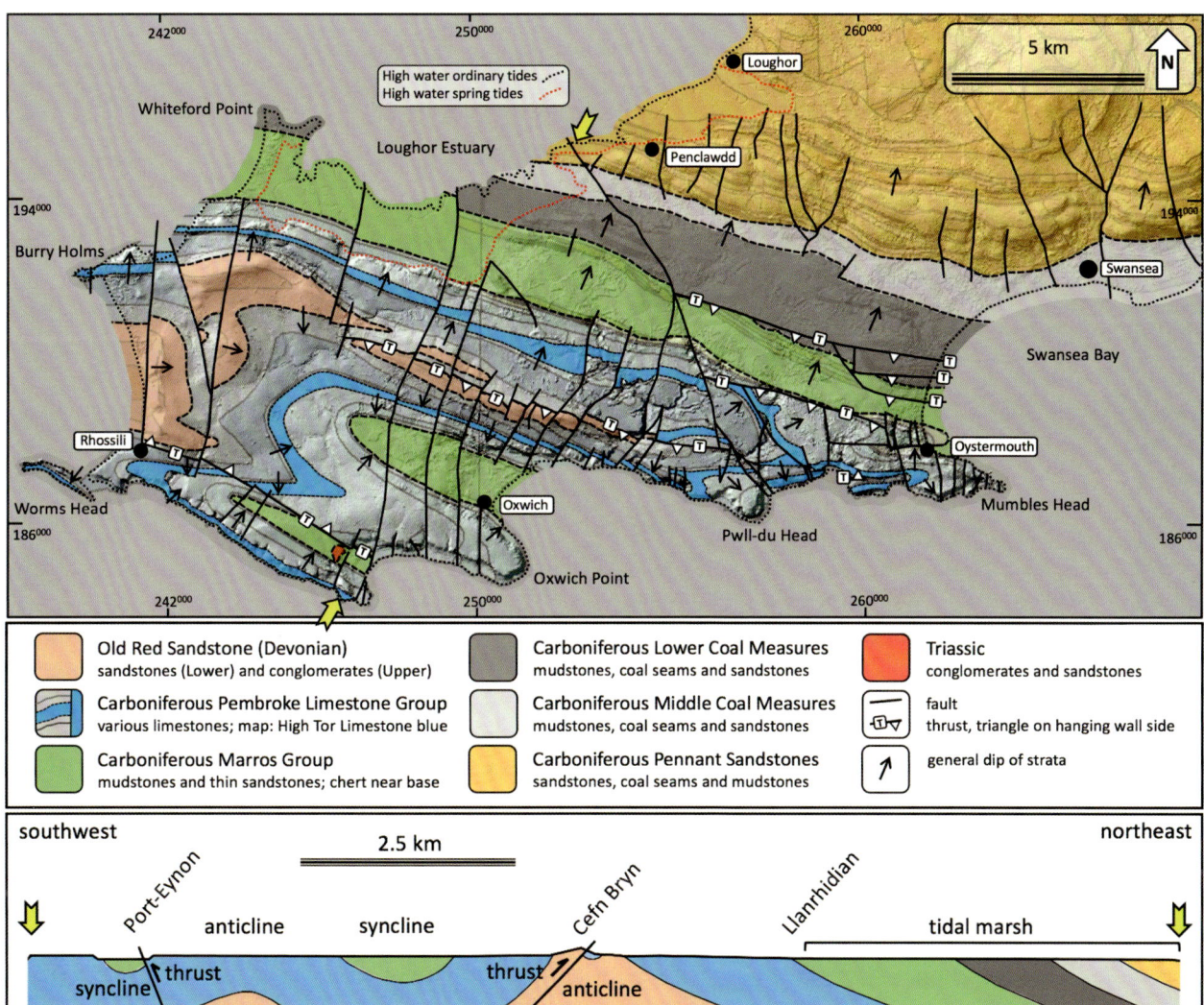

Figure 3. Geological map of Gower redrawn simplified over a base shaded relief image*. The cross-section reaches between the high-water limits of ordinary tides and is marked by the two yellow arrows. It is simplified, showing true dip directions and only the Cefn Bryn thrust fault and the Port-Eynon back-thrust fault. The small outcrop of Triassic rocks in Port-Eynon is centred in the syncline there but is just out of the plane of section.
*(Provided by BGS © UKRI and derived in part from NextMap Britain elevation data provided under licence to BGS from Intermap Technologies).

The geological structure of Gower is well shown by following the blue-ornamented High Tor Limestone layer from Burry Holms in the west (Fig. 3). It traces east and then southeast along the north flank of Cefn Bryn, dipping northeast under the younger rock layers. In a complicated fashion it then wraps back westwards in the vicinity of Pwll-du Head and strikes northwest along the southwest flank of Cefn Bryn where it now dips towards southwest. Northwest and west of Oxwich it zig-zags and ultimately forms the coastal cliffs of southwest Gower. This irregular pattern is because the rock layers are folded, as is best seen in the cross-section (Fig. 3). The relatively soft Marros mudstones (ornamented green) can be seen to occur in 'down folds', synclines, forming the low-ground embayments at Oxwich and Port-Eynon. The 'up folds', anticlines, intervene, with the most prominent one bringing the oldest rocks of Old Red Sandstone up to form the high ground of Cefn Bryn.

Gower rocks are cut and offset by numerous fractures. These are faults, the most prominent set mostly trending somewhere just east of north, across the folds just considered. These faults will figure throughout this study, for many are mineralised or guide significant valleys. This set is omitted from the cross-section (simplicity and laziness of the author), but two faults are shown. These strongly relate to the folding of the rock layers. The mountain-building Earth-crustal crunch, a plate-tectonic collision far to the south in Europe and known as the Variscan Orogeny, here pushed the layers from the south-southwest northwards. This compression, while causing shortening of the crust by folding also led to fractures and thrusting of rocks upwards. Thus the Cefn Bryn thrust fault (Fig. 3) has allowed older (ORS) rocks to be shunted upwards and northeastwards over younger rocks in a fold. The fault at Port-Eynon is a 'back-thrust' on which rocks have been forced up southwards, backwards, over younger layers that are also folded. The map reveals some tight folding with thrust faults in the east, towards Mumbles Head and northwards into Swansea Bay. The folded limestones and back-thrust exposed in Caswell Bay (SS592876), between Mumbles Head and Pwll-du Head, constitute a renowned demonstration of Gower geology that can readily be viewed from the beach, for as long as the tide allows.

We will in subsequent chapters consider all of the rocks depicted in Figure 3, but it is worth pointing out here how the main rock units will be involved. We should note immediately how the map is misleading, because it suggests only solid rocks at the surface. But Gower is substantially blanketed in superficial deposits that hide the solid geology beneath, most widely inland by glacial debris and at the coasts by sand and marshes. Actually, only some few percent of the ground surface is of exposed rocks, although much of the covered rock is easily identified, especially the limestones. The spectacular coastal cliffs of Gower are almost entirely of Carboniferous limestones while the Carboniferous Marros and Lower Coal Measures rocks are mainly poorly exposed, because they are relatively soft and were more easily eroded, so tending to form subdued lower ground. The Middle Coal Measures and Pennant Sandstones are not well exposed, but they form prominent ridges that speak of the underlying tough rocks, while myriad abandoned coal mines speak of their former seams of importance.

Momentarily returning to the beauty of Gower, most visitors will be familiar with the fabulous vistas of the land and coast southwest of a line from Oystermouth trending northwest to Whiteford Point (Fig. 3); let us call it south and west Gower. This area is dominated by the Carboniferous limestones and prominences of Old Red Sandstone, and certainly this essay mainly centres on aspects of this beautiful tract. The chapter on landscapes, while ranging far and wide off Gower, deals primarily with south and west Gower. The chapter on the glaciations is all-inclusive, necessarily treating northeast peninsular Gower and even the North Gower of the original Lordship[7], whence cometh the ice. Unsurprisingly the chapters on caves and sinks and springs are restricted to dealing in the Carboniferous limestones of peninsular Gower, with some territorial pride in now excluding involvement of the counterpart limestones of the north crop. The chapter on storms, dunes and landslides is essentially about the mobile elements of the coast and less to do with the solid rocks.

The reader understandably may infer from all this that south and west Gower is *the* interesting part to consider, but that would be missing much. North Gower too is outstandingly beautiful where relics of our coal and other past industries rest alongside even more scenic beauty. Our salt marshes and their tidal pills are exceptional, as are views from the sandstone ridges along and across the Loughor Estuary and into the hills far to the north.

Overview and introduction Notes

[1] Retired as George Herdman Professor of Geology at the University of Liverpool. Studied mainly physical processes of volcanic activity, including geophysical fluid dynamics, geochemistry and structural geology. Primarily a field geologist with many seasons of research in Pembrokeshire, Snowdonia, the Lake District and Glencoe, as well as on numerous volcanoes worldwide. Dived on ancient volcanoes off northwest Scotland and on active volcanoes off coasts of Iceland and New Zealand. Recently focused on debris flows and avalanches, involving laboratory experiments and field studies of the 1980 eruption of Mount St Helens, USA. Latterly described avalanche phenomena in craters on the Moon, but there only via astonishingly detailed photographs.

[2] LiDAR stands for Light Detection and Ranging, in this case a remote sensing method that uses pulsed laser light to measure precise distances to the ground, normally from an aircraft.

[3] A truly fascinating exposé of the human-life history and place settings revealed by the Welsh language of Gower is in 'Gower and what it means' penned by Philip Stephens (Stephens 2021).

[4] An outcrop (noun) is where a particular rock unit is at the surface; hence the unit 'crops out' (verb) at the surface. The units we discuss, loosely speaking, are at the surface along the 'north crop' and the 'south crop' – strictly the northern outcrop and the southern outcrop.

[5] Karst is the name given to distinctive limestone ground that has been subjected to long-term dissolution by slightly acidic water. Clints, upstanding blocks of limestone, and grikes, the slots between the clints, are a characteristic karstic weathering feature of exposed limestones.

[6] The reader is directed to 'images' on the web under 'Carboniferous crinoids, corals and shells' and also to the Burren Geopark website for additional relevant explanations.

[7] The Lordship of Gower, established in the Middle Ages under Norman rule (since 1116), includes the tract north of the neck of peninsular Gower, extending between the Loughor, Amman, Twrch and Tawe rivers.

Return from Whiteford.

2 LANDSCAPES CUT IN TIME: PENEPLAINS AND PLATFORMS

2.1 Introduction

A striking feature of the landscape of southern Wales is its stepped succession of broadly planar surfaces. Although the terrain in general appears subdued, rounded and undulating, the shores are characterised by platforms ending at cliffs and inland views can show wide areas with summits of similar heights. These 'surfaces' record the interplay of erosion with uplift or subsidence of the ground, generally relating to sea level, but a comprehensive explanation of them has proved elusive. Where there is a stepped succession of surfaces at different heights with rather steep slopes between them, it appears that, on geological timescales, long periods of comparative stability with erosional planation (levelling) were separated by short pulses of uplift.

This chapter involves a few tricky concepts, which, hopefully, are adequately explained. The aim is to provide an entirely new understanding of the landscape development, which necessarily involves the modern geology and physics of plate tectonics and deep-Earth processes. It turns out that Lundy, the island 50 km southwest of Gower across the Bristol Channel and once an active volcano, marks both the cause and the extent of uplift that affected southern Wales. Further, it was the opening of the North Atlantic Ocean between Britain and Greenland that ultimately led to regional subsidence, before the glaciations of the Quaternary Period dramatically affected global sea levels and embellished our vistas on Gower. All this takes some explaining. Hopefully the prospect of a new understanding will sustain the reader through the heavy-going sections of this chapter, but, if not, there is a summary at the end.

We first consider issues concerning land level relative to sea level, because this is central to understanding the landscape development. Then we discuss key topographical features of peninsular Gower and, necessarily, its hinterland and offshore areas. This wider view is essential because the presently surrounding sea has had little to do with the longer-term landscape development.

We look closely at the processes of coastal planation, with special reference to Gower beaches, before working through new ideas concerning the causes of land uplift and subsidence relative to changing sea levels. This is where Lundy provides crucial insights into the timing and controls of the geological mechanisms. These new ideas, possibly, are the most fascinating. The stepping of our landscape has been described and debated for over 100 years, but it is only recently that a satisfactory complete explanation has been discovered and here it is being applied to Gower. In following sections, we figure out the influences of sea level changes in shaping our land and coast before and during the currently ongoing Ice Age[1]; the iconic Pleistocene 'Patella Beach' is reinterpreted here as recording an exceptional 'super-storm'.

Detailed images of the land surface and sea floor that have only recently become available have proved crucial in much of this work. There is a final summary, with diagrams that might be useful for reference at times in reading this rather lengthy and involved chapter. By the way, this chapter is about the landscapes cut into rock over tens to hundreds of millions of years; features of more recent superficial deposits due to glaciers, rivers, slopes and wind are discussed later.

2.2 Absolute and relative levels

Nothing in or on our planet stays still. The whole dynamic Earth and its oceans constantly adjust their shape as they are pulled about by changing forces of gravity. The changes are partly due to movements of our heavenly neighbours, especially the Moon, and also because inner and outer masses of our planet – the core, mantle and crust – slowly move around.[2] It would be ideal to speak of a fixed point in Earth to which levels of land and sea could be perpetually and precisely related. However, because nothing stays still, such a point can only be notional, although that is fine for our purpose. Thus, notionally, we can think of a 'fixed' centre of our planet so that we can distinguish *absolute* uplift and subsidence of land from *absolute* changes of sea level that reflect changing volumes of on-land ice. Most of what we will discuss regarding landscape evolution concerns the *relative* changes between land and sea levels that govern the position of the coast, which in turn influences the base level to which rivers and their valleys tend to adjust. But, of course, we will also need to address the *absolute* changes because they govern what happens relatively (see box).

metres. Conversely, during previous full-on glacial episodes sea level was much lower than now and hence the coasts were far away.

Adding a little to this grasp of relative rise and fall of sea level and shifting coastlines, we need to know that during glacial-interglacial cycles absolute levels of the solid surface of the Earth change in response to the shifting loads of both ice and seawater. Melting away of ice (deglaciation) removes its load and causes land to rise buoyantly, albeit very slowly, which may displace seawater as is happening now in the northern Baltic region. Related to this melting, any added depth of seawater loads and depresses the seabed. Hence with global warming one certainly should not anticipate a uniform depth of flooding of all Earth's coasts.[3]

We will be concerned with absolute uplift of the ground surface that results mainly from deep-Earth processes with heating beneath the Earth's crust. Uplift due to removal of load in deglaciation proves to be of less concern in our southern Wales area. Absolute subsidence will be related to lessening deep-Earth influence and reduced crustal buoyancy according to cooling (density increase). Subsidence due to glacial loading was not a major control here.

While absolute sea level rather obviously varies according to the amount of water locked away as ice that is grounded on land, it also varies with changes of ocean-basin depth, which in turn relate to the temperature and density of the ocean-floor rocks and hence their age.[4] Our story of landscape evolution inevitably involves the interplay of geological forces and climate.

It is important to appreciate that now is not typical of most of Earth history. We live in the Quaternary Period, which started some 2.6 million years ago and features unusual development of thick and extensive ice caps at both poles with consequently lowered global sea levels; ice on land means less water in the oceans. This period has involved cycles of cold glacial and somewhat milder interglacial conditions. 'Now' is relatively mild interglacial *or* post-glacial time, depending on whether or not it turns cold again in future. If all ice on Earth melted now, sea level would rise many tens of

2.3 Views on views

In upland regions, groups of irregularly shaped hills viewed in a wide panorama and into the far distance commonly have summits at similar heights that together define an extensive level or gently sloping surface (e.g., Fig. 4). Of course, the hills individually are rounded, and the landscape is variously cut by ravines and valleys, but it is interpreted that the defined surfaces are remnants of ancient subaerial peneplains. Each peneplain is considered to represent a 'final' (near equilibrium) surface that was levelled by deep weathering and erosion of the rocks, mainly by streams and rivers, during long-lived crustal stability. The existence of the peneplain surfaces at various heights above modern sea level is thought to reflect periodic crustal uplift that caused rivers and their head-water catchments to be 'rejuvenated' (disrupted equilibrium), with consequent renewed active down-cutting and hence erosional planation to lower base levels.

*Figure 4. **(Top)** Winter view towards south over southern Wales, from near Plynlimon (SN815878) across numerous summits that are 5-45 km away. Seen together these hill tops define an apparent plateau surface – the prominent dark skyline – at around 470-510 m above sea level. In the distance (faintly seen in the centre between the wind turbines), 65 km away, is the escarpment of the Brecon Beacons reaching 850 m. **(Inset map)** During the Late Devensian glaciation, the southern part of the Welsh Ice built thickly in the area of this top view, in the vicinity of the red oval (ice limits redrawn from Glasser et al. 2018). The ice flowed across the Brecon Beacons to the south, some of it towards Gower. **(Bottom)** View north from the Brecon Beacons escarpment (SN829213), with Fan Brycheiniog at 802 m and Fred on the left. The distant blue-green skyline is the same apparent plateau as in the top image, at 470-510 m, with the obvious partly forested plateau in the middle ground at 400-420 m. (Photo locations are marked by crosses on the inset map).*

In lowland regions, wide panoramas also can reveal extensive surfaces similar to those of the uplands. Most notorious are the 200-foot, 400-foot and 600-foot 'platforms', now, less sweetly, 61 m, 122 m and 183 m OD[5]. The so-called '61 m platform' (Figs 5-7) is spectacularly well developed on the Gower peninsula and also in south and west Pembrokeshire, e.g., on Skomer and Ramsey islands. The identification specifically as 200-foot or 61 m for that particular

*Figure 5. Early October mists at dawn, viewed from Cefn Bryn (SS495897). The mist lies at about 60 m OD, thus picking out the irregularities in Gower's inland '61 m platform' surface. (**Top**, view to east, to beyond Swansea Bay) The sun is over Porthcawl and the distant skyline is a composite of many summits, rising from around 300 m behind Port Talbot (centre) to over 500 m above the head of the Vale of Neath (extreme left). In the foreground, mists just fill the cwms cut into the platform. In the left-hand middle distance, silhouetted against the distant range, the dissected ridge north of Swansea Bay – Cefn Coed, Town Hill, Kilvey Hill – rises irregularly from 170 to 190 m. This was previously taken to be part of the 'surface' at 183 m. (**Middle**, view to southwest) The distant profile above the mist is the southern constructional feature of the Paviland Moraine, which is built upon the coastal platform (Chapter 3). The nearer mist fills in between the low flank of Cefn Bryn and the partly eroded north-eastern part of the moraine, surmounted by the hamlet of Berry at 70 m. (**Bottom**, view to west) Beyond Reynoldston, the mist fills the basin formerly occupied by the Paviland ice lobe and now draining north to the coast via Burry Pill (Chapter 3). The distant skyline is Rhossili Down with tops at 193 and 185 m, also previously considered as part of a planation surface at 183 m.*

level on Gower is understandable, but, as originally recognised by TN George (1938), this platform actually slopes gently. Although previous researchers have tended to focus on specific levels (e.g., Fig. 7), the existence of a continuous gentle slope is more important.

Inspection of the coastal fringe around south-western Wales widely reveals a sloping plane surface ranging down from around 75 m OD or more inland, to 40 m and in places 20 m or less before being truncated by modern cliffs. This slope is generally less than 1°, only locally 1.5°.

The notorious '61 m platform' of Gower is better considered as a remnant part of an originally more extensive, gently sloping, planar surface. Apart from the present cliffed coastline, this *sloping* planar surface is the most distinct feature in the Gower landscape (Figs 5 and 6). As discussed below, it proves to be the most recently planed-off of all of the extensive levels, but, importantly, there is evidence that the lowland profile was eroded down close to this level long before this most recent planation. In other words, the existing gently sloping coastal planar surface was formed on a landscape that had already been deeply cut across the various rocks of southern Wales. We shall see that many hundreds of metres thickness of rocks had been eroded off over tens of millions of years, long before the planation that formed the existing gentle coastal slope.

*Figure 6. (**Top**) View east from the westernmost top of Worms Head (SS384877). The ridge of the Inner Head of the Worm is at 45 m, the main headland rises gently, at around 1.5°, from 45 m to 80 m at Rhossili, while the summit of Rhossili Down is at 193 m. This low-tide view reveals the modern 500 m-wide marine platform also sloping gently seawards. (**Middle**) View east from Tears Point (SS409870) showing Fall Bay to Port-Eynon Point. Most of the cliff edge is at 60 m although towards the farthest point the surface slopes gently to 45 m, as at Worms Head. The broadly planar inland slope is mantled by low mounds that are the Paviland Moraine (**arrows**), including a western remnant surmounted by the house sited on the skyline. (**Bottom**) View west towards Tears Point and Worms Head from the cliff above Paviland Cave (Goat's Hole; SS436859). The continuity of the eroded planar 60 m surface here is striking, as is the outer slope of the overlying Paviland Moraine (western remnant) to the right. Views such as this gave rise to the notion of a '200-foot (61 m) platform'.*

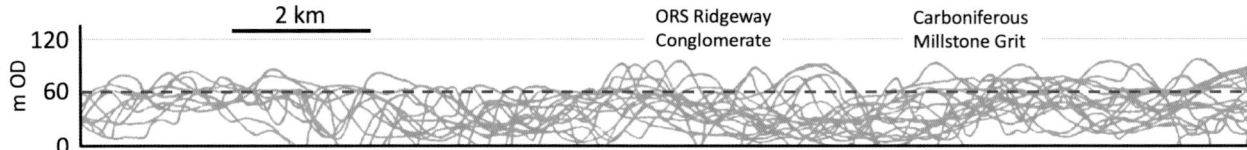

*Figure 7. Superposed west-to-east profiles across a belt of country west of Tenby in south Pembrokeshire, viewed from the south (after Goskar and Trueman, 1934, figures 2 and 4). The diagram represents an area that is roughly 16 km from north to south and it was intended to show that despite variable incision there is wide planation to a surface at about 61 m (**dashed red line**). Above 61 m OD the groups of relatively steep-sided profiles relate to resistant hard rocks (**labelled**) whereas some other departures relate to a gentler slope rising inland. Importantly, however, other recurrent levels are visible in the plot, e.g., at around 50 m, 30 m and 10 m, and the diagram masks the fact that, although irregular in detail, there is a general slope southward down towards the coast from about 75 m, as there is on Gower.*

The distinct coastal sloping surface is readily reconciled with marine planation, which is the cutting of a planar surface by wave action at a shoreline of the sea (Section 2.6). The adjacent higher surfaces notionally at 122 m and 183 m OD, however, are less distinct and less certainly attributed. They inevitably have been degraded by subaerial weathering and erosion, including the effects of glaciation, and close inspection typically reveals numerous variations about any level so that faith must play a part in identifying surfaces with any specific altitude. It turns out, as explained below, that only the coastal-fringe slope is due to marine planation. The higher surfaces are much older and were formed subaerially.

Inland and at levels above about 300 m, the extensive plane surfaces that define plateaux in wide and distant views (e.g., Fig. 4) can in places be proved to be due to subaerial peneplain development. This is confirmed by deep chemical weathering that affects uppermost rocks (e.g., Fig. 8). Crucially, such deep weathering also occurs at lower levels, down to 65 m in Pembrokeshire (Battiau-Queney 1984), which necessarily implies that *the lowland terrain profile also involved subaerial peneplain development, <u>before</u> its marine planation.*

The weathering typically involves the uppermost several metres of rock, which have been altered and softened by the breakdown of silicate minerals and some dissolving of quartz (silica). This leaves insoluble clays with distinctive red-brown hydroxides of iron while tough quartzites are reduced to deposits of quartz sand (locally quarried). This alteration typically occurs in warm and wet climates, which we know existed here for tens of millions of years before the start of the Quaternary Ice Age about 2.6 million years ago (see stratigraphic-age chart Figure 9). Thus, the main landscape peneplain levels are truly ancient.

Sediments produced by the ancient peneplain erosion accumulated in offshore basins around Wales (e.g., Tappin et al. 1994) and these show that development of the upland landscape dates back to tropical Paleogene times (Fig. 9). Records of possible earlier developments, in late Cretaceous times (from about 80 Ma), would today be cryptic at best. Upland surfaces would have been continuously modified during Paleogene (66-23 Ma) and Neogene (23-2.6 Ma) weathering and down-cutting erosion, so early patterns are likely to have been erased completely.

*Figure 8. (**Top**) View northeast across the north crop of the South Wales Coalfield, around the upper Swansea Valley (vicinity of SN8517). The highest summit on the left, at 725 m OD, comprises Old Red Sandstone strata forming part of the north-facing Black Mountain and Brecon Beacons escarpment. Mid-right are the quarries in Carboniferous limestones at Penwyllt and the top-right skyline around Carreg Cadno comprises tough quartz-rich Marros Basal Grit. The view shows that extensive surfaces are substantially independent of the underlying various rock layers. Whereas all of the rock strata dip gently south at about 15°, the right-hand skyline surface slopes at only 6° (lines show the apparent slope difference). Furthermore, sloping terraces, evident in the lower flank of Fan Gyhirych (**arrows**), trace semi-continuously across the middle ground and these also slope more gently south than the underlying strata.*

*(**Bottom**) Around the summit of Carreg Cadno robust rock outcrops (e.g., SN871156) show polished surfaces with striations recording southerly directed flow of the Late Devensian Welsh Ice from over the escarpment to the north. Ice movement was away from the camera and parallel to the stick.*

*(**Hidden**) North of the summit of Carreg Cadno, near Pwll Byfre (just over the skyline; SN876166), the Marros Basal Grit shows deep subaerial weathering indicating that the surface up there is part of an ancient peneplain. Thus, the Carreg Cadno skyline is in part an ancient peneplain, albeit somewhat shaved by Welsh Ice of the Late Devensian glaciation.*

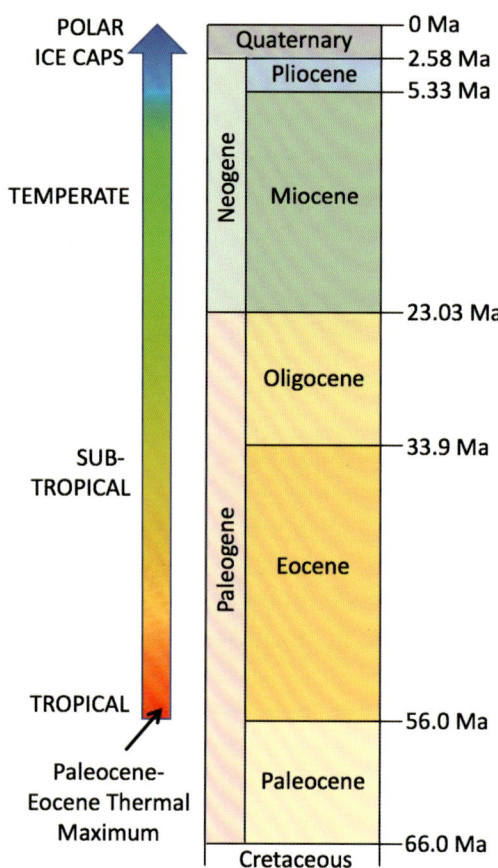

Figure 9. Subdivisions of geological time and the general climatic cooling during the interval of interest, from about 56 Ma. Ma means millions of years (mega annum), conventionally 'ago', i.e., relative to 'now'.

Later we shall see that Quaternary landscape modifications are a beautiful and intriguing 'late dressing' of the previously evolved main surfaces of Gower. However, here it is important to stress that the landscapes were never cut into a 'blank medium' and that the bed-rock geology of Old Red Sandstone and Carboniferous strata necessarily influenced the outcomes of erosion. As TN George (1938) pointed out regarding differential erosion of the limestones: "... every bay, inlet, gully, mere, cave or sound is eroded along a joint or a fault or a fold or a soft bed ...", and this applies to the other rocks too, as we shall see. Quaternary processes cut much of the present cliffed coastline, modified valleys and some skylines, and left the extensive drapes of glacial and periglacial deposits, all upon the frame of the underlying bedrock.

2.4 Imagined ancient landscapes

It is not at all obvious to see, but the coastal land surface on Gower widely lies just beneath the original level of an ancient and uneven erosion surface (an unconformity)[6] that had been cut across and deeply into the folded and faulted Old Red Sandstone and Carboniferous strata. The folding and faulting started in Late Carboniferous times, about 300 million years ago, and deep erosion across Gower developed mainly in Early Permian times, by some 290 million years ago (Chapter 1 and box).

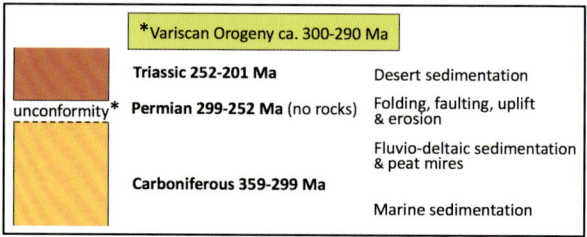

Evidence for the ancient erosion surface is in vestiges of Triassic red sediments that occur within deep hollows locally preserved in the underlying limestones, for example at 35 m OD at Port-Eynon and at the high-tide level in Fall Bay (Fig. 10). Similarly, fine-grained red ochreous (hematite- and goethite-rich) sediments commonly occur in fault fissures on south Gower[7] and they have also recently been recognised infilling a fault void intersected inside Tooth Cave, at about 55 m OD (Section 4.8). The red sedimentary deposits are the youngest rocks preserved on Gower, at around 220-210 million years old, and they record generally arid desert environments prone to flash floods. It is entirely reasonable to deduce from these vestiges that the Triassic red rocks and their basal ancient erosion surface once occurred more widely somewhere 'not far' above the present land surface[8] and subsequently became planed-off. But then what was the relationship with the higher surfaces on Gower that culminate at around 183 m? What was the original areal extent and thickness of that mainly red Triassic cover?

LANDSCAPES CUT IN TIME

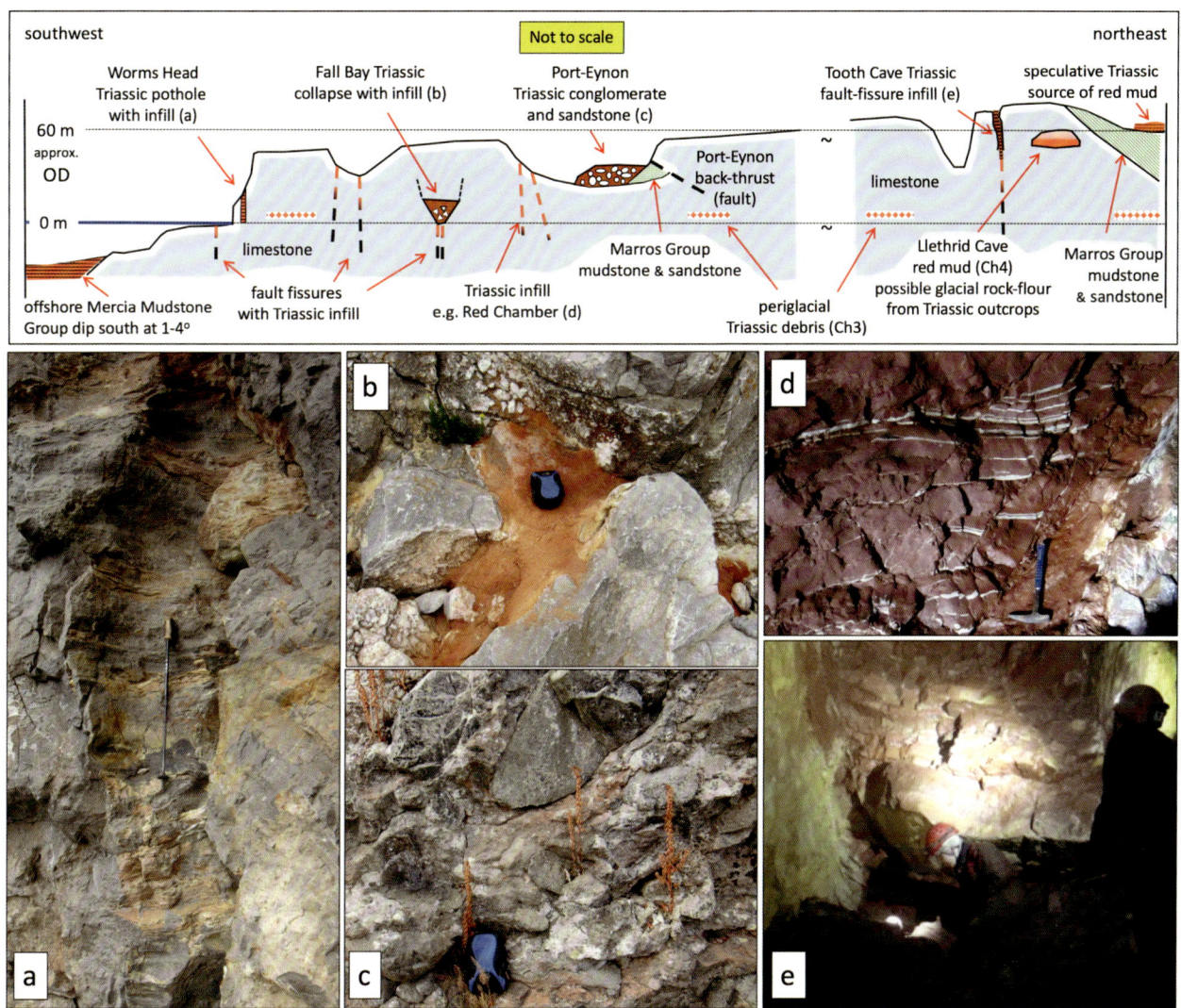

Figure 10. Schematic cross section across southwest and central Gower showing the main relics of Triassic sedimentary strata and a key to the photographs a-e below.
(a) Section of an irregular near-vertical pothole about 1 m in diameter (stick is ~1 m). Vestiges of pale red-brown-yellow calcareous sandstone and siltstone (micrite) adhere to the limestone wall, the semi-circular form of which clearly registers dissolution; there is no evidence for opening on a fault void. This pothole, on the westernmost face of the Outer Head of Worms Head (SS384876), is one of very few examples on Gower of Triassic-age karstic cave development.
(b) Red sediment infill between a jumble of angular boulders of limestone that show breakage across original mineralised fault surfaces. Some fault-vein mineralisation nearby appears in situ and the rock exposure, some 25 m wide at the high-tide level in Fall Bay (SS413873), appears to represent a collapse that would originally have formed a surface depression (a doline). Camera case is 12.5 cm.
(c) Poorly sorted rounded to sub-angular cobbles, boulders and blocks of limestone with red sandstone matrix constitute the main Triassic outcrop on Gower, at Port-Eynon (SS466854), and are interpreted as recording deposition from flash floods in a mainly desert environment.
(d) Ochreous siltstone lines the fault-fissure that is now a sea cave which extends into a mine at Red Chamber (SS426867), 700 m southeast of Mewslade Bay. The white bands mark natural chemical leaching along original bedding with offsets due to faulting. Hammer is 33 cm. (continued...)

Figure 10 continued. (e) Underground, red-brown-yellow sandstones and siltstones fill a fault-fissure, some 2 m wide, which has been intersected by the main flood-bypass cave passage in Tooth Cave (SS531909). These show diverse structures that record progressive infilling by sediment washed in from the contemporary Triassic surface far above (Section 4.8).

It is intriguing to visualise that precursors of the prominent Gower peaks of tough Old Red Sandstone rocks that reach up to about 180 m – the long ridges of Cefn Bryn and Rhossili Down, and the more rounded Llanmadoc Hill and Hardings Down – must also have stood proud in Permo-Triassic times, more than 200 million years ago. We know that these hills have been substantially degraded, most recently in Quaternary times by frost and ice (Chapter 3), and we think that they are likely to have been weathered down during the general Paleogene to Neogene peneplanation of southern Wales (see below), but now we might wonder how much they reflect originally persistent peaks of a Triassic desert landscape. Is the 183 m level a really ancient tough-rock-surface relic? And was this ever completely buried?

The closeness of parts of our present landscape to the Permo-Triassic erosion surface has led some researchers to view the ancient surface as a sort of template somehow influencing later developments, but it is unclear how this might work, especially if there was a thick cover. We have very little direct evidence as to whether there was any considerable thickness of overlying Triassic and younger Jurassic rocks, or even Cretaceous strata, that might have buried the Permo-Triassic template. There is, however, some indirect evidence.

Triassic sandstone fragments are moderately abundant in Quaternary debris all along the south coast of Gower, especially in the vicinity of Worms Head in the west and near Rotherslade in the east (Chapter 3). These must have been eroded from Triassic rocks originally exposed to the north. Also, red mud in older parts of Llethrid Swallet Cave might record glacial grinding of inland Triassic rocks (Chapter 4) that are no longer known there (Fig. 10). Furthermore, a Lower Jurassic ammonite found in Quaternary deposits at Horton (Smith et al. 2002) confirms that some younger Mesozoic remnants existed and were reworked from somewhere to the north.

Triassic-rock cover is continuous nearby in the sea floor off the south Gower coast (Tappin et al. 1994). Bathymetric images showing seafloor-exposed Carboniferous strata (Fig. 11) suggest that the contact with that cover is topographically highly uneven, as it is on land. Triassic strata could lie as little as 650 m from the shore at Worms Head, between -10 m and -20 m OD, while they must lie deeper than about -35 m OD some 2-3 km south and southeast of Port-Eynon Point. Given that coastal exposures of Triassic sediment in hollows and cracks are common (Figs 10 and 11), it is probable that the original uneven unconformity surface existed not far above the seafloor rocks and is hidden close to where modern sediments now lap onto the limestone. Lower Jurassic strata overlying the Triassic occur 2 km offshore to the south and Cretaceous rocks are preserved farther southwest (Tappin et al. 1994, figures 19, 30 and 42).

The fragmentary remains suggest that there originally was Triassic and Jurassic cover on Gower. Those strata are widely preserved on top of the older rocks to the east in southern Wales, in the Vale of Glamorgan and around its gently sloping coastal fringe, e.g., at Sker Point and round to Barry (Fig. 12). It seems simplest to envision that a similar cover existed on Gower but was eroded away. Later in this chapter we will discover that the present landscape and the seabed to the south of Gower were deeply cut through a great thickness of uplifted rocks. In this light, the present proximity of the modern surface to the original Permo-Triassic erosion level, at least for the coastal sloping platform, may simply be a coincidence that has little to do with any template.

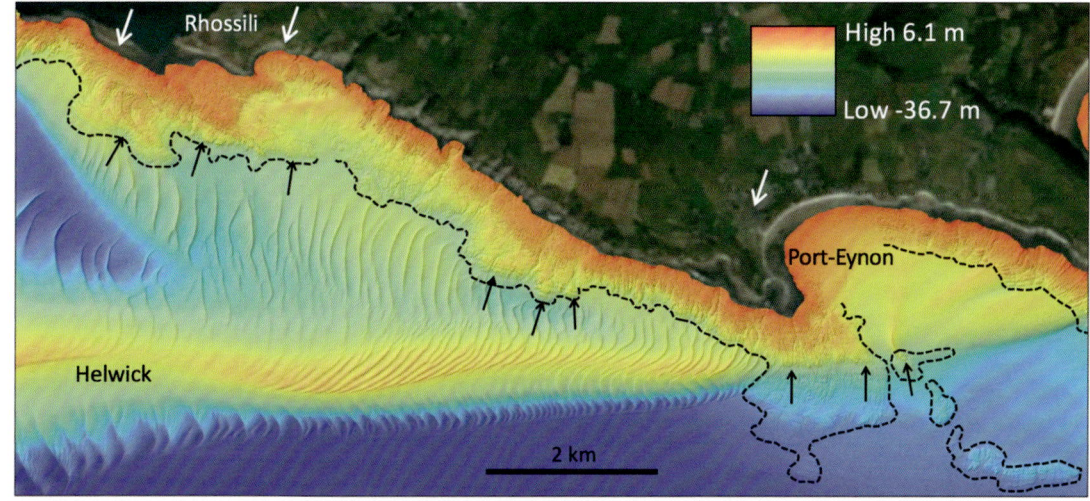

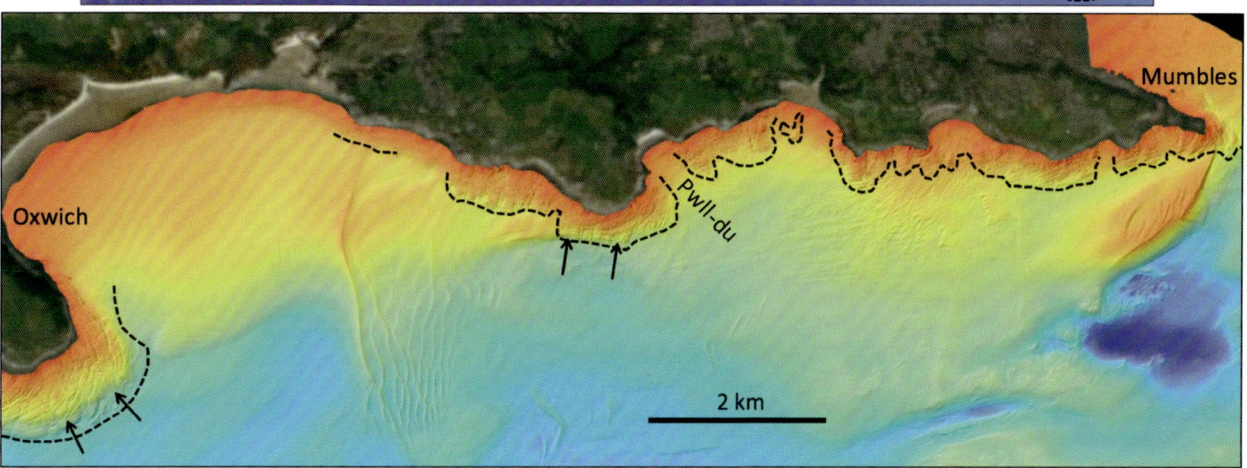

*Figure 11. Coastal bathymetry off south Gower, from Worms Head, top left, to Mumbles Head, bottom right. **Dashed lines** delimit seafloor exposures of folded Carboniferous strata and so exclude any substantial Triassic cover. Such exposures south of Port-Eynon extend down to 35 m below sea level (-35 m OD, coloured dark blue). The continuity here of the erosional etching of the limestone surface emphasises the fact that modern sea level obscures part of an ancient-massif landscape (see below). **White arrows** show the locations of significant on-land remnants of Triassic sedimentary rocks. **Black arrows** indicate the base of submarine cliffs. Depths are relative to the level of the lowest astronomical tide: LAT. (Base image multibeam bathymetry, at 2 m x 2 m resolution, from The Admiralty Marine Data Portal, kindly processed and provided by Gareth Carter). The beautiful forms of the Helwick sand bank and associated subtidal dunes are discussed in Chapter 6.*

The Carboniferous limestones and the Old Red Sandstone prominences of Gower must have formed a substantial upstanding massif early in the onset of Triassic sediment accumulation here. This is evident from the 1-4° southwards dip of the Triassic Mercia Mudstone (Fig. 10) occurring offshore at depths ranging from -10 to -35 m OD (Fig. 11); the lowermost strata project northwards directly into the older rocks, which thus would only become buried progressively. It can be fun to scramble along the Old Red Sandstone conglomerates on top of Rhossili Down and imagine the fields, commons and coastal limestones below to the south and east as a ruddy desert landscape of wadis with bluffs of red-brown sandstone and grey limestone, with surrounding aprons of debris grading farther south down into plains of silt.

But there is more still to imagine. There is evidence for a nearby latest Triassic seashore. It has been recognised that the fault-fissures containing red sediment, and calcite with iron oxide/hydroxide mineralisation, show no evidence for limestone dissolution to produce the cavities. Fault movement opened the spaces for those infills of sediment and minerals (Wright et al. 2005). Flash floods in the desert led to water washing sediment underground, but it did not, at the levels exposed, form any water table where caves might form. Indeed, such generally arid environments with little vegetation are not conducive to development of 'aggressive' water that would dissolve limestone. However, the pothole exposed on Worms Head (Fig. 10a) does require that a water table existed above it when it formed; it is a record of karstic weathering with cave formation in the limestones. So how is this resolved? What might have changed?

Potholes and caverns containing Triassic sediments have long been known in the Carboniferous limestones in and around the Mendip Hills and in South Glamorgan, predominantly exposed in quarries. These occurrences have been the focus of research because the sedimentary infills contain fossils of vertebrates, most importantly of early dinosaurs (Robinson 1957; Whiteside and Marshall 2008). Most famous amongst the fossils is that of the 2-metre-long bipedal herbivore *Thecodontosaurus antiquus*[9], amongst the oldest known. For our interests, both the dinosaur diet and associated organic microfossils (palynomorphs) indicate a climate shift from arid to one favouring vegetation, while the cave developments indicate raised water tables.

Whiteside and Marshall (2008) have elegantly provided detailed evidence that the raising of the water tables must reflect incursion of the sea towards the limestone massifs, where land plants and animals lived above water-filled (phreatic) caves. Seawater must have prevented rainwater from simply draining out. Thus, the Gower Triassic pothole of Worms Head records the Late Triassic transition to marine conditions that heralded the shallow Jurassic seas. We would be justified in imagining that our later view from Rhossili Down would be from an island in the sea.

2.5 Beneath the sea

In order to understand the uplift and planation history of southern Wales it is necessary to consider here, briefly, the coastal bathymetry along south Gower in particular and in the Bristol Channel in general (Figs 11 and 12). Offshore beyond the gently sloping edge of the tidal platform there is a moderately steep slope cut in rock, mostly limestone, between -15 m and -25 m below OD (Section 2.12). Beyond this there is generally a broad and smooth 'plain', very gently sloping, at less than 1°, to -80 m OD and beyond (Fig. 12). Much of the sea floor deeper than -25 to -30 m OD around southern Gower must comprise Triassic strata (Mercia Mudstone Group) that dip at about 1°-4° broadly southwards (Fig. 10 section). In places the offshore slope appears as a continuation of the on-land slope, with valleys continuing into deep water as at Pwll-du Bay and in Carmarthen Bay.

The offshore surface is cut into rocks generally with only thin and variable superficial cover that includes peat, sand, gravel and till, although moraines, sand banks and sand waves are locally prominent. Its smooth profile is in places broken by islands formed of relatively hard rocks, most notably Lundy, 50 km southwest of Gower (Fig.12). This island, mostly of granite reaching up to 140 m above sea level, stands proud of adjacent softer seabed rocks at -40 m to -50 m below sea level and is topographically similar to those instances in southwest Pembrokeshire where the broad subaerial planar surface is punctuated by steep profiles of resistant rocks (technically inselbergs), e.g., on Ramsey Island (Carn Llundain at 136 m) and east of St David's Head (Carn Llidi at 181 m).

Thus, the offshore sea floor is quite like the land surface, and, although it was worked over by wave and current action more than the coastal lowlands, previous deep subaerial erosion is graphically evident. Lundy island holds the key to understanding all of this, as explained later, but first we should go to the seaside...

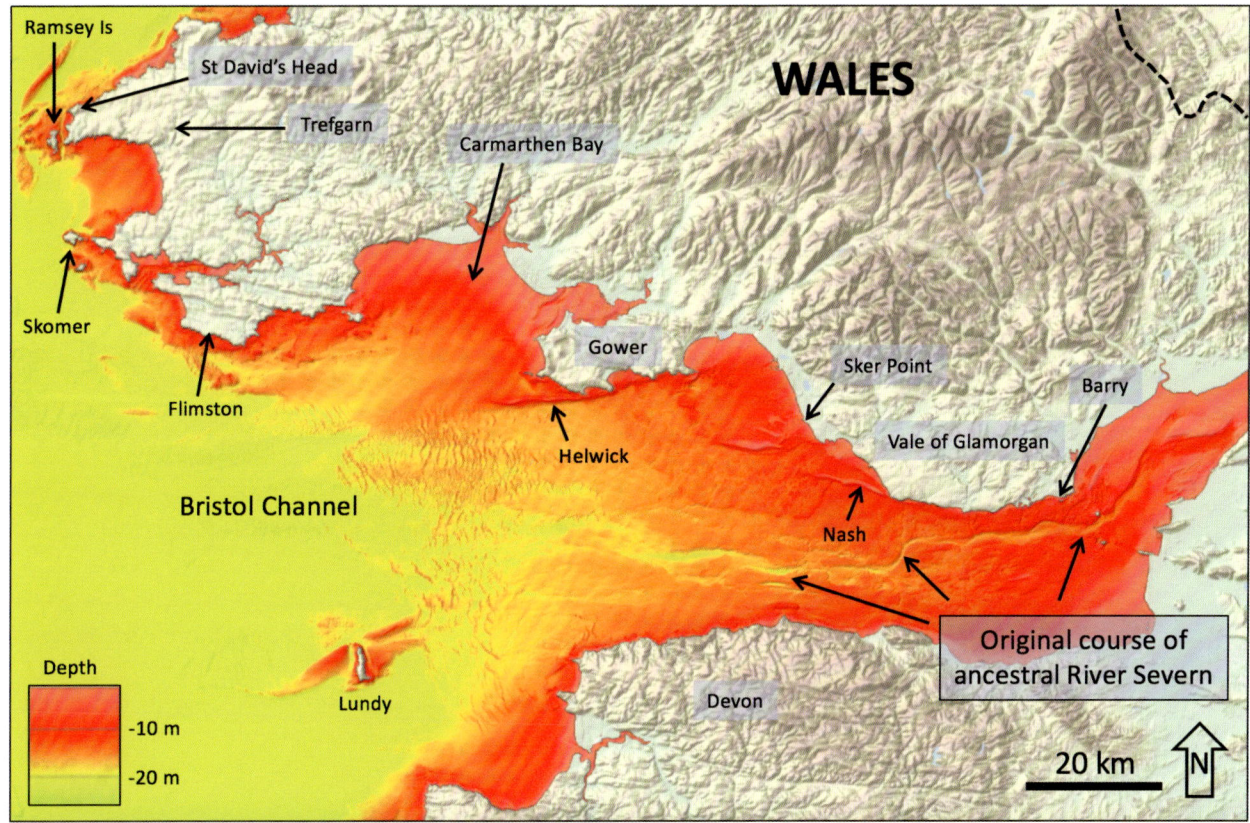

Figure 12. Bathymetry of the Bristol Channel area with coastal locations mentioned in the text. Although widely notched by cliffs generally less than 60 m high, the coastal fringe of south-western Wales, from St David's Head to Barry, widely appears to continue into the seafloor slope to the south. Arcuate seafloor features southeast of Gower, between the Helwick and Nash sand banks, are glacial moraines; north-south repeated features are subtidal dunes. (Base image from EMODnet[10] 2019).

2.6 Coastal platforms, modern and ancient

Here we see how gently sloping platforms are cut in rocks by prolonged seashore wave action. After the last glaciation, which peaked with ice on Gower some 23,000 years ago (Chapter 3), mean sea level globally rose fairly rapidly to near its present height relative to the land by about 7,000 years ago (Fig. 13). Since then, it has slowly risen perhaps a few metres, not much, and *globally* it continues to rise *now on average* by 3.2-4.2 mm/year, at an increasing rate.[11] It is important to note that because our planet Earth deforms according to mass-loading under gravity, the 'global rise' of sea level does not translate into any uniformly distributed rise. Especially where on-land ice has melted away, the relative sea level typically has fallen as the land rebounds after removal of the load (Miller et al. 2005, 2011; Tamisiea and Mitrovica 2011).

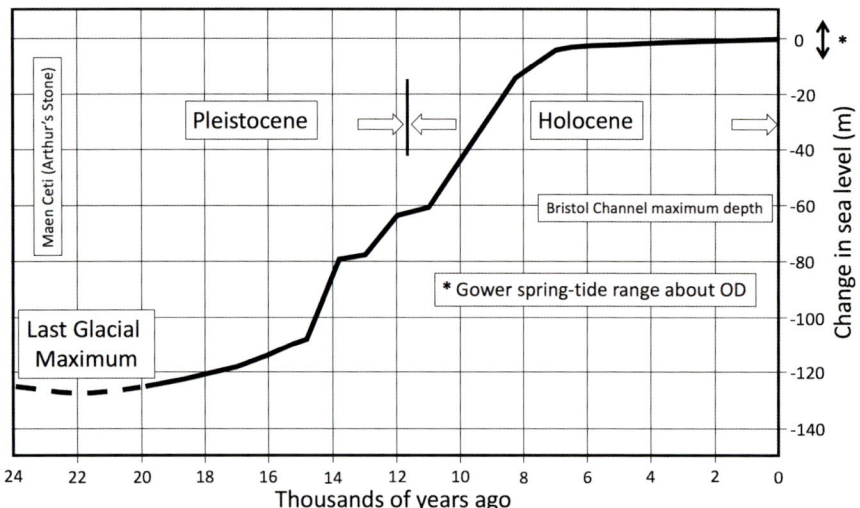

Figure 13. Post-glacial sea level rise (after Fleming et al. 1998). The steep steps relate to major melting episodes affecting grounded ice sheets. Details of this rise in terms of icesheet melting and coastline advance are graphically illustrated at:

https://www.psmsl.org/train_and_info/geo_signals/gia/palaeoshoreline_webpage/HTML/HOME.htm

With the second largest tidal range on Earth occurring on the Bristol Channel coasts, the gently sloping tidal rock platforms along southern Gower are especially spectacular. These tidal surfaces together with the cliff-edged sloping platform above them (Figs 14 and 15) constitute the most iconic landforms on Gower: truly outstanding natural beauty! However, as is so often the case, none of what we see has a simple explanation. Although we can observe the modern planation processes at work, as described here, it proves incorrect to assume that the modern shore has been cut in one go. As we shall establish later, the planation progress is slow, and the present shore has been re-visited by waves on several occasions. But let us do the obvious bits first.

Figure 14. The modern rock platform forming the low-tide causeway to Worms Head is 500 m wide during low spring tides and, judging from the building waves on the left, continues to slope gently for a similar distance offshore. This wave-cut surface is closely similar to the gently sloping cliff-top platform above. (continued...)

Figure 14 continued.

*(**Top**) Oxwich Point from a distance appears as a plateau but is actually a moderately straight-crested, northwest-southeast trending ridge at around 85 m OD dropping gently to cliff edges at about 60 m. The old surface cuts smoothly across the limestone strata (here dipping 20°E), just as the modern beach platform does.*

*(**Middle**) Viewed westwards parallel to the cliffs (image above), the surface at about 60 m OD is strikingly flat, with slight undulations like the modern platform, although from here it rises 25 m rather smoothly northwards inland over a distance of 600 m.*

*(**Bottom**) Drone view of the strikingly planar southern slope down to the cliff edge on Oxwich Point. Port-Eynon Point at about 45 m is visible in the distance. (Photo courtesy of Andy Freem).*

*Figure 15. (Page opposite). **(Top)** Three Cliffs Bay with Pobbles beach in the foreground (SS540877), viewed west towards Great Tor with the 61 m platform sloping landwards up towards the foot of the '183 m' ridge of Cefn Bryn. In the foreground the large-boulder strewn ledge is part of the Patella Beach platform cut in the coast during an exceptional Pleistocene high stand of sea level, at around 10 m OD, some 123,000 years ago (Section 2.14). **(Middle)** Viewed north from above the Patella Beach near Pobbles, the general surface slopes up to 75-90 m in the middle distance, while in the far distance (right of Pennard Castle) the skyline is the ridge supposed to form part of the '122 m' level at Three Crosses (SS574943). (Drone photo by Andy and Antonia Freem). **(Bottom)** Viewed east from Oxwich (SS503864) the surface smoothly undulates. From Southgate, by Fox Hole, as far as Deep Slade (Hunts bay) the platform is at 65-70 m OD and from there it rises gently south-eastwards to 97 m on the summit of High Pennard (Pwll-du Head).*

The commonly used expression 'marine abrasion platform' does not register the important dissolving away of the carbonates that make up limestones. Gower's entire rocky coast, from Oystermouth round to Llanrhidian, is limestone and dissolution is plainly evident as a process of the planation in the high-energy zones of the incessantly rising and falling, big tides (Fig. 16 below and next page).

Figure 16. (Previous page and above). At the shore of Deep Slade (SS562867), below Hunts Farm, the spiny and pocketed limestones show clear evidence of dissolution in the surf zone. Many temporary pools show scalloped fringes that record most aggressive dissolution around water margins. Aerated, bubble-rich surface waters, sea spray and naturally acidic rainfall contribute to this gradual corrosion. Atmospheric CO_2 is captured by breaking waves so that the surface water can be weakly acidic and hence aggressive, while seawater in general is mildly alkaline. **Above right** *shows fluting due to spray-runoff at about 10 m OD near Ravenscliff Cave (SS5464687304). Scale is 10 cm.*

Dissolution is also evident in the diverse hollows and grooves that form at various scales reflecting fluid-dynamic processes and molecular-scale effects, some patterns being quite bizarre. Microrills, which are mm-scale, close-spaced and variably sinuous grooves (Fig. 17 below), form by capillary flow of water beneath wet sand and are in effect a hidden corrosion of the solid rock (Section 6.4).

Scallops and longitudinal or transverse ribs (Fig. 17 opposite) are formed by corrosion that reflects water-rock contact interactions in which vortices of water are more 'aggressive' – i.e., tend to dissolve more limestone – where flow is fastest.

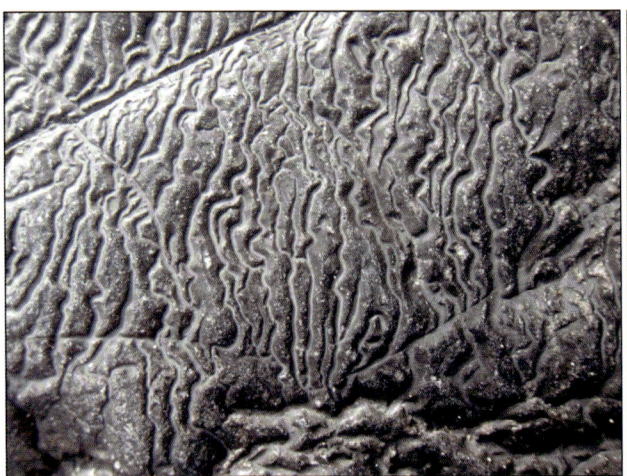
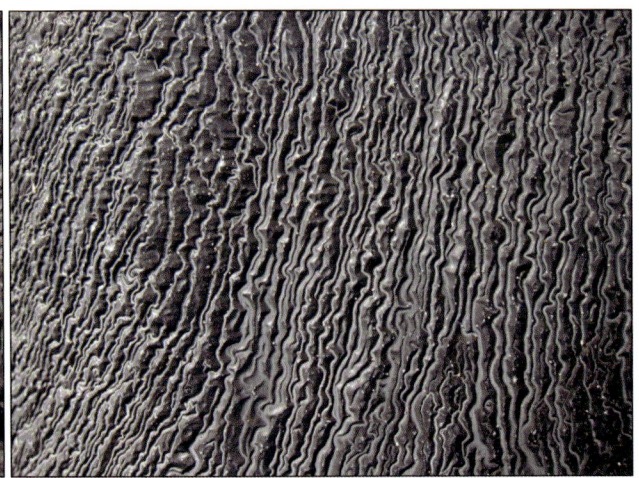

LANDSCAPES CUT IN TIME

Figure 17. (Page opposite). Beautiful and fascinating millimetre-scale grooves, known as microrills, formed by limestone dissolution that occurs beneath wet sand, here at the top of the beach in front of Little Tor (SS526877). These two examples are exposed on a narrow platform that is periodically exposed when the beach sand is removed during storms, although they most probably formed by capillary-flow of water beneath relatively long-lived sand dunes (Fig. 20 and Section 6.4).

*(**Above left**) Solution scallops and fine ribs on a persistently exposed rock platform at the top of the beach at Prissen's Tor, Broughton Bay (SS425936). Prevailing sea-water flow from top to bottom here forms broadly spoon-shaped intersecting hollows with the steeper slope on the up-current side. These scallops record the variability of limestone dissolution around decimetre-scale turbulent-flow instabilities, while the finer ribs are also due to dissolution around fluid-dynamic instabilities, in this case millimetre-scale flow-parallel vortices formed in thin flows.[12]*

*(**Above right**) Roughly recording a slow rate of limestone dissolution, the subdued and corrugated hollow to the right of and parallel to the scale is the remaining half-part of an originally cylindrical hole drilled for quarry blasting in the early to mid-19th century (top of the beach at Prissen's Tor; SS425936).*

Marine abrasion (Figs 18 and 19) obviously differs from the dissolving of the limestones. Surfaces are polished smooth as waves dash lifted stones against proud surfaces and the backwash drags them off again. The surprisingly loud rumble and clattering of boulders, cobbles and pebbles dragged out in backwash is dramatic testimony to their collisions and the damage being done. Where the beach platforms on Gower cut into drift containing hard, quartz-rich cobbles and boulders (mostly periglacial deposits), for example from local Old Red Sandstone or derived from the north crop of the South Wales Coalfield (see Chapter 3), these are manifestly abundant in the beach accumulations and are likely to be the most effective agents in the abrasion. Where the erosion has cut deeply, pebbly upper reaches of the beach can give way to sandy slacks exposed during low water.

Figure 18. At the shore of Deep Slade (SS564866).

(Top) Abrasion by cobbles and pebbles produces polished and rounded limestone rock surfaces.

(Lower left) Abrasion undercuts the spiny, pocketed dissolution surfaces.

(Middle right) The relative effects are clearly shown where stones thrown up into pockets are trapped and then mill deep smooth hollows into the dissolution surfaces.

(Lower right) Detail from the cobble beach (shown lower left), with tough quartzites and quartz-pebble conglomerate out-numbering the rather smaller grey, less durable limestone fragments.

Figure 19. Viewed to the west-northwest, the broad beach platform between Oxwich Point and Horton (SS503849) illustrates the contrasts between predominant limestone dissolution (darker surfaces) and prevalent abrasion. The pebble- to boulder-sized agents of the abrasion have been piled to the northwest of the pale abraded surface by southerly storms, but the next westerly gale will have continued the work in the foreground.

The loose sands of Gower beaches widely rest thinly on the modern rock platform, or on glacial or post-glacial deposits, as seen exposed after some storms. The sandy beaches lack any counterpart in the ancient marine platform above, for obvious reasons. They can be considered as parts of the modern beach platform in a so-called 'dynamic equilibrium' with the constantly changing waves and tides, although, perhaps surprisingly, they mostly do not derive from cutting of the rocky shore. Instead, the beach sands mostly originated by reworking of glacial tills and outwash sourced generally from upland Wales to the north, or from the Irish Sea area, and also by reworking of alluvial deposits of the ancestral River Severn out in what is now the Bristol Channel.

During Pleistocene times, the original Severn valley axis deepened westwards to beyond a level that is now more than 80 m below OD (Fig. 12), but as post-glacial sea levels rose (Holocene; Fig. 13) the glacial outwash and sediments of the ancestral River Severn were reworked into sandy beaches that migrated towards the present positions. The offshore Helwick 'banner' sand bank[13], extending from Port-Eynon Point to far southwest of Worms Head and constantly reworked by the tides, is of similar origin (Fig. 11 and Section 6.6).

The ancestral Severn and its tributaries widely eroded quartz-rich rocks while reworked glacial material also became relatively enriched in quartz, because clay particles were winnowed away. Consequently, Gower beaches are dominantly of quartz-rich sand, with the hard quartz grains abrading the softer rocks of the limestone platforms beneath currents and waves. Thus, the sandy beaches and dunes of Gower, e.g., Figure 20, do not directly reflect the limestone of the adjacent platform being cut today. Pebbles, cobbles and boulders of the high-energy coastline, however, can predominantly comprise limestone, with some reworked glacial erratics, and these clearly aid in platform erosion and in becoming rounded themselves must yield limestone sand and silt grains. Such fine-grained liberated limestone particles, with large surface-to-volume ratios, are particularly susceptible to dissolution in the surf zone near to where they form, which, again, contributes to the fact that the sands do not directly reflect the rocks eroded around Gower beaches.

Sand dunes are very much an attractive backdrop to many Gower beaches, forming a distinctive part of the landscape. However, their formation, migration and disappearance, in times ranging from immediately post-glacial, through early human occupancy, catastrophic medieval storms and into living memory, are worthy of a separate later account (Chapter 6). Now we should get back to explaining the less mobile land features of rock.

Figure 20. View east across Nicholaston Burrows (SS522879) towards Little Tor showing re-configuration of the vegetated dune sands and drainage of Nicholaston Pill that has occurred in approximately 170 years. Vegetated dunes evidently surrounded Little Tor in ca. 1850 although these were gone by 1894, when mapping by the Ordnance Survey showed a close similarity to the modern configuration. The limestone-dissolution grooves (microrills) shown in Figure 17 most probably formed under the relatively stable sand of these earlier dunes (see also Section 6.4). While the coastline here is limestone, the beaches and dunes are predominantly of quartz sand. (Top image John Dillwyn Llewelyn; National Library of Wales archive).

2.7 Lundy: a vital insight

Much of the landscape history discussed so far has involved changes of *relative* levels between the land and the sea. Pioneer geomorphologists over a century ago recognised and attempted explanations for these changes, long before understanding of the ups and downs of global (eustatic) sea level, discussed later, and before the advent of plate tectonics and the dynamic-Earth model[14]. Even with these latter revelations, it is only the fuller appreciation of deep-Earth processes found in the past few decades that has enabled creation of a robust(ish) 'theory of everything' for our purposes; the hidden land-surface mover has been revealed and tests of its existence and function have been passed satisfactorily, it seems, so far.

Somewhat ironically, a clue for our understanding of the changes of relative levels has always been in our sight, 50 km away, until it rains. Lundy island (Fig. 12), mostly composed of igneous (magmatic, originally molten) rocks, is especially important to understanding the landscape evolution of Gower and its surroundings, because the measurable absolute age of its origin and its regional setting provide direct evidence of the timing and causes of considerable regional uplift and erosion and, ultimately, subsidence. We must now temporarily divert for a far-ranging journey in time and space...

Lundy rocks are parts of an ancient volcanic magma-plumbing system; magma is molten rock. This system originally linked the molten-rock source at great depth in the Earth to magma-filled chambers at shallow levels and to eruptions that built a volcano at the surface – Lundy Volcano. The island is mainly of granite cut by narrow sheets (dykes) mostly of basalt, and nearby related intrusions beneath the sea floor include a large igneous-rock complex and regionally extensive dykes. The Lundy granite must have formed at a *bare minimum* depth of 1 km below the contemporary surface[15] and age-dated samples show that the volcano was active in Late Paleocene times, around 58 million years ago (59.8 ± 0.4 – 57.2 ± 0.5 Ma; Charles et al. 2017).

The obvious erosional removal of the volcano and other rocks originally overlying the granite at Lundy is clear direct evidence that there has been very substantial erosion since about 56 Ma, although it probably initiated earlier while the volcano was active. Lundy provides perhaps the most graphic evidence for the Bristol Channel seabed and coastal southern Wales having been deeply eroded down, by at least 1 km, long before Quaternary times and any glaciation or marine planation then (Fig. 21).

Lundy is a southern outlier of a vast volcanic region that was originally centred in the area of western Scotland, north-eastern Ireland and east Greenland, but extended more widely. This huge area of numerous volcanoes was formed from magmas derived by partial melting of hot mantle rising from deep within the Earth, during the interval from 62 Ma until about 52 Ma (Fig. 21) and peaking between 61 Ma and 58 Ma. The rising hot and buoyant mantle, known as a 'mantle plume', had two important physical consequences for our understanding of the landscape and seascape development.[16]

Firstly, there was dome-like dynamic uplift over the rising plume. The slowly upwelling mantle pushed the Earth's crust upwards over an area more than 2000 km in diameter, with the greatest uplift above the plume axis, which lay beneath northwest Britain and east Greenland. The amount of uplift diminished towards the plume edge, which, for our sector of interest, lay south of southern Wales and Lundy.

The second consequence was that magmas from the partially melting mantle plume were added beneath (technically 'under-plated') and intruded into the Earth's crust, while also forming substantial volcanoes at the surface. The hot materials added beneath and into the crust contributed to its overall buoyancy and involved forming granite of relatively low density, all of which contributed substantial uplift. Thus, Lundy records considerable Paleogene uplift that increased northwards. It does not, however, directly explain the following subsidence and we must look even farther afield to understand how that happened.

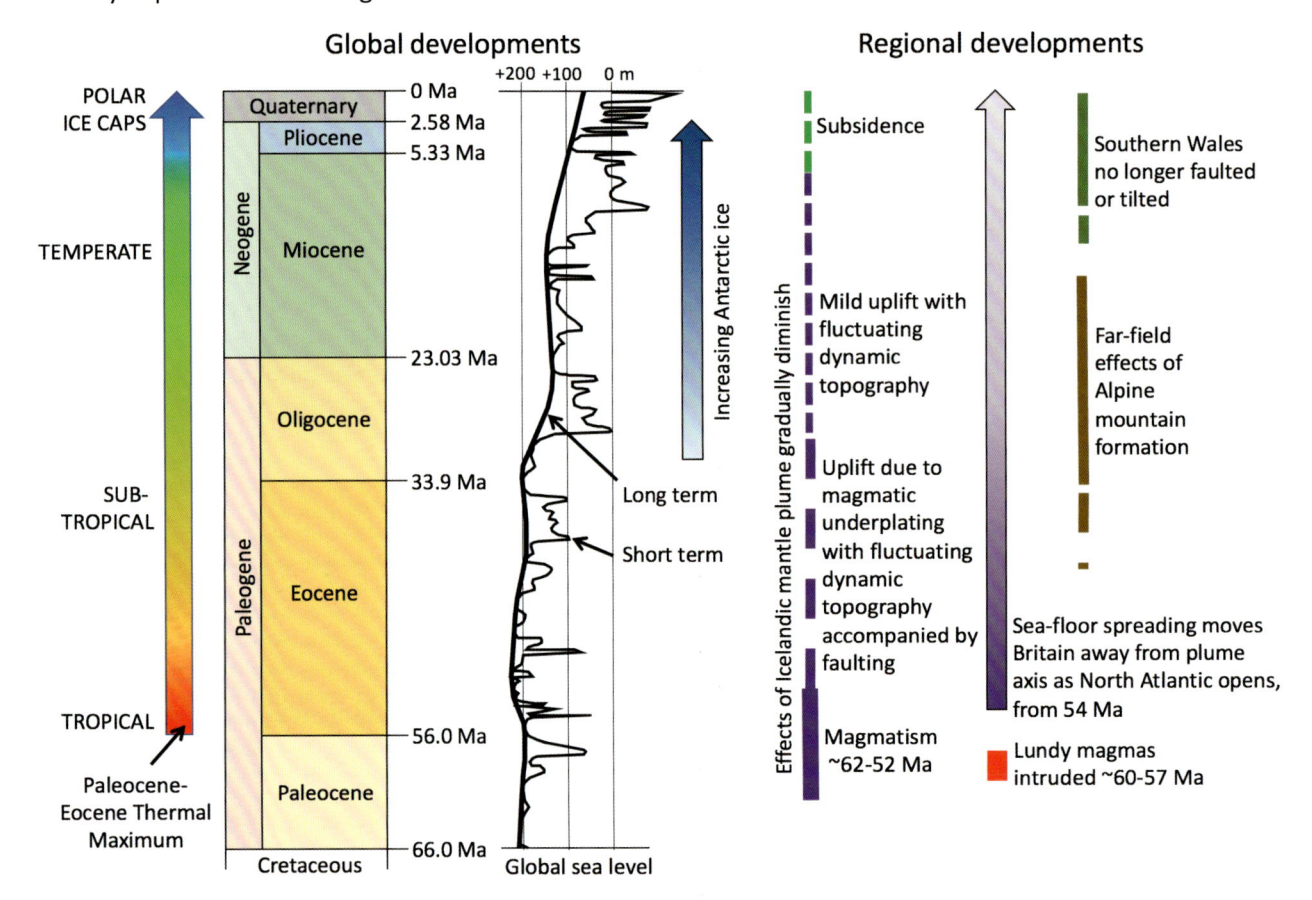

Figure 21. Inter-related developments that influenced the topographic evolution of southern Wales. Late Paleogene through Neogene times are regarded as 'Icehouse', when climatic temperatures progressively diminished. Through this time the global sea level fell; the long-term trend mainly reflects the increasing average depth of the oceans as seafloor crustal temperatures diminished, and the short-term spiky lowerings of sea level directly reflect fluctuations in on-land ice volumes and cooling-contraction of ocean waters. The long-term trend assumes ice-free conditions and shows that sea level would on average be some 70 m higher than now without ice on Earth. The global sea-level curves are after Haq et al. (1987). Other authors have produced slightly different curves and while the absolute amplitudes of the short-term variations may be too great (e.g., Miller et al. 2011), the general form and frequencies displayed by the original influential plots are fine for our purposes.

The mantle plume that in Paleogene times lay beneath northwest Britain and east Greenland is the same huge feature that now underlies Iceland on the Mid-Atlantic Ridge. It is simplest to think of mantle plumes as very long-lived and anchored deep in the Earth, widely believed to be at, or close to, the surface of the Earth's core some 2900 km down; they move their anchor very little whereas the Earth's crust is continuously mobile above them as tectonic plates. These plates, especially where heated and stretched in domes over plumes, as here, are susceptible to being torn apart along extending rifts. After the growth of Lundy Volcano, the North Atlantic Ocean started to open by rifting and formation of oceanic crust, making new sea floor between Greenland and Britain.

Because of seafloor spreading after about 56-53 Ma, the crustal plate on which Britain is situated gradually drifted towards the edge of the plume and thus gradually lost its original dynamic uplifting support (Fig. 21). Further, the hot magmas that had been added to the Earth's crust here cooled and crystallized so as to become denser and less buoyant, which also contributed to subsidence. These plume-related uplift and subsidence processes are now considered by far the most important for understanding the Paleogene and Neogene landscape evolution of Britain. Important advanced-research ('difficult') papers that explain this include White and Lovell (1997), Jones et al. (2002), Al-Kindi et al. (2003), Jones and White (2003), Tiley et al. (2004), Davis et al. (2012) and Schoonman et al. (2017). Here is the first time this research has been applied to understand the Gower landscape.

Previously, the far-field effects of the mountain building that formed the Alps and Pyrenees, some 1000 km away, were considered as possibly having caused widespread uplift in Britain. Alpine deformation started in Eocene times and climaxed after about 34 Ma, during Oligocene to mid-Miocene times (Fig. 21). It did cause folding of strata in southern England and some movement on faults in the Bristol Channel area (e.g., Sticklepath-Lustleigh and West Lundy Fault Zones; Tappin et al. 1994), but it could not account for the place, scale or timings of the vertical motions that affected western Britain, e.g., farther north along the eastern Irish Sea where there was 1.5-3 km of uplift with attendant erosion (Jones et al. 2002). From Late Miocene times onwards (Fig. 21) there has been only en-bloc (all together, coherent) vertical movement and general tilting of southern Wales and the Bristol Channel area.

Now we should understand that our region was dramatically uplifted for a duration sufficient for deep erosional planation, e.g., removal of Lundy Volcano and more, and that subsidence necessarily followed. A profound remaining effect due to the mantle plume influence having been greatest in the north is that western Britain, including Wales, was gently tilted, mainly southwards. Part of this tilt results from there having been less Paleogene crustal thickening towards the south, but even today there is some tilt attributable to the buoyancy of hot mantle from the Iceland Plume (Figs 21 and 22).[17] Gower is just a small part of the regional picture and fits into it precisely. Next, we should try to understand when the sea played its part.

2.8 Relations with sea level

We know that when Lundy Volcano formed, in Late Paleocene times, the climate of Earth was much warmer than now and there was no polar or other significant accumulation of ice; this has been regarded as 'Greenhouse' time and it ended with the Paleocene-Eocene Thermal Maximum (Fig. 21). Melting of all the ice on Earth *now* would raise *average* sea level to about 70 m OD (~61 m from the melting of Antarctic ice and ~7 m from Greenland ice), and during the Last Glacial Maximum, 26,500-22,000 years ago when ice reached onto Gower, sea level *globally* was about 125 m lower than now (Fig. 13). Evidently, we live in 'Icehouse' conditions with our modern sea level still some 70 m below the ice-absent maximum.

Very simplistically, ice-volume-related variations of *absolute* sea level give us some 200 m to play with for relative change, whereas mantle-plume-related variations of the *absolute* land level could have approached x7.5 greater for relative change of sea level in southern Wales. Conceivably this was up to 1.5 km, given the 1.5-3 km of uplift to the north in the east Irish Sea area (Jones et al. 2002). The deep erosion of our area can only be explained by involving considerable absolute land uplift.

Supported by the underlying mantle plume (edge) and by the buoyancy of the thickened and relatively warm crust, our area of southern Wales and the Bristol Channel sea floor to the south must have been elevated *many hundreds of metres* higher than any sea level of that time. Conditions then were subaerial, tropical to subtropical and wet, so that weathering and erosion could proceeded quite rapidly, with the products being washed away by rivers into local basins or into the distant sea. We can reasonably anticipate that Lundy Volcano and southern Wales were subjected to 'efficient' erosional processes of peneplain formation from Paleocene times onwards, because of the uplift. Thus, as we found earlier, our landscape levels are fundamentally ancient; Quaternary modifications by ice and seawater were relatively superficial.

Now for another twist. An additional attraction of the plume-magmatism-related hypothesis in explaining the topographic evolution of southern Wales lies in the rather difficult physics of fluid dynamics. Explained simply, any hot mantle that rises buoyantly in a plume and spreads out beneath the rigid outer layers of Earth[18] can be considered, in geological timescales, as a fluid flow. The overall form of the buoyantly rising mantle would be something akin to an enormous bizarre mushroom, with a head more than 2000 km in diameter and, in the Iceland Plume case, with radially directed fingers rather than a continuous circular outline (Fig. 22). Suffice it to say that continuously flowing fluids like the hot mantle in a plume do not move steadily through the cooler mantle, but rather form series of waves or blobby pulses, similar to those in an ornamental 'lava-lamp'. Thus, from Paleocene times until the significant drift of Britain off the Iceland Plume, uplift attributable to the plume can be expected to have occurred in pulses. This has been confirmed where sediment accumulation reflecting uplift-related erosion has been found to occur in pulses (White and Lovell 1997). Thus, the stepped nature of our upland landscape, which suggests unsteady uplift, now can be reconciled with unsteady mantle flow.

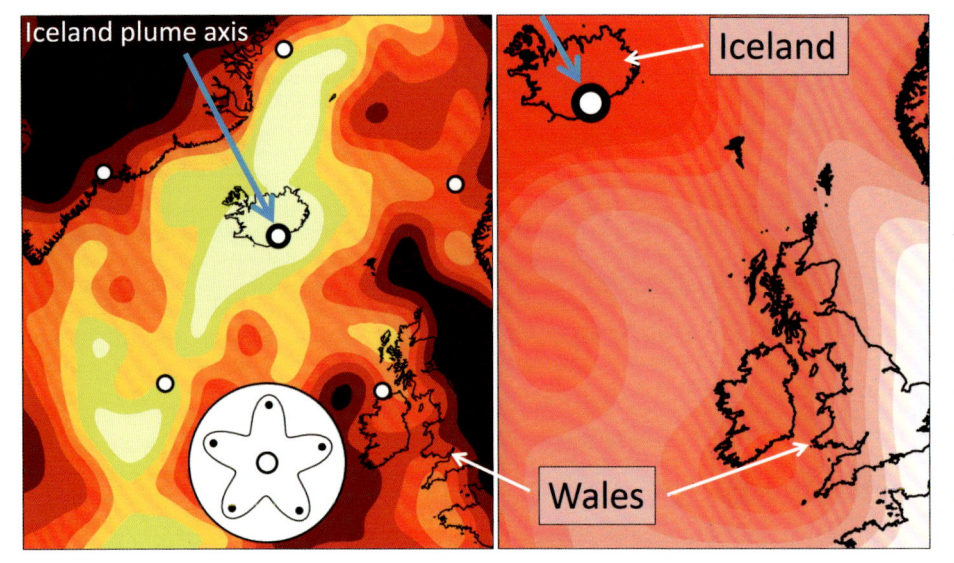

*Figure 22. (**Left**) Southern Wales today lies near the end of a southeast-directed 'hot finger' emanating from the Iceland Mantle Plume. South Wales is 1500 km from the plume axis. The inset **black on white pattern**, relating to the five small circles on the map, is the schematic top-view form of a mantle plume with radially spreading 'fingers'. The feature is defined by relatively high mantle temperatures, which diminish with radial distance and hence also relate to a decline of dynamic and buoyant uplift.*

*(**Right**) Shades of red show variable crustal uplift due to the underlying relatively hot and thus buoyant, slowly flowing mantle. The uplift decreases towards the east under Britain. Both the previous active uplift and the residual uplift now relate to gentle southwards and south-eastwards tilting of southern Britain. The original volcanoes of western Scotland and the north of Ireland are deeply eroded at the present surface, whereas the modern London Basin is 'simply' tilted eastwards. (The maps depict computed geophysical models of mantle viscosity and Earth's gravity, respectively taken to represent temperature and dynamic topographic uplift. They are azimuthal polar projections centred on Iceland, originally published by Schoonman et al. (2017) with graphical modifications kindly supplied here by Hannah Galbraith-Olive and Nicky White of Cambridge University.)*

2.9 When and why the sea came up

Absolute sea level, i.e., relative to our notionally fixed centre of the Earth (Section 2.2), varies according to changes of (1) total ocean-basin capacity, and (2) total sea-water volume.[19] This measure is called global sea level, also known as eustatic sea level. Provided that the absolute level of some land area remains constant (which it rarely does), global sea-level rise and fall respectively cause shorelines to migrate up and down, onto and off that land. Absolute sea level fell during Late Paleogene and Neogene times, in parallel with global 'Icehouse' cooling (Fig. 21). The long-term falling trend mainly reflects increasing ocean-basin capacity as seafloor-rock temperatures diminished. Seafloor spreading diminished during this time, so there was less relatively buoyant warm oceanic crust flooring the ocean basins.

The short-term, spiky trend (Fig. 21) mainly registers the total volume of ocean waters, with the 'glacio-eustatic' falls of sea level directly reflecting increased on-land ice volumes ('grounded ice'), as well as contraction of ocean water due to general cooling. There are additional more subtle controls that need not concern us (e.g., Rowley and Markwick 1992; Miller et al. 2011). Ice-volume changes have been one of the primary controls of absolute sea-level change during the so-called 'Icehouse world' of the past 33 million years.

Antarctic ice accumulated significantly from about 33 Ma and substantially from about 14 Ma, being then a major driver of global sea-level fall (Fig. 21). In Pleistocene times, from about 2.6 Ma, this effect was added to by the growth of northern polar ice caps on land. This is recorded by rock debris dropped from ice bergs in the North Atlantic and Arctic oceans (Jansen et al. 1988). Through all of this time, for our area of interest, we can anticipate lessening of both the amount of plume-related uplift and the crustal buoyancy, which would result in subsidence of the land from the maximum level attained. So now we must consider the 'relative sea levels' as distinct from global absolute levels.

From Late Miocene into Pliocene times the *absolute subsidence of our land surface overtook the global absolute lowering of sea level, so that the land experienced a relative rise of sea level* and hence submergence of the lower part of the subaerial series of peneplains. Consequently, in Late Pliocene times, marine sediments were deposited within estuaries occupying former river valleys and shorelines briefly were high on the former subaerial surfaces.[20]

A Late Pliocene palaeogeography of Gower is suggested in Figure 23. The limited records of marine sedimentation and planation as far up as 75 m above present sea level show that there was only brief influence of the *maximum* sea advance in modifying the higher parts of our coastal scenery. While marine planation is profound at and around the 61 m level, no really obvious notch or cliff line was formed at or just below 75 m OD (Fig. 23) and weathered or loose soils and sediments occurring there were not completely scoured away.

After that Late Pliocene maximum high stand of the shoreline, relative sea level fell sharply during Pleistocene times. *Ice-related global sea-level fall then overtook the diminishing plume-related subsidence, switching the previous relationship.* Today we can see records of sea level having been far higher than present, perhaps up to 75 m higher, because we remain influenced by existence of substantial land-based ice in Antarctica and Greenland. Although glaciers have retreated from Gower and sea level has risen, we are still in the global 'Icehouse' conditions of the Quaternary Period, albeit in a relatively warm interglacial-like stage, known as Holocene.

We should now see that the overall form of the regional surface, including the major broad valley of the ancestral River Severn (Fig. 12), was formed subaerially before Early Pliocene times (5-4 Ma). That surface then became partially submerged and erosionally planed along coasts, up to what is now 75 m OD, in Late Pliocene times.

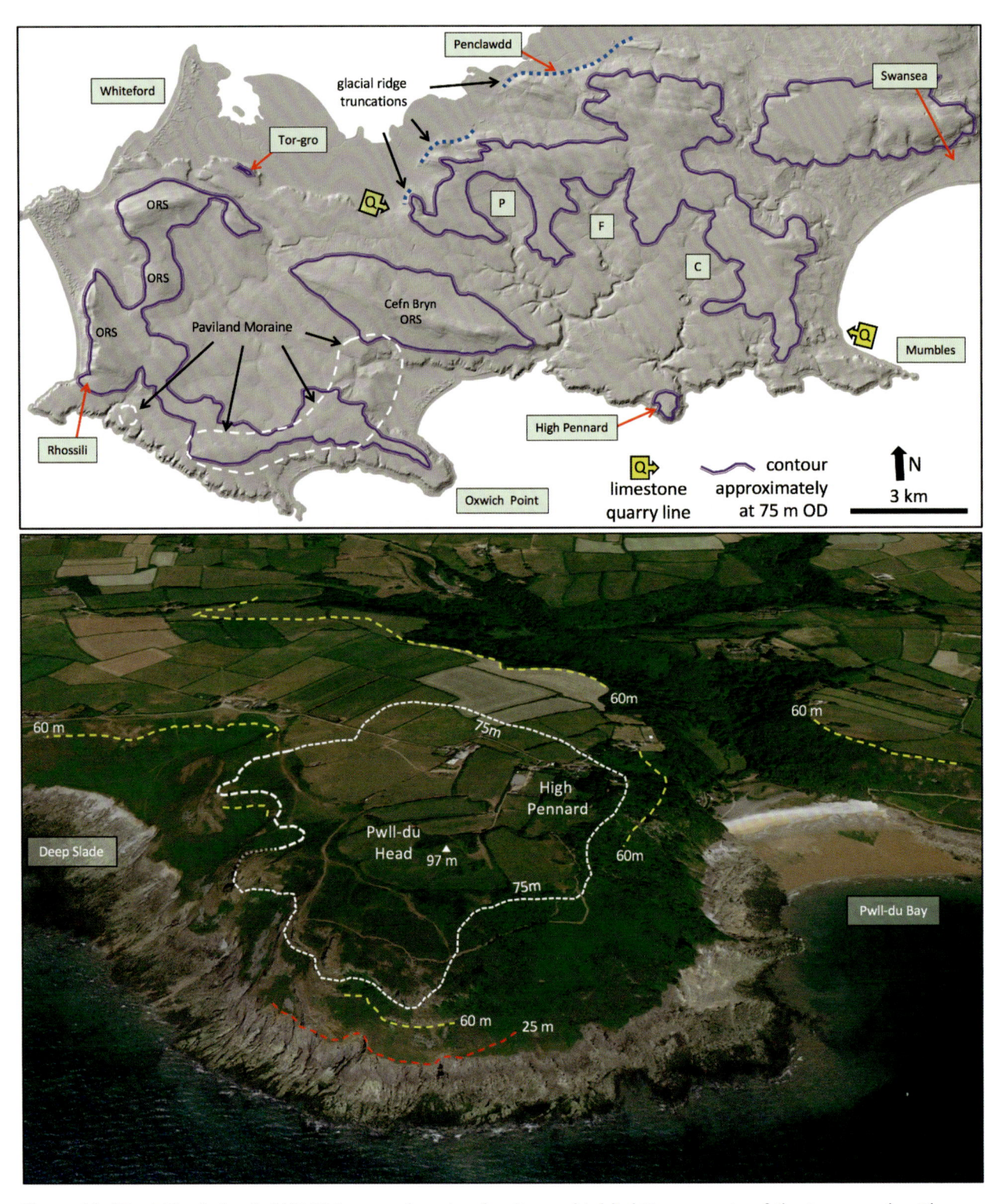

*Figure 23. (**Top**) Shaded-relief LiDAR image of peninsular Gower highlighting aspects of the topography. The **purple contours** at approximately 75 m OD indicate roughly the higher parts of the surface that perhaps were beyond the reach of marine planation and were lowland parts of the subaerially weathered and eroded peneplain. One might imagine this as a short-lived archipelago of islands lacking cliffed shorelines. (continued...)*

Cefn Bryn and the western summits of Old Red Sandstone (ORS) probably were prominent landscape features for considerable geological time.

*The '**limestone quarry line**', **Q-Q**, indicates the string of numerous old and more recent quarries in the upper parts of the Carboniferous limestones succession; it roughly marks the top of the limestones beneath the overlying mudstones and sandstones to the northeast, which form catchments draining into the limestones; **P** is the Pengwern catchment, **F** is Fairwood and **C** is Clyne. The Late Pliocene marine planation surface is only well developed south and west of the quarry line, on the limestone. The few limestone outcrops possibly not planed by the sea are High Pennard, the ridge of Oxwich Point, Tor-gro and an area west of Mumbles. The Paviland Moraine (from Shakesby et al. 2018), with the remnant proved farther west (this study), formed upon the marine-planed surface; it is a late addition to the topography. Similarly, the truncations of the sandstone ridges of northeast Gower (**blue dotted lines**) are late sculptures due to glacial ice coming from upland Wales. (Base data © Natural Resources Wales).*

*(**Bottom**) Oblique Google Earth view north over High Pennard and the seaward end of Bishopston Valley. Like the purple contour of the image above, the 75 m OD **white contour** distinguishes high ground that may have remained beyond the reach of waves during the Late Pliocene marine planation of the southern Wales coastal fringe. On the high ground above the general marine planation level, here at 60 m OD, there is no really convincing horizontal notch or cliff line that might have registered a still-stand of sea level as it rose to about 70-75 m. Topographic steps in the higher ground appear to relate to the variable durability of the limestone strata. Most of the summit above 75 m OD, as well as the eastern slopes, are of the Oxwich Head Limestone, which is relatively robust and on Gower forms some of the prominent headlands. However, there is evidence that both Anglian and Devensian glacial ice crossed the summit here and that could have shaved off some subtle features, respectively about 450,000 and 23,000 years ago (Chapter 3).*

2.10 The headless valley problem

Here we re-visit an old problem: the problem of the slades. To sustain the reader in this discussion, it may help to know that resolving this issue bears crucially on understanding the time of development of some valleys and caves on Gower. It is perhaps worth introducing the problem via the writing of TN George (1932, p. 315), who was grappling with the age and origins of the various platforms and plateaux on and around Gower. He wrote, in classic scholarly prose: "It is manifest, therefore, that the minute form of the coast of Gower was impressed upon it at a very early date by agents of denudation compared with which the erosive forces of later times have been insignificant." He then continued rather desperately: "The problem need not be considered at length, as no semblance of a solution is apparent; but it may be indicated that the chief difficulty rests in an explanation of the deep gouging necessary to produce the steep cleanly-cut valleys, rarely more than a few hundred yards long, that run from the plateau level to the shore; while at the same time the plateau itself remains largely free of any notable modification by denudation subsequent to its initial planation."

Slades are the short dry valleys that cut into the flat-topped cliffs; good examples occur between Oxwich Point and Horton (Figure 24), and also at Ram Grove (SS430867; Chapters 3 and 4). They characterise the limestone coast while their platform hinterland, although widely blanketed by superficial deposits (drift), has sinkholes that testify to subterranean drainage.[21] Closely similar features also occur in the steep sides of the more extensive limestone valleys, e.g., in the Bishopston Valley inland from Pwll-du (Fig. 23). On the platform, short streams that are only active during storms typically disappear rapidly into the limestone and do not flow in the slades.

Dry valleys and sinks are characteristic of 'karst' – the name given to distinctive limestone ground that has been subjected to long-term dissolution by slightly acidic water[22]. 'Pavements' of variably regular round-edged blocks (clints) separated by deep open cracks (grikes) normally characterise karst at the surface, but on Gower these are mostly buried by glacial deposits.

Because the water exploits fissures to form underground drainage, ultimately via potholes and caves, it is difficult to understand the dry valleys in terms of normal surface erosion. They have been attributed to collapse of former caverns, but here there is little direct evidence for this. Common fracture control of the valleys, via faults and joints, suggests that freeze-thaw weathering in Quaternary times could be involved, but, as we will see, the dry valleys existed before the Ice Age.

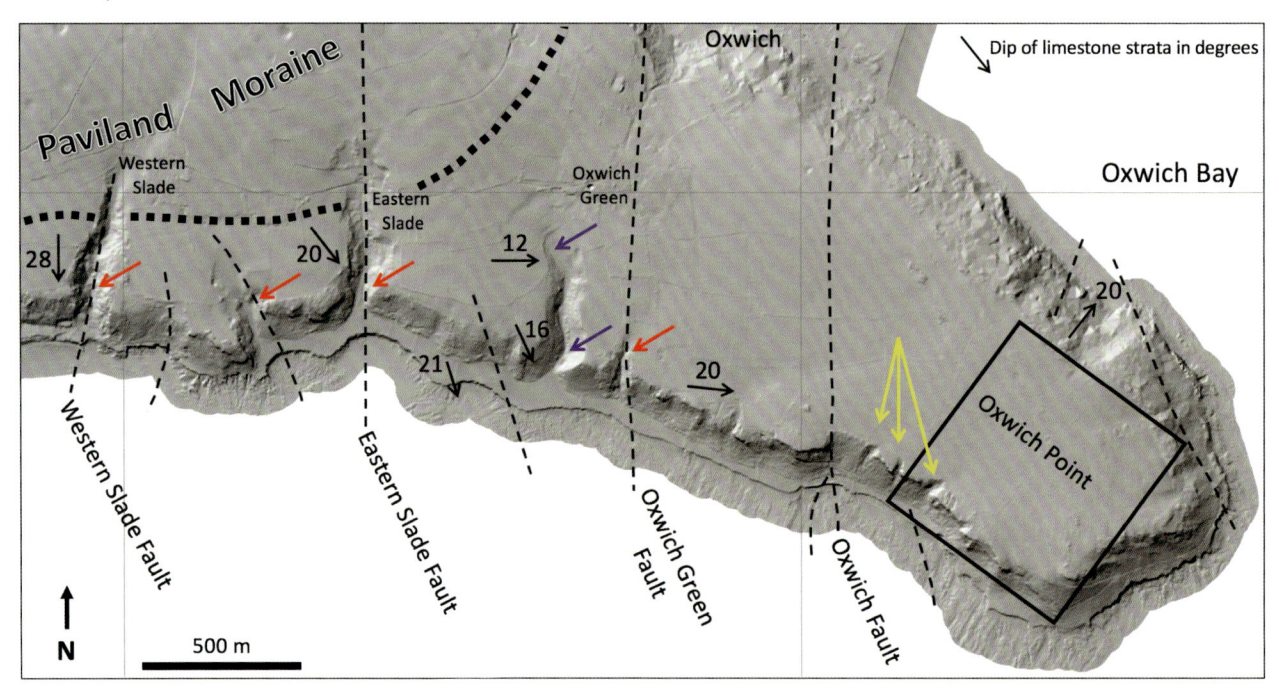

*Figure 24. Shaded relief LiDAR image of Oxwich Point. (Base data © Natural Resources Wales). The box (lower right) marks the area of the middle and bottom photographs of Figure 14. The platform above the cliffs widely has a thin veneer of Anglian till, although the feather edge of the Paviland Moraine extends towards Oxwich Green (**dotted-line** limit from Shakesby et al. 2018, figure 5). The south-coast limestone cliffs are fringed by a low apron of debris that extends upwards into the lower parts of short dry valleys, the 'slades', and towards the coast this apron is truncated at a low cliff on top of the beach platform. **Red arrows** mark slades formed on faults, the **purple arrows** mark a sinuous slade with no obvious fault control, and the **yellow arrows** mark pronounced headless notches possibly formed on bedrock fractures (joints). Pockmarks in the superficial deposits (e.g., top left) are kettle holes where buried ice blocks melted (Chapter 3).*

The problem that concerned TN George was in the evident cutting of 'deep' valleys into the platform cliffs where there appears no headwards continuation via which some agent of the erosion could have been delivered. He recognised that the platform was planed-off "at a very early date" – in fact the Late Pliocene marine high stand described above – but he could not place the dry-valley erosion. In resolving this it perhaps helps to consider the slades as one type of limestone dry-valley erosional features in a range of types. At one end there are long and branching valleys that deeply cut the platform for several kilometres, some largely dry (e.g., Green Cwm) and some with semi-persistent streams (e.g., Ilston Cwm). Then there are the 'true' short slades (e.g., at red and purple arrows in Fig. 24, and Deep Slade east of Southgate in Fig. 15), and at the other extreme there are the isolated cliff-edge notches (e.g., at yellow arrows in Fig. 24). The longer valleys clearly have existing upstream rainwater catchments that are not on the limestone (Fig. 23 Top), and we consider their formation elsewhere in relation to the dry-valley caves (Section 4.7). Many of the slades clearly developed on structural weak zones in the limestone, commonly N-S faults or joints (i.e., fractures with no offset), but it appears that some are not. Similarly, not all slades are completely headless. Importantly, and of course not available to be seen by TN George, LiDAR imagery allows us to see that some slades have significant headwards continuation into the platform, albeit perhaps subtle and masked by drift (Fig. 25). Furthermore, some slades have offshore continuations (see below).

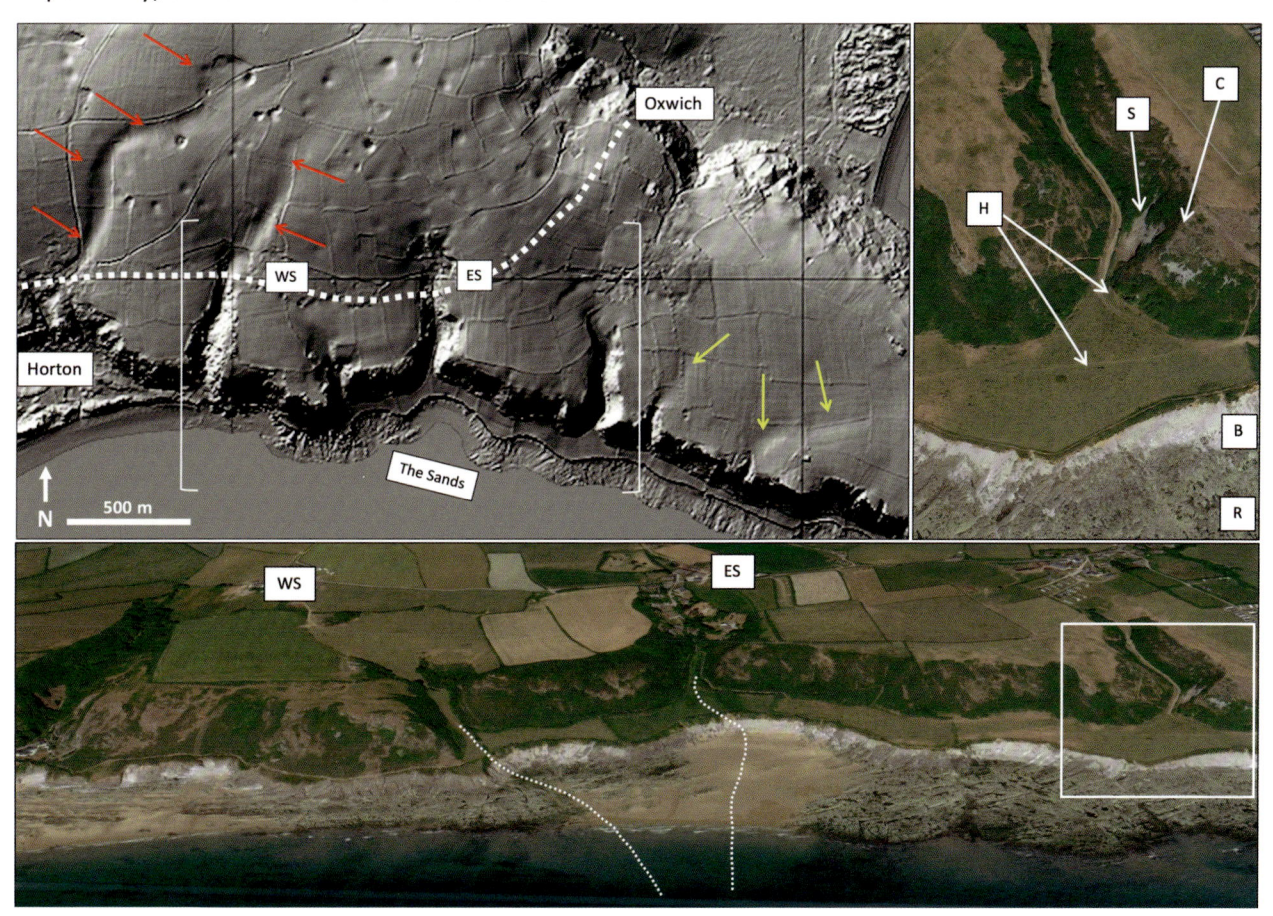

*Figure 25. (**Top left**) Strongly shaded relief LiDAR image of the Oxwich Point to Horton area (SS487854). (Base data © Natural Resources Wales).* **Red arrows** *show the continuations of valleys within (beneath) the Paviland Moraine (**dotted-line** moraine limit from Shakesby et al. 2018 figure 5) and the* **yellow arrows** *indicate subtle continuations beneath thin (weathered) Anglian till. Pockmarks indicating kettle holes are prolific in the moraine, appearing less prominent towards its edge and scarce beyond the supposed limit.* **White bracket**s *indicate the limits of the bottom image.* **WS** *and* **ES** *mark the hamlets of Western and Eastern Slade.*

   *(**Bottom**) Google Earth oblique image centred on The Sands (SS487854). The two **dotted lines** mark traces where the slade valley rock floors (buried) originally continued across the present shore. The other slades, on either side, show no trace through the modern rock platform, indicating that they were not cut so deeply. The* **white box** *marks the location of the top-right image.*

   *(**Top right**) This slade apparently lacks fault-control, as evident from the rock platform beneath it. Screes (**S**) beneath cliffs (**C**) on the dry-valley side pass down into a fan of slope-reworked scree plus soil, described as head (**H**). This deposit is exposed at the shoreline where it rests on limestone breccias and reworked till (debris flow deposits), in turn underlain by silty sands, cemented sandstones, and ancient beach deposits (Patella Beach; Section 2.14). This section is variably developed just above the modern storm beach (**B**) or rock platform (**R**).*

The simplest solution to the perceived problem of the slades is that they are remnants of minor karst valleys that existed *before* the Late Pliocene marine planation. Thus, the upper reaches of those valleys that originally were more substantial, e.g., at Horton, Western Slade and Ram Grove, were shaved off by the marine planation to leave only the subtle vestiges that TN George, understandably, could not see. One might visualize an original degraded limestone-cliffed massif with breaks and ravines of various scales recording karstic weathering, commonly marked along faults and joints. This massif would have been clear to see from the banks of the ancestral River Severn, with the Old Red Sandstone peaks behind. Thus, the slades are ancient, the bare-rock elements of which probably were cut by Early Pliocene times (5-4 Ma), like the ancestral Severn valley. This has some serious implications for the age of some Gower caves, as we shall see next.

2.11 The perforation of Great Tor

Great Tor (SS529876) is one of the iconic coastal features of Gower and it presents critically important insights into the landscape evolution. Viewed with the benefit of understanding the considerable antiquity of the slades, and with the services of a drone-borne camera, the striking promontory shows several key features (Figs 26 and 27). Great Tor is perforated by a plexus of modest-sized cave passages that all lead into open air and must have formed before its erosional isolation. The passages show features typical of phreatic development, having formed fully flooded with fresh water beneath a water-table surface that existed at least several metres above them. The scale of the passages suggests a moderately extensive original cave system, now almost all lost to erosion.

The Great Tor phreatic cave passages (known as Leather's Hole) must have formed *before* cutting of the slades and their flanking cliffs, which inevitably would cause them to become drained. The slades formed before the Late Pliocene marine planation 4-3 million years ago (Section 2.10), so they must be at least 5-4 million years old, formed by Early to Mid-Pliocene time. Thus, the phreatic passages in Great Tor must have formed long before 5 million years ago.

The time of existence of an ancient water table above phreatic caves is a recurrent issue in developing an understanding of the dry-valley caves of Gower. Leather's Hole in Great Tor tells us that there was an ancient limestone massif with karstic features and underground drainage long before the slades reached their current depths, which, in turn, was long before the Late Pliocene cutting of the coastal platform surface.

Although we know of late Triassic pothole formation (Fig. 10) in the limestone massif that would form Gower, there is no evidence that the slades or the caves in their flanks, as described here, existed so long ago. After the formation of caves in Late Triassic times, the region was covered at least by Jurassic rocks, like the Vale of Glamorgan, so cave development could not resume in the Carboniferous limestones until they were exposed again by erosion (Fig. 2 and Section 2.4).

Ancient Leather's Hole and many other Gower caves existing at the same time would have been flooded with water again when the sea reached up to 70-75 m OD during the Late Pliocene marine transgression[23]. The evidence from Pembrokeshire and High Pennard (Fig. 23) is that the sea did little erosion at its highest stand; subaerially weathered, soft material was not stripped off and there was no high mark, e.g., any notch or small cliff, sufficient to survive the subsequent Pleistocene sculpture and dressings by ice. Hence the maximum marine incursion over Gower may not have been long-lived, but re-flooding of caves like Leather's Hole at 35 m OD would have been protracted.

Following the Late Pliocene marine transgression, Pleistocene lower sea levels cut the cliffs and platforms at and beneath the present beach level (Section 2.12). The Great Tor promontory, like High Pennard, must have been influenced by the Anglian ice sheet that reached across the Bristol Channel about 450,000 years ago, when melt water might have penetrated the caves before and after any complete freezing. Clearly, the promontory form was largely complete by 123,000 years ago, because its east flank is notched by the Patella Beach high stand at about 10 m OD (Fig. 26; Section 2.14).

The passage walls of Leather's Hole are ornamented with rather decrepit calcite flowstone and they are strangely etched near the entrances (Fig. 27). A sample of the flowstone yields a U-series radiometric age of 426,250 ± 14,380 (Section 4.10), while the etch patterns may relate to times when Lundy island was not visible from Gower.[24] The cave was excavated by Colonel ER Wood (1819-1877), who found remains of fox, wolf, hyaena, red deer, horse, woolly rhinoceros and mammoth; one imagines that the latter were brought in there by scavengers, in bits.

Figure 26. Great Tor juts out between Tor Bay (behind) and Three Cliffs Bay (foreground), and its summit lies not far beneath the Late-Pliocene-cut '61 m' sloping coastal platform. At 35 m OD Leather's Hole is a remnant of a branched cave system that initially formed beneath a fresh-water water table. The cave passages must originally have been connected with fresh water within a continuous body of rock, long before the bays formed on either side. The bays are of the same ilk as the slades discussed above; Three Cliffs Bay is the outlet from Green Cwm and Ilston Cwm, while Tor Bay might have been classed as a slade had its inner reaches not been overwhelmed by blown sand (Section 6.2). The cave must pre-date the deep cutting on either side and thus it registers subaerial weathering with karst development before Early Pliocene time, substantially more than 5 million years ago. The view also shows a sea-level-cut notch of the Ipswichian (MIS 5e) 123,000-year-old Patella Beach at about 10 m OD. (Drone photo courtesy of Andy and Antonia Freem).

*Figure 27. Leather's Hole, Great Tor (SS533873). **(Top)** Eastern entrances and view from there towards the east over Three Cliffs Bay. The view emphasises the antiquity and erosional isolation of the cave when one realises that it formed below a water table of a fresh-water cave system that once must have continued extensively within solid rock. **(Bottom)** The passages, initiated more than 5 million years ago and perhaps as much as 10 million years ago (Section 4.10), are 'decorated' with calcite speleothem ('flow-stone') and a peculiar karstic dissolution of the limestone near the entrances, perhaps reflecting wetting from atmospheric mists. (Top left drone photo courtesy of Andy and Antonia Freem).*

2.12 Cutting the modern coast

We know that in Quaternary Pleistocene times there were many dramatic fluctuations in ice volumes – the so-called 'glacials' of maximal ice and 'interglacials' of diminished volumes – such that 'glacio-eustatic' sea-level changes dominated coastal developments (Fig. 21). The fluctuations were synchronised according to subtle cyclic variations in Earth's orbit around the Sun and the tilt of its rotation axis, which caused variations in solar influence and consequently variations of temperatures and ice volumes at the poles (Fig. 28).[25] Thus, the southern Wales fringe and the Bristol Channel area were the sites of a number of varied marine incursions and withdrawals, with associated trimmings of the surface. These are not simple to unravel.

Only once during Pleistocene times did our present land area become obviously modified by sea level higher than today. This particular high stand is recognised globally (technically known as Marine Isotope Stage 5e) and it reflects a brief mild interval – the Ipswichian interglacial – at around 123,000 years ago. On Gower it formed the widely developed coastal notch and narrow platform at around 10 m OD, with an associated 'Patella Beach' deposit, as described below (Section 2.14).

The especially cold intervals (glacials), on the other hand, were characterised by especially low sea levels and generally freezing conditions, with glaciers directly impacting Gower twice. First was the Anglian glaciation, about 450,000 years ago (interval of 478-424 ka), when ice reached over Gower, across the Bristol Channel area to Somerset, Devon and Cornwall, scouring over Lundy. Second was the Late Devensian glaciation with the so-called Last Glacial Maximum some 23,000 years ago, when ice just reached onto and locally over Gower (Chapter 3). Thus, the obvious Pleistocene features on peninsular Gower are one high-level notch with 'beach' and two sets of glacial and periglacial deposits.

We need to look at the cliffs and offshore again to consider what else might have happened in Pleistocene times. Given that the Ipswichian interglacial involved sea level rise to around 10 m OD, with cliff erosion, it might be anticipated, perhaps naively, that the three interglacials that followed the Anglian glaciation, with temperatures higher than now (Fig. 28), could have had sea levels near to or coincident with the present coast of Gower. This view is supported by various analyses of absolute (global) sea level fluctuations, which show the post-Anglian through Ipswichian spell as having the highest levels (e.g., Spratt and Lisiecki 2016). Possibly this was the time when the main coastal cliffs were cut.[26]

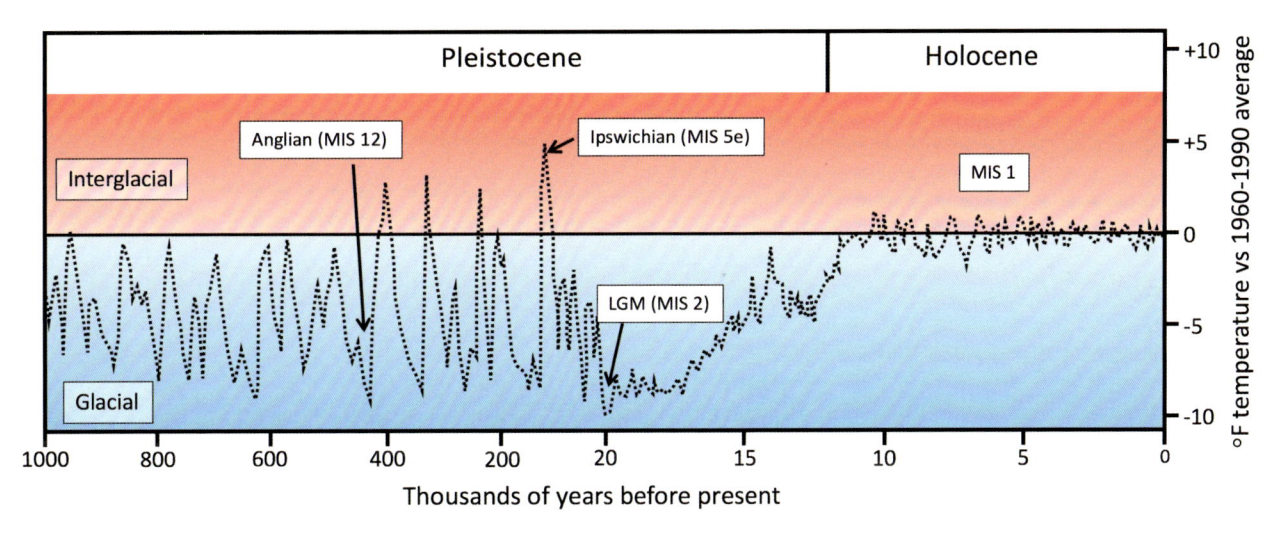

Figure 28. Global temperature fluctuations of the last 1 million years; note the timescale change at 20,000 years. The Pleistocene trace is a hybrid of Antarctic and Greenland data modified to represent global conditions (Lisiecki and Raymo 2005; Jousel et al. 2007). Global temperature estimates over the ~12,000 years of the Holocene are from Marcott et al. (2013). LGM is Last Glacial Maximum; MIS is Marine Isotope Stage.

Offshore surveys of the seabed reveal that the tidally exposed rock platform of our present coast locally continues gradually for over 1 km to about -15 m OD, at which depth it steepens down to about -25 m. Beyond this the rock surface slopes more gently again, into deeper water and locally buried by sand. Thus, there is a cliff line submerged offshore (Figs 11, 29 and 30).

The submerged platform and cliff, and some seabed limestones beyond (Fig. 11), show erosional patterns etched into the limestone bedding, faults and joints, just as is seen on the tidally exposed parts. Furthermore, the submerged cliff shows gullies and narrow valleys practically identical to the slades that cut the on-land limestone platform (described above). Thus, there are (at least) two sets of cliffs: one with its top edge at about -15 m OD, and the other, more familiar to us, along the southern coast of Gower typically rising on average to around 60 m OD. The close similarity of the limestone erosion in the two sets of cliffs and platforms (Figs 11, 29 and 30) emphasises the fact that modern sea level obscures the history of the submerged surfaces; what we see now is just the upper part of an ancient massif.

Although it is possible that the submerged cliff between -15 m OD and -25 m OD was cut during one or more Pleistocene interglacial episodes, it seems conceivable that it is part of the first slope that was cut by the subsidence-related rise of sea level and marine planation that reached about 70 m OD in Late Pliocene time. In either case, viewed from near the ancestral River Severn during the latest Pleistocene low sea level, one would see the steep edge of the proto-Gower limestone massif standing proud, with two sets of cliffs interrupted by ancient gorges and valleys and topped by two uniformly gentle slopes. Behind this would be prominent hills of Old Red Sandstone, Coal Measures and Pennant sandstones.

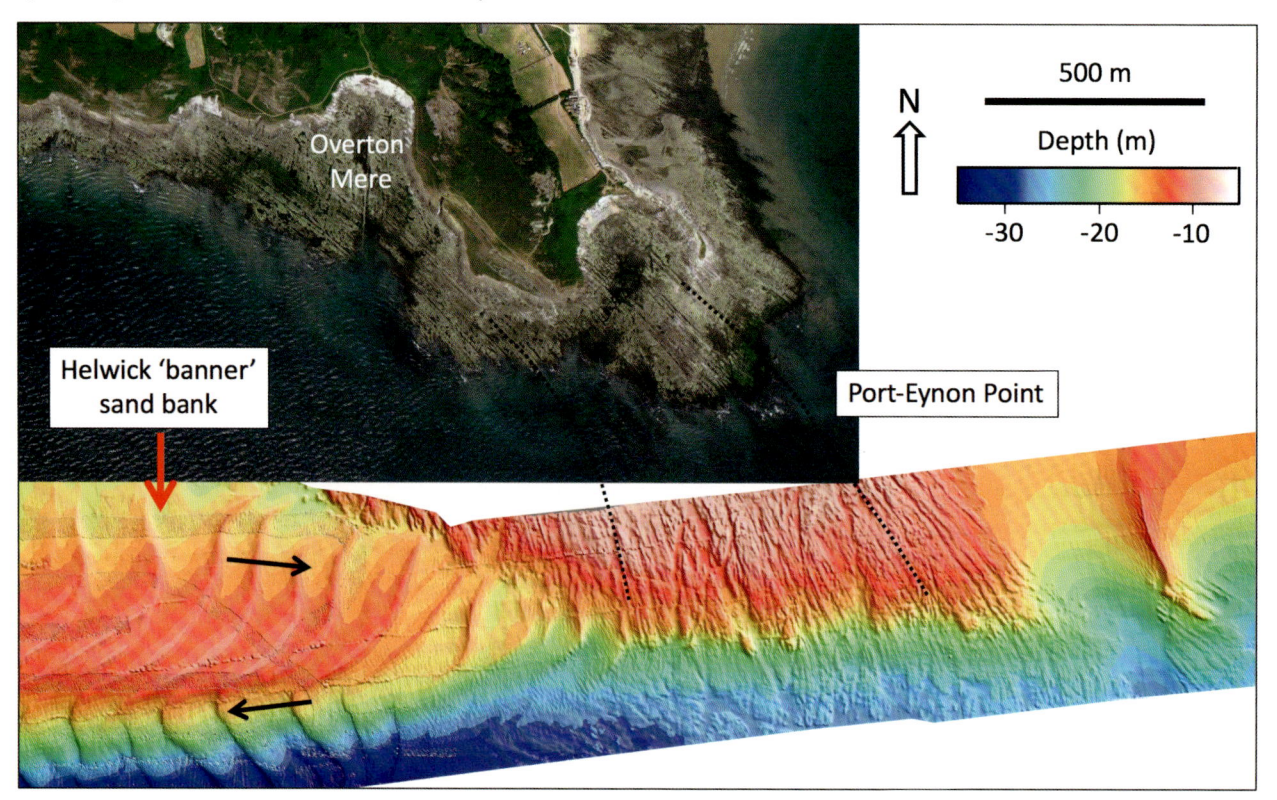

*Figure 29. Google Earth view of Port-Eynon Point combined with a seabed survey produced by high-resolution multibeam echo-sounders. The two images are placed together at the same scale to render subaerial-to-submarine continuity (**black dotted lines**). The coloured relief image shows the easternmost end of the Helwick sand bank resting across the submerged cliff, at around -15 to -25 m depth, and the edge of the rock platform above it, which is about 1 km wide off the point. The panel shows a change of bedding strike from NW-SE on land to around N-S offshore (close and parallel continuous curved features). A prominent 'grain' of somewhat irregular eroded faults and joints on the shallow rock platform cuts the bedding and trends mostly close to N-S. Note the scale of the sand waves on the Helwick bank. The bank is generally 20-25 m tall and the sand waves register strong eastward currents on the north side during incoming tides and westward currents on the south side during ebb tides (**arrows**). (continued...)*

LANDSCAPES CUT IN TIME

Figure 29 continued. Low tide at Overton Mere (bay) emphasising the existence of two marine platforms. The limestone bedding here dips northeast at 45° and the horizontal trend of beds – the bedding strike – is parallel to the coast in general. (Offshore survey published by Schmitt and Mitchell (2014), with supplementary detail kindly supplied by Neil Mitchell. Helwick bank overall is 13.5 km long and up to 2.7 km wide (Fig. 11); it is considered in detail in Section 6.5).

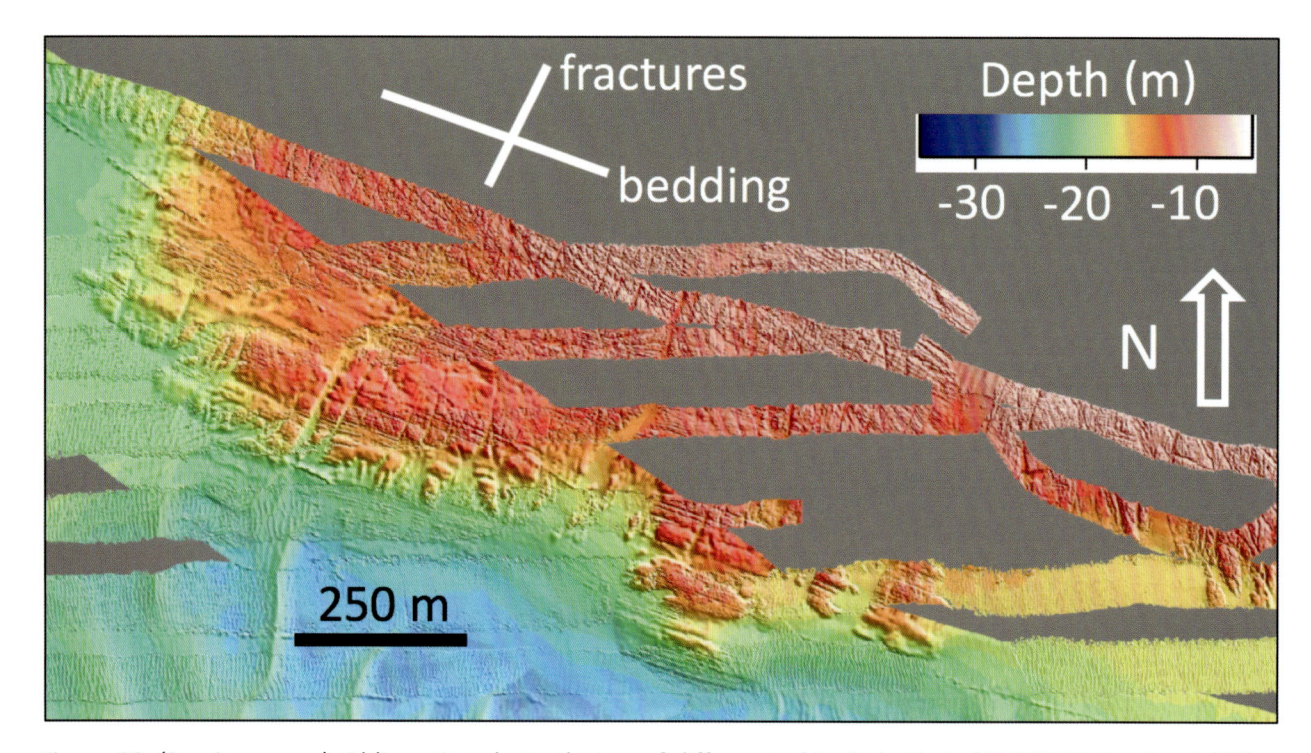

Figure 30. (Previous page). Oblique Google Earth view of cliffs east of Foxhole Slade (SS437859; Paviland Cliff on extreme left) showing various fracture sets cutting across the limestone bedding, as seen offshore. The upper slopes are widely mantled by various superficial deposits of screes and reworked glacial debris.

*(**Above**) Rock platform up to 1 km wide due south of the coast between Paviland and Longhole caves (SS4485), showing an irregular margin with a relatively steep drop from about -15 to -25 m, which is the submerged cliff. The erosional relief of the submerged cliff and platform is similar to the on-land cliffs and platform where the slades are well developed but partly obscured by superficial deposits (e.g., Fig. 24). The fractures occur in sets, mostly trending a few degrees on either side of north; they dip east or west or vertically, as evident in the cliffs behind. The image utilises two echo-sounder surveys; the ribbons are data collected along individual boat tracks. Diverse scales of sand waves beyond the rock platform reflect tidal currents mainly running west to east before the surveys. (Survey image kindly supplied by Neil Mitchell).*

2.13 Loads of ice

Let us step aside for a moment to consider some Earth physics that has been a complication in understanding platform landscapes. The key technical term is 'isostasy', which considers the level of the ground according to the loading and buoyancy of the Earth's crust 'floating' on the denser but slow-flow-adjusting mantle beneath, known as the asthenosphere. During ice-sheet advance on land, the ground surface absolute level responds to the new load of the overlying ice. Ice loading causes the Earth's crust to be pushed down over a wide area, by displacing underlying (asthenospheric) mantle material. And when in deglaciation the ice load is removed that crust slowly rebounds buoyantly, in a timescale of thousands of years. When this occurs near coastlines the rebound can form so-called 'raised beaches'; an earlier beach is raised above sea level as the ice-sheet load diminishes.

Ice loading and rebound complicate interpretation of relative sea-level variations because they tend to mask global absolute changes; it is difficult to gauge the load due to ice that has long-since disappeared and the buoyancy of the crust on the mantle has to be inferred. We will, however, not be troubled to unravel the potential intricacies of marine planation versus ice advance and retreat here.

Although southern Wales was within or near ice limits during two glacial episodes, for *most* of Pleistocene time the Bristol Channel area and Gower were far from the major icesheet loads that advanced and retreated farther north; these areas remained above sea level with mainly cool temperate or periglacial subaerial landscapes. Retreat of the few hundreds of metres maximum thickness of Anglian ice that loaded on and around Gower, and the greater thicknesses farther north and west, will have caused some isostatic rebound, but this would not have been clearly registered locally. Similarly, retreat of the thin edge of the Devensian Welsh Ice that reached Gower and Swansea Bay, with only several hundreds of metres thickness farther north (e.g., Lambeck 1993, 1995), would have registered little effect in the valley of the ancestral River Severn relative to the impacts of the marine flooding that would follow (Shennan et al. 2006). In passing it is perhaps interesting to note that the crust beneath Wales, at up to 36 km, is the thickest beneath Britain (Davis et al. 2012), and so, like the Welsh peoples in general, hard to put down and ready to rise up!

Happily, amongst the many complicated ups and downs of Pleistocene relative sea levels, there was only one clearly registered rise to above the present level. It has a fascinating and long-studied record on Gower, and, for our purpose, it helps to distinguish the ages of other landscape changes and indeed cave developments. It resulted from an interglacial mild-climate episode around 123,000 years ago (MIS 5e or Ipswichian times; Fig. 28). Melting of polar ice caused absolute sea level to rise to about 10 m OD, where it produced a coastal notch and where the iconic 'Patella Beach' deposit was laid down. It is perhaps helpful here to warn that in the following section the 'Beach' deposit is reinterpreted as significantly post-dating the notch. It is not in itself a true beach deposit recording the high stand, as has been thought, but instead it represents an exceptional climatic event. The deposit and its relationships are described first, referring to its original epithet 'Patella Beach'; a summary and the synthesis leading to the new interpretation of it then follow.

2.14 Rough times on Patella Beach

Patella Beach on Gower refers to cemented fossiliferous deposits and an underlying gently sloping platform or a notch carved in the modern limestone cliffs. It was named by TN George, in his 1932 paper, for its typical content of common limpet shells (Patella vulgata). The deposits are everywhere distinctive and quite unlike any other beach sediment of the peninsula coast, irrespective of their cement and obvious antiquity (Figs 31-33). Patella Beach deposits are preserved at numerous localities along the entire rocky south coast, from Mumbles Head to Worms Head, and are composed of various pebble-, cobble- and breccia-conglomerates with matrix of coarse sand or gravel. They are characteristically fossiliferous, in places strikingly so (Fig. 31), with abraded and fragmented shell debris dominated by robust forms of limpet, periwinkle and dog whelk, respectively Patella, Littorina (with Melarhaphe neritoides) and Nucella species. These are creatures that inhabit shores between the regularly exposed high-tide mark and the shallow waters just below the low-tide mark, i.e., the littoral zone.

Patella Beach deposits are invariably cemented onto a rock surface of limestone. This surface commonly shows ancient features of coastal erosion identical to those of the modern marine platform (described above), with gullies, small potholes and various slopes including benches and notches (Figs 26 and 32). Patella Beach deposits occur mostly at around 8-10 m OD but range up to 15 m OD and down to 0 m OD. In some places remnants are traceable sloping seawards over 15 m on and within the present beach platform (e.g., near Tears Point, SS404872; Fig. 32 middle left), and some remnants occur inside coastal caves (Section 4.3). Most of the deposits are within the range of modern tides and storms and widely have resisted wave erosion for the past five to six thousand years since sea level reached back up to them. In some places, however, e.g., around

Pwll-du Bay and Oxwich Point (Fig. 32 top left), the limestones under Patella deposits show relatively deep karst erosion with metre-scale pinnacles akin to inland-exposed ancient surfaces far from sea level (e.g., on cliff edges where superficial deposits are lacking). Evidently the deposits reach high levels where the ancient marine platform transitioned upslope to subaerial karst.

Patella Beach deposits and the erosional features are our local record of the interglacial high stand of sea level that is marked world-wide and occurred around 123,000 years ago (Pedoja et al. 2011; Kopp et al. 2019). The high-stand records are **well known at broadly similar levels across southern Ireland, in the western Mediterranean, and in the Bahamas and Bermuda** (e.g., Hearty and Neumann 2001; Hearty and Tormey 2017; Polyak et al. 2018). They record a brief relatively mild interval during Late Pleistocene times (MIS 5e; Ipswichian), **spanning the interval 128,000 to 118,000**, when there was less polar ice than now.

The Patella Beach of Gower is not a 'raised' beach, as it has been called. True raised beaches, such as those that are well developed around western Scotland, result from real raising of the Earth's crust as it rebounds following deglaciation and removal of huge loads of ice up to 2 km thick. Raised beaches form by isostatic uplift; instead, Patella Beach deposits formed during a global (eustatic, absolute) rise in sea level. Many geologists will recollect the 'old chestnut' examination question: Explain the differences between isostatic and eustatic...

As noted at the outset, Patella Beach deposits are unlike any modern beach deposit on Gower and indeed they differ widely from most in coastal Britain. One obvious difference is in their content of angular limestone fragments. In comparison, today's beaches show more 'maturity' in terms of the daily tidal shoreline reworking and attrition that causes fragments to become rounded. Even at Gower coastal sites of former limestone quarries, as at Pwll-du and Oxwich (Chapter 6), there is little such angular material on the beach. It is easy to understand our modern beaches of well-rounded pebbles and cobbles formed by the persistent mechanical reworking (e.g., Figs 18 and 19), and that rounding manifestly is 'fast' considering that quarrying occurred only in recent centuries. So, what does one make of the relative angularity of so many fragments in the Patella Beach deposits, and their mixture together with other shapes and sizes of debris?

*Figure 31. (Page opposite). Patella deposit lithology. Patella Beach was named (George 1932) for its content of thick-shelled common limpets (Patella vulgata; **white arrows**), although there are also abundant various periwinkles (Littorina littorea and rudis, and Melarhaphe neritoides) and dog whelk (Nucella lapillus), with myriad fragments including other species. Top, middle and bottom left are vertical sections and bottom right is a plan view. Top is from the southwest shore of Oxwich Point (SS504849), middle and bottom are from the east shore of Deep Slade (SS564866). It is common, as here, for there to be gradations from shelly gravels into layers rich in a range of rock-fragment types, predominantly rounded to angular limestone, but also assorted fragments derived by reworking of the earlier, Anglian, glacial deposits. The mix of large particle sizes, the absence of fine sand or silt or mud matrix, the varying angularity versus rounding of rock fragments, and the robust nature of the surviving shells are characteristic. Locally the larger and more tabular rock fragments show 'jamming fabrics', here arranged sub-parallel to one another and at a steep angle to the bedding, as in the inclined stack in the middle image, just above '1 cm'. Mostly the deposits are firmly cemented by calcite.*

LANDSCAPES CUT IN TIME

49

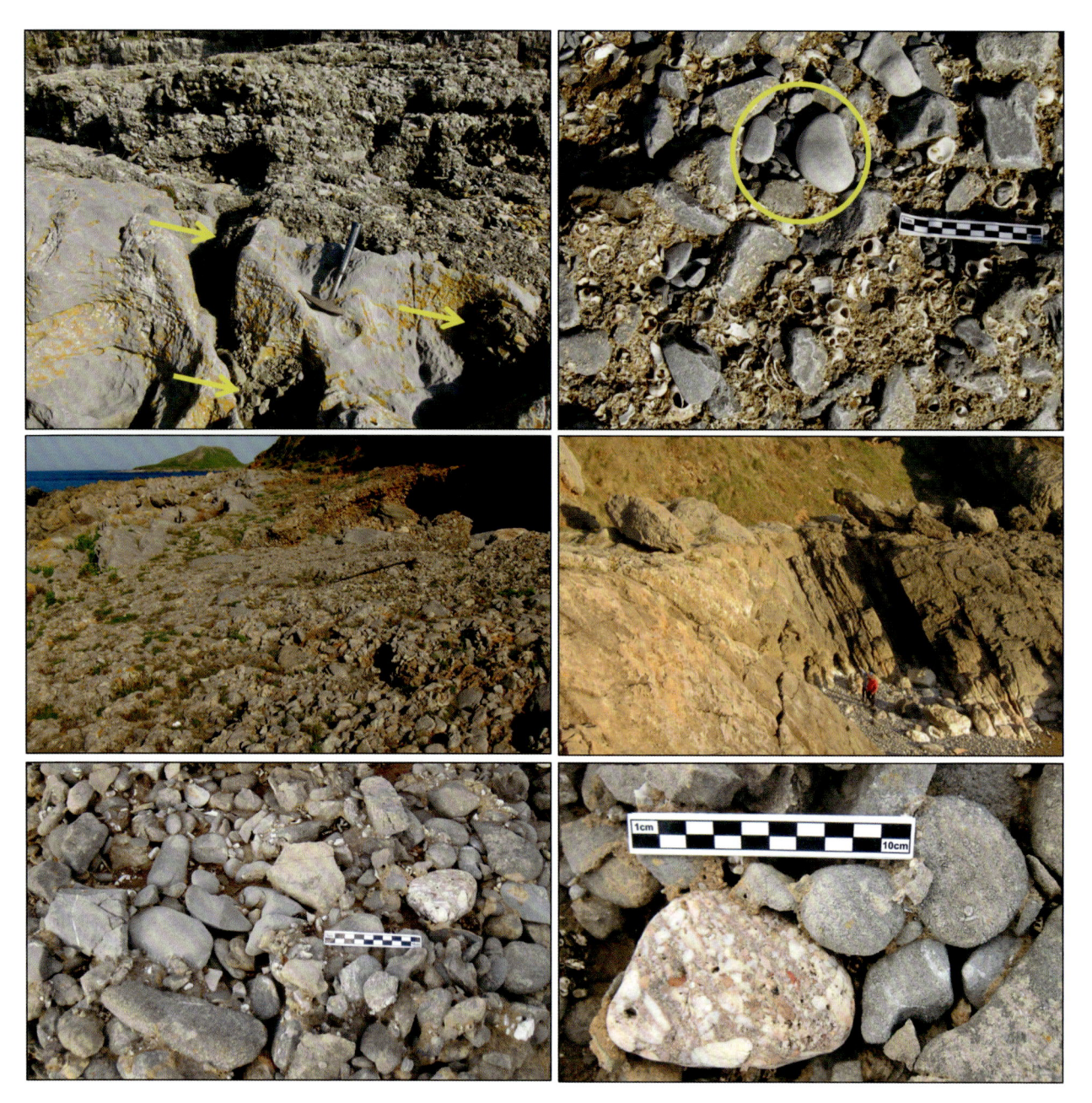

*Figure 32. Patella deposit lithology and relationships with substrate. (**Top**; southwest shore of Oxwich Point) Deposit firmly cemented onto an inclined limestone surface and also (**arrowed**) within deep, karst gullies and holes. This confirms the pre-burial existence of the karstic weathering. The limestone fragments are less rounded than the modern beach pebbles, two of which are **circled**. The deposit is especially rich in limestone fragments near its top and adjacent to a contemporary vertical cliff against which it was banked. (**Middle left**) Extensive 7°-sloping Patella deposit west of Tears Point (SS407871), where it rests on and fills between steeply dipping limestone beds. (**Middle right**) Cliff-backed platform between Pobbles and Shire Combe (SS541877; persons for scale on the modern beach). The boulders on the platform are from above and not simply related to the platform or Patella deposit[27]. (**Bottom**) Weakly cemented Patella deposit on the platform (middle right image), with both angular and well-rounded limestone pebbles and an Old Red Sandstone conglomerate fragment derived by local reworking of Anglian glacial deposits. The mixture of angular with very well rounded fragments and the wide range of particle sizes are distinctive attributes.*

*Figure 33. Patella deposit stratigraphic relationships. Devil's Kitchen (a-c; SS547872) and Watch-house Bay (d-e; SS548872) are indicated, with the headland of Shire Combe beyond. (**a**) Pebbly Patella Beach deposit is cemented over and between steep limestone beds (hammer is 33 cm). (**b**, plan view) The 'beach' deposit is penetrated and gradationally overlain by coarse sandstone with included slightly rounded fragments of Triassic red sandstone (**arrows**). (**c**) The coarse sandstone is over 10 m thick, cross-bedded and banked steeply against near-vertical limestone cliffs of an ancient gully (not shown). (**d**) Overlying the sandstone, or directly resting on the 'beach' deposit, there is profoundly angular breccia of limestone partially cemented and with an orange-brown silty matrix. (**e**) The limestone breccia above the Patella deposits occurs both beneath and above discontinuous layers and lenses of this extremely heterogeneous, indistinctly stratified sediment comprising diverse pebbles, cobbles and boulder set in orange-brown fine-grained matrix. This is reworked and redeposited glacial till, always stratigraphically above (younger than) the Patella deposits where they both occur. The large fragments include red and brown sandstones, quartz-pebble conglomerate, white quartzite and greenish sandstone. All but the red and brown sandstones derive from the north crop of the Carboniferous succession and must have been delivered as erratics to somewhere nearby on Gower by Late Devensian ice (Chapter 3).*

Remembering that the Patella (Ipswichian) interval was bracketed by freezing glacial conditions (Fig. 28), it is most probable that the angular limestone fragments within Patella deposits were derived from the periglacial material that existed on the coast that the Patella sea advanced over. Rather obviously, the *later* highly angular limestone breccias that lie *above* the Patella Beach deposits (e.g., Fig. 33d) were formed by frost shattering of exposed rocks, i.e., by the intense thermal and consequent mechanical stressing typical of periglacial freeze-thaw landscapes, typically forming screes; technically, they are 'thermoclastic' breccias (temperature-broken). Thus, we may infer that the angularity of the material *within* the Patella Beach deposits records advance of the sea on a coast where there was abundant *previously existing* periglacial angular material, which is unlike today's beaches.

The considerable mixture of fragment sizes and the extremely variable extents of rounding, with local imbrication fabrics and only gradational layering, strongly suggest that the Patella deposit was dumped rather quickly with little or no wave or current reworking. It is remarkably uniform and yet extremely 'immature' relative to our modern beach sediments. This begs a reinterpretation.

Quite by coincidence in this study, while examining possible evidence for the 'Killer Wave of 1607', which is attributed to a storm surge coincident with a high spring tide in the Bristol Channel (Horsburgh and Horritt 2006; see Chapter 6), it was found that the Patella Beach deposits higher on the rocky shore include many enormous limestone slabs that had been flipped over by some 90° to become more stable in resisting wave impacts. The Patella-age boulders are considerably larger than the nearby 1607 suspects (Fig. 34) and are interpreted here as consistent with impacts of exceptional waves. This begins to make sense vis a vis the other Patella deposits where the extreme immaturity and apparent rapid dumping would be consistent with storm deposition.

*Figure 34. Exceptional Patella deposits. On the southwest-facing flank of Tears Point, near Worms Head (SS407871), Patella Beach deposits include thick, slabby limestone boulders up to 2.5 m in diameter (**red arrows**) that lie 'flipped over' at roughly 90° to the in situ underlying beds from which they are derived (pole in top right and lower left panels is 1 m). Fossiliferous Patella Beach deposit (**labelled P**) extends widely across the sloping 'platform' here and clearly rests upon and between the boulders confirming their similar age. (continued...)*

*Figure 34 continued. The parallel sloping orientation of the slabby boulders, facing southwest, records powerful storm waves that turned them so succeeding waves would dash upon them with least tendency to move them again. The **bottom-right** panel shows three relatively small (ca. 1 m diameter) flipped-over slabs, which are lower on the coast section here and are attributed to the devastating historical storm surge of 1607 (Chapter 6). Evidently the latter event was trivial relative to the power of the Patella-time storm waves recorded higher up.*

There are some interesting historical complications that need sorting out before considering a new interpretation of the Patella deposits, and they originate in the extensively excavated Minchin Hole cave (SS555868). TN George (1932) distinguished there a marine shelly deposit above and thus, he thought, younger than his Patella Beach. It has smaller and different abundances of shell species, and it includes remains of land mammals and birds; he named it Neritoides Beach (after abundant small littoral-zone gastropods, now Melarhaphe neritoides). He found that this deposit locally overlay an 'Ossiferous Breccia', which, as the name suggests, is (was then) rich in bones, here of diverse land mammals and birds together with limestone blocks in an earthy matrix. George (op. cit.) considered the breccia to be a non-marine slope deposit representing a significant interval of time. Since the breccia overlies his Patella Beach, he inferred two beaches and hence potentially two high stands of sea level[28].

Some confusion stemmed from this interpretation of two beaches, Patella and Neritoides, but Sutcliffe et al. (1987) found that laterally the Neritoides layer grades down into the Patella layer while elsewhere also occurring under and partly interstratified with the earthy breccia. Thus, the bone-rich earthy breccia is essentially a wedge-form partly within the Neritoides deposit, which forms a gradational top on the Patella layer.

So, after all, there is only one fossiliferous unit. It has non-marine mammal and bird remains towards its top, widely within marine shelly deposit and also locally in a land-derived breccia (albeit partly an in-cave-collapse deposit). Thus, there was no switching of marine-terrestrial-marine conditions, but instead the shoreface (littoral) sea creatures lived alongside the coastal land mammals and birds, under the same mild temperate conditions.[29] We are now left only to conjure the means by which the marine shelly deposits came to have remains of non-marine faunas mixed into them. Again, this is not typical of a simple beach deposit and the key features together require a reinterpretation.

Key features of the Patella deposits (Fig. 35) are:

(1) They are remarkably uniform in their unusual character across the entire south Gower rocky coast, at between 0 m and 15 m OD. They widely occur on or beneath cliffs at the top of the broad rocky shelf slope that extends down to -25 m OD. Where they occur on cliffs at the back of sandy beaches, the sand there is mostly just a thin cover on a solid shelf. The deposits always rest directly on the limestone with no intervening other material and locally they extend far up into karstic subaerially weathered ground.

(2) The deposits have a typical thickness of up to about 2 m with no sign of any protracted break within them. They are of similar composition whether at the bottom of deep pockets in the limestone or near the top. The abundance of fossil shells relative to included limestone fragments is generally the most variable feature.

(3) The deposits locally show evidence of *en masse* deposition in their imbricate and steep jamming fabrics of large platy fragments; these fabrics respectively show that there was shear and bed-parallel compression within a rapidly depositing thick layer.

(4) On the coast between Shire Combe (SS541876) and Deep Slade (Hunts Bay; SS562868), there is a subtle gradation up into deposit with smaller shells and incorporated bones of land mammals and birds (the Neritoides layer). Here the Patella layer locally contains remains of red deer (Cervus elaphus). The entire fossil content is indicative of a mild climate. **The land faunas have been interpreted as recording**

"a more genial climate than that of the present day" (report by AS Kennard in George 1932).

(5) The Patella deposit top typically shows evidence that it remained subaerial; it widely shows abrupt grading up into wind-driven sand, locally with abundant land-snail shells (e.g., Caswell Bay[30], SS594873) or has land-derived earthy breccia in its top. It shows no sign of reworking by water and is widely overlain by thin soil and/or periglacial debris (Fig. 33d and e).

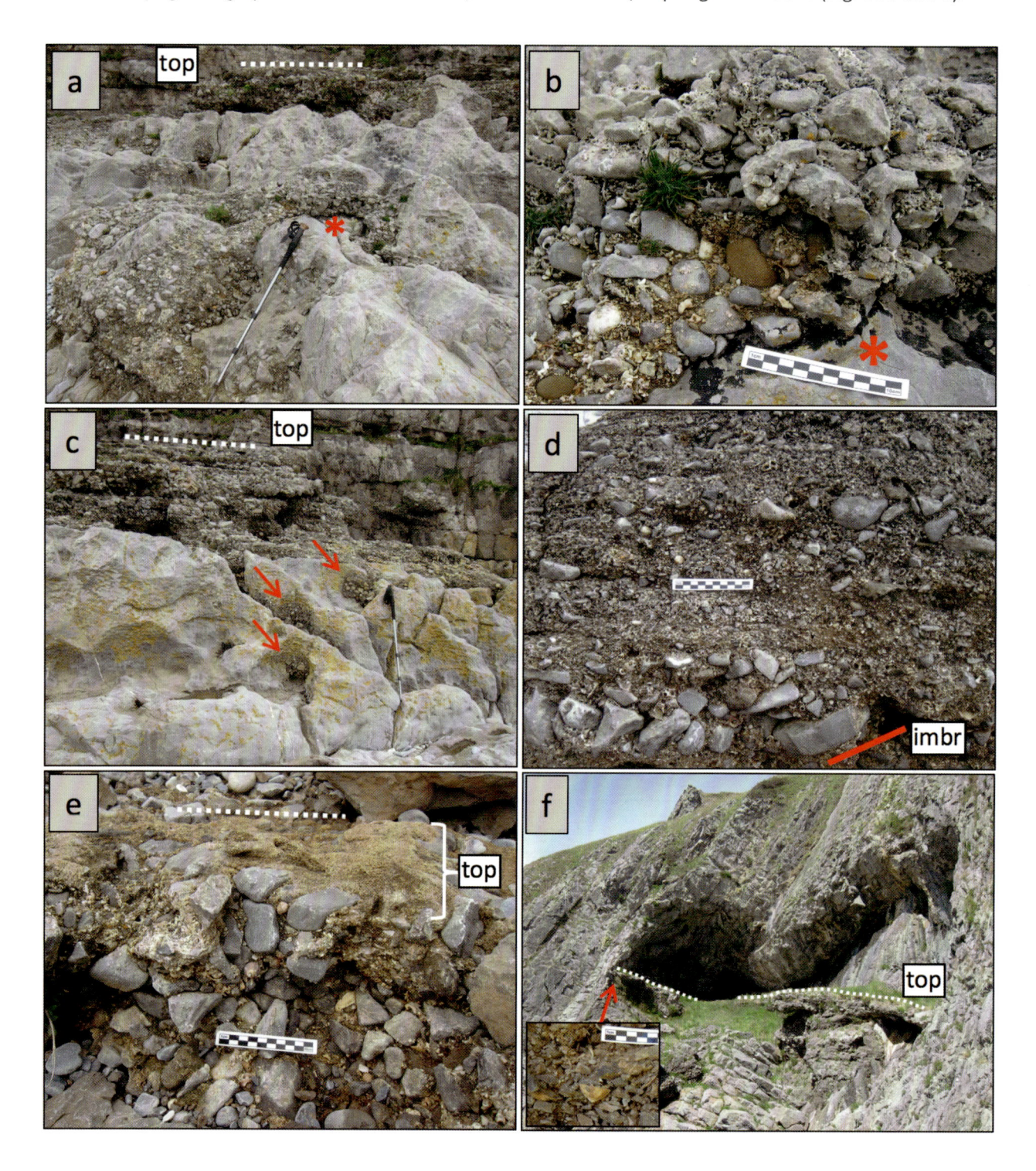

*Figure 35. (Page opposite). A new view of the Patella deposits. In photos (a) and (b) the **red asterisk** marks the same place. (a-e) Southwest shore of Oxwich Point, SS504849; (f) Ravenscliff Cave near Southgate, SS5464687304.*

(a) The 2-metre thickness of Patella deposit here penetrates and buries a deeply karstified limestone surface. The deposit has no other material at its lowest contacts and shows no break other than fluctuations of limestone pebble content throughout.

(b) Predominantly rounded various rock fragments on the lower left side are succeeded by a concentration of angular and subangular fragments towards top right.

*(c) Vestiges of similar deposit adhere to original karstic surfaces (**arrows**), which, like those shown in (a) are <u>unlike</u> the modern beach platform where pockets are variously infilled by rounded fragments where milling-erosion occurs. This karst is typical of the ancient surfaces exposed far above sea level where unmodified by ice.*

*(d) Stratification is locally developed and mostly defined by variation in abundance of rock fragments that show various degrees of rounding and range from platy angular to blocky well rounded. Locally the large fragments, where packed together, show inclined sub-parallel alignment, known as imbrication (**imbr**). This is characteristic of shearing within a rapidly depositing layer, here showing shear from left to right.*

(e) The top of the deposit is sharply gradational into coarse shelly sandstone that in other places grades into wind-blown stratified deposits (e.g., Fig. 33b and c).

*(f) Here a sparsely fossiliferous layer predominantly of reworked scree breccia forms the deposit top, which is overlain by weakly cemented deposits of wind-blown sand that have been excavated within the cave. The top layer on the left contained a single Patella shell (**centre of inset**)[31] and is a remnant of a bank that sloped down into the cavern and was deeply covered by wind-blown sand. The excavation through the sand layer yielded the 'Hippopotamus fauna' that characterises the warm Ipswichian interglacial episode (Sutcliffe et al. 1987).*

The simplest interpretation of the Patella deposits, including the uppermost transitional (Neritoides) layer, is that they are the product of a single exceptional storm with associated storm surge that caused debris to be catastrophically dumped beyond the reach of normal storm waves. The evidence that the dump remained unreworked and subaerial after deposition suggests that sea level at the time was several 10s of metres lower than the high-stand notch at 10 m OD, i.e., at some considerable distance seawards of the present shoreline. 'Normal' storms of today commonly form waves that run up more than 10 m above OD and we know from tragic experience that exceptional storm surges can raise sea level in a wide area to more than 8 m above the normal high tide level.

The large overturned limestone boulders at Tears Point (Fig. 34) attest to an exceptional storm, although those boulders are small compared to some moved in storms. On the west coast of Ireland in the winter of 2013-2014, for example, huge boulders weighing 50-100 tonnes were transported tens of metres by waves that ran out high above high tide level (Cox et al. 2018). So, the wave power may not have been absolutely exceptional, but the singular occurrence suggests that the Patella event was a freak, although others could have occurred later during the further fall of sea level.

Isolated coastal debris ridges due to exceptional storms with associated storm surges are well known and quite common (e.g., Cox et al. 2018). They arise irrespective of coastal slope. The debris is swept up and dumped all together near the farthest reach of large waves that surge onto coasts rather like tidal bores. Consequently, the debris ridges are typically isolated from the fair-weather shoreline by a bedrock surface and are not connected with any beach deposit. Such waves, which would arrive in succession throughout the duration of the storm peak, would tend to strip both the normal shoreface and the run-up zone of loose material, as well as stripping ledges on cliffed coasts.

In this light, it is perhaps simple to imagine how the Patella deposits came to have chaotic mixtures of rounded pebbles, including reworked glacial (Anglian) material, along with highly angular thermoclastic scree fragments and blocks of limestone from the inundated coast, and of course along with ripped off littoral-zone shelly faunas. The Ravenscliff Cave occurrence (Fig. 35f) neatly preserves a bank of reworked scree breccia (with littoral-zone fauna) that faces into the cave where it must have buried land mammal remains before storm-driven sand filled much of the remaining space.

Coasts inundated by run-up waves must also take the less energetic return flow of flood waters, which tend to erode and drag down land materials including, in our case, unfortunate mammals, snails and birds. This would account for the 'Neritiodes' finer grained top of the event deposit, with its content of non-marine faunal remains. Hell must have broken loose in Minchin Hole during the storm, with collapse of internally banked earthy debris and parts of the cave roof. It may be imagined then, after the surge, that exceptional winds could blast coarse sands across the exposed platform and against exposed cliffs.[32]

Now, at last, it should be evident why it is considered here that the Patella *deposits* do not constitute an ancient beach. The notches and platforms at around 10 m OD do register the Ipswichian MIS 5e high stand of sea level, but the deposit was formed some time afterwards during the fall of sea level. It is then something of a coincidence that the Patella storm waves widely dumped their debris load at the notches and platforms previously formed during the high stand. The debris is now reinterpreted as the 'Patella superstorm surge deposit'.

It is, of course, difficult to know the local weather patterns of the Ipswichian interglacial interval around 123,000 years ago. From evidence of robustly dated notches, elevated boulders, storm-sediment banks and corals marking sea levels in the Bahamas and Bermuda, Paul Hearty and colleagues deduced that the maximum high mean sea-level reach was short lived, at about 118,000 years, towards the end of the Ipswichian interval. They found that it occurred just before a rapid fall of sea level that was coincident with "oceanic reorganisation" that led to "massive deep low-pressure systems" with associated storm surges that added >6 m to tide levels, which they referred to as "super perfect storms" (Hearty 1997; Hearty et al. 1998; Hearty and Neumann 2001; Hearty and Tormey 2017).

Today we commonly suffer storms that previously impacted western North Atlantic shores, so we might wonder whether western Britain may have been similarly susceptible in Patella Ipswichian times. The system of **westerly mid-latitude winds of today, variably directed by the jet stream, would have been similar then.** Evidence from the huge flipped boulders near Tears Point (Fig. 34), and from the Patella deposit in general, suggests that indeed there was at least one storm of exceptional magnitude that occurred when sea level was falling.

Deterioration of the Ipswichian interglacial 'mild' climate to the ultimate truly cold glacial conditions, with related dramatic sea-level fall, is especially well marked on Gower by the faunal transition reflected in remains found in the so-called bone caves, including Ravenscliff Cave, Minchin Hole and Bacon Hole. Populations with hippopotamus, straight-tusked elephant, narrow-nosed rhinoceros, fallow deer and sea birds, give way to those with brown bear, woolly mammoth, reindeer and humans, plus the fabulous woolly rhinoceros, which became locally extinct by about 35,000 years ago, and no sea birds (Sutcliffe et al. 1987; Sutcliffe 1995; Dinnis et al. 2019).[33]

Finally, we step away and briefly consider the wider coastal scene during Patella high-stand times. The sea penetrated up a flooded Pennard Ria from a Pennard Bay (now Three Cliffs Bay) to Parkmill and beyond (Fig. 36). The Patella deposits and notches tell us that most of the bays and promontories around 123,000 years ago were virtually as now, from Mumbles Head right around through the tidal breach of Worms Head. Soon, with sea level falling rapidly, it would be freezing again. Eventually, some 100,000 years later, the Welsh Ice advanced onto Gower for the Last Glacial Maximum (Fig. 28).

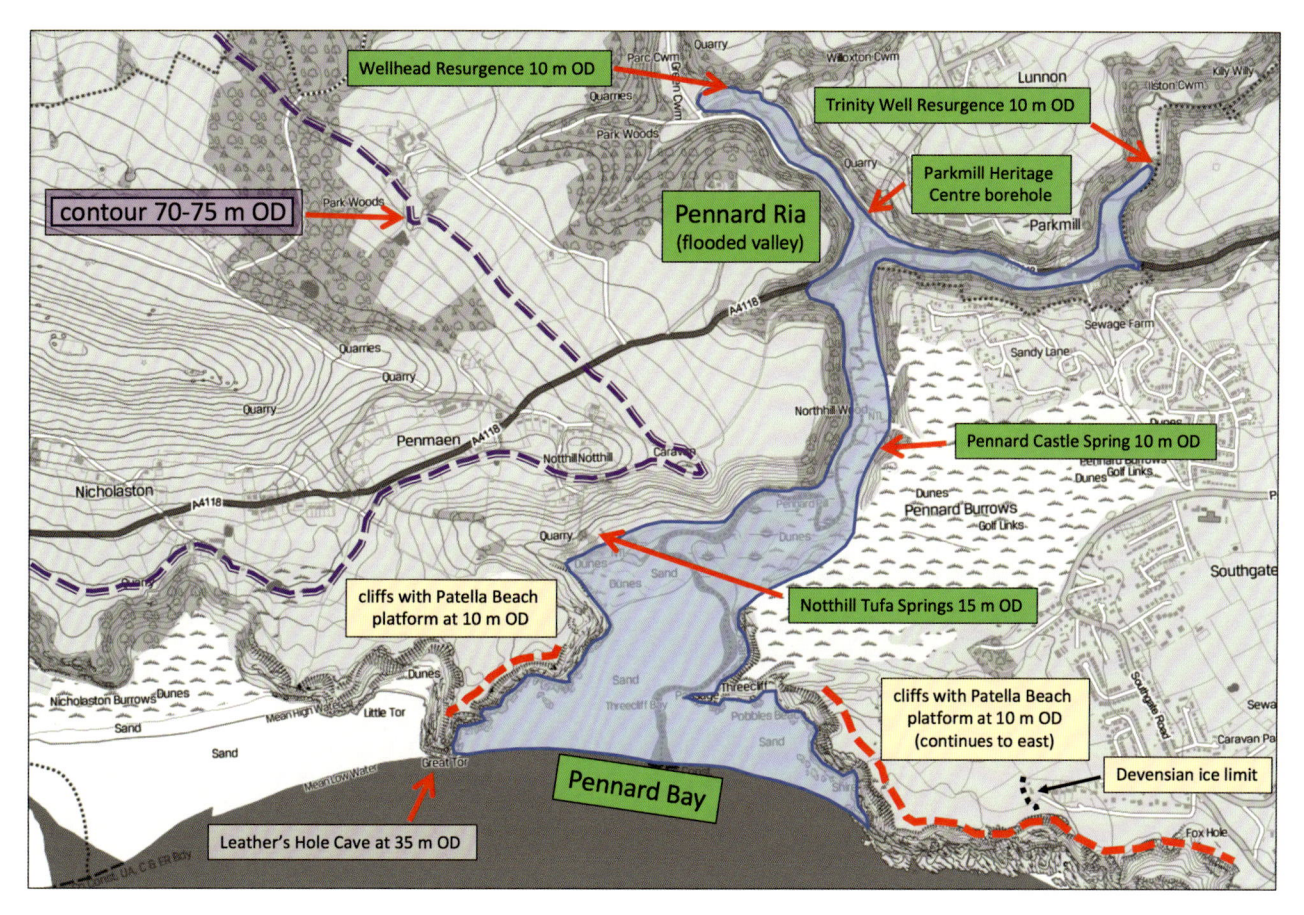

*Figure 36. Coastline and hinterland around Three Cliffs Bay (SS536879) showing aspects of the Ipswichian high sea level around 123,000 years ago. Pennard Bay then was the entrance to Pennard Ria (**blue**) when sea monsters could have swum into the lower reaches of Green Cwm (to Wellhead) and Ilston Cwm, and enormous land snails browsed happily nearby in the relative warmth with hippos, rhinos and elephants. The **purple dashed line** shows the tentative Late Pliocene high-stand shore at 70-75 m OD (see also Fig. 23). Like the other Gower valleys including the 'slades,' some form of this valley existed long before Late Pliocene times. (Map © Natural Resources Wales).*

2.15 More storms and cement

At last the sea returned to make Gower a peninsula. In Heatherslade Bay (SS553871) and Watch House Bay (SS549872), between the high and low-water marks (at 0 m OD), there are patches of cemented beach deposits (Figs 37 and 38). The constituent pebbles, cobbles and boulders are dominated by rock types that were delivered to Gower by glaciers during the Last Glacial Maximum, which here was about 23,000 years ago (Chapter 3). During that glaciation the sea coast was far out beyond the Bristol Channel area and after it, during deglaciation, marine conditions returned fairly rapidly at first and then only gradually up the broad and shallow valley of the ancestral River Severn (Fig. 12).

The Heatherslade conglomerate is no more than about 1 m thick and between the stones is sand with abundant small shell fragments, the whole being robustly cemented together with a calcite (calcium carbonate) cement (Figs 37 and 38). At 0 m OD the cemented conglomerate has at least locally been able to withstand modern storms. Apart from its distinctive rock types and the cement, it is physically closely similar to the modern storm beach and its base is not at a significantly different level or slope. Like modern deposits with rock fragments that have any flattish shape, the fabric is imbricated, that is, with the shapes aligned upwards away from the direction of waves and strong currents that would flip them over if they faced the other way. Also indicative of storm activity, the shelly debris is intensely fragmented.

*Figure 37. Cemented beach deposit in Heatherslade Bay (SS553871) with roughly 40% limestone fragments (**l**), some with calcite veins, mixed with some 60% comprising rock types typical of the Late Devensian, Last Glacial Maximum till; freshly broken, these are various white and buff quartzites (**q**), quartzitic conglomerate (**g**), brown sandstone (**b**) and green sandstone, probably Pennant (**p**). These are bound in coarse cemented sand containing abundant fragments of thin shells, which grades up into cemented medium sand (**s**). Scale is 10 cm.*

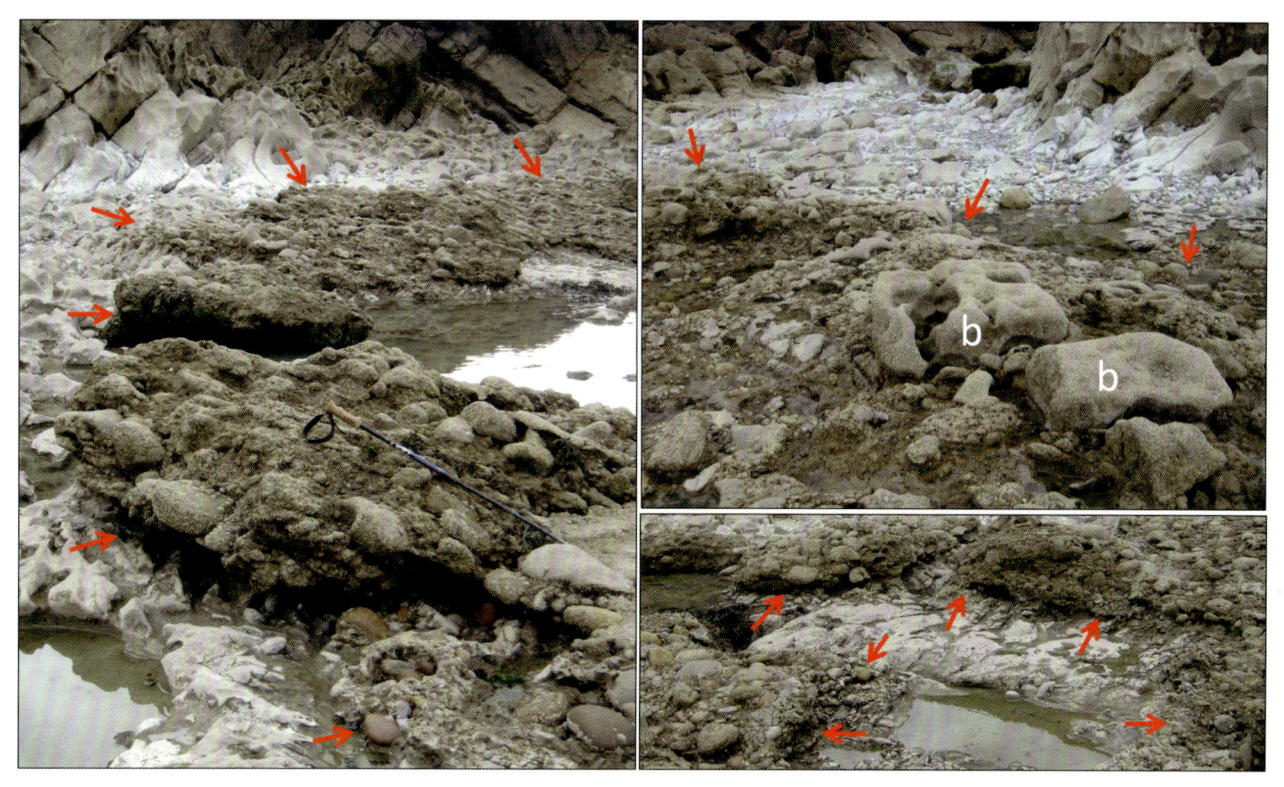

*Figure 38. Cemented pebble-to-boulder conglomerate at 0 m OD in Heatherslade Bay; **red arrows** mark the base of the deposit, which is cemented onto beach-platform limestone. The beach platform slopes gently seawards from an abrasion-polished base to a small cliff. The top of this cliff is at about 5-7 m OD and is a seaward sloping platform on which there are remnants of the Patella Beach. **(Left)** A crude imbricate fabric of seaward-dipping cobbles and boulders lies sub-parallel to the stick and is steeper than the basal contact. **(Top right)** Large blocks of limestone **(b)** are embedded in the conglomerate, and the whole deposit here is closely matched by the modern beach visible in the background. **(Bottom right)** A 'window' through the cemented conglomerate reveals the underlying platform at a level and slope indistinguishable from the modern surface.*

We know that the Heatherslade cemented storm-beach deposit formed after the last Pleistocene glaciation, during and since when the sea was mostly nowhere near this level. As the climate improved significantly around about 12,000 years ago (Fig. 28), marking the start of Holocene times, the Bristol Channel then gradually flooded until the sea reached close to present levels 7,000-5,500 years ago (Fig. 13). Pennard Valley would again have been flooded by the sea, this time as far as Parkmill (Fig. 36). But what of the cement?

None of the modern active beaches show cementation, but one might sensibly look for it in the earliest storm beach of Pwll-du (SS573871), for example, where successive pebble-cobble banks have been stabilised and infiltrated with sand and organic-rich acidic fresh water that could dissolve and reprecipitate carbonates (Fig. 150). Another possible analogue is within the scree and sand banked against the ancient rocky shore over springs at the back of Three Cliffs Bay. Here, at the Notthill springs (Fig. 36 and Chapter 5), carbonate-rich waters actively deposit tufa along their stream beds and are likely to percolate through the sediments and form a binding cement there.[34]

It is possible that the Heatherslade deposit was from a big storm when sea level first reached *near* here, and the resulting advanced storm-beach deposit was then buried under sand so that run-off and seepage water from the cliffs could deposit carbonate to bind the whole together. In this scenario the deposit would register that last increment of sea-level rise, perhaps just the last couple of metres. The Heatherslade cemented conglomerate should be younger than the 'fossil peat' layers that are exposed in Gower beaches (e.g., Swansea, Oxwich, Port-Eynon, Broughton and Whiteford bays). Peat layers at Port-Eynon and Broughton record woodland and fresh-water sedge fenland and give radiocarbon dates of between about 5,500 and 3,790 years before present (Philp 2018 and pers. comm.). These must have formed a little earlier than the Heatherslade conglomerate during the Holocene re-advance of the sea up the Bristol Channel.

The platform with cliffs at the back that became notched by the Patella (Ipswichian) high-stand level obviously must predate Patella times. Heatherslade conglomerate rests on that platform, so we see that the present shoreline is reoccupying previous Pleistocene seashore levels. The Heatherslade deposit confirms that coastal planation is rather slow, here trivial in the past 3,500 years. Evidently then, most of the cliff and platform development of the Gower coast must have occurred during the 2 million years between the Late Pliocene and the Late Pleistocene, probably mostly after the Anglian glaciation (Section 2.12).

2.16 Summary

The landscape evolution inferred in this chapter is summarised in Figures 39 and 40. Here are a few general remarks. Two comparatively recent scientific advances have enabled this new inference of the landscape evolution. First, through the work of countless researchers, our understanding of the dynamic Earth has advanced to the point when we can appreciate the detailed effects of plate tectonics in terms of both upheaval and climate.[35] In our Gower case we see how mantle-plume activity and opening of the North Atlantic Ocean caused uplift and then subsidence that interacted with ice-related changes of sea level.

The second advance is in the ground imagery that is now available. LiDAR images that see through vegetation in great detail, Google Earth views, and detailed inshore bathymetry now facilitate analyses where previous workers could not see. Although new facilities provide new information, they do not preclude user error, so the new inference is not *fait accompli* but rather a string of (hopefully logical) deductions yielding a new hypothesis in need of testing (Figs 39 and 40).

The ground that would become Gower peninsula was eroded down progressively for tens of millions of years before its partial marine planation and the following Quaternary onset of glacial and interglacial times. Major subaerial erosion was consequent on regional uplift caused by upwelling of deep mantle. Melting of that mantle resulted in formation of a vast volcanic field that was active from about 62-52 Ma and heralded the opening of the North Atlantic Ocean. Lundy Volcano tells us that the uplift must have resulted in erosional removal of at least 1 km thickness of rock down to near the present level of the Bristol Channel floor.

Rocks that would form Gower presented a relatively rugged massif profile in the limestones, where karst terrain with caves had developed and gradually became dissected by steep-sided dry valleys, many following faults. Only later came the subsidence and marine planation that formed the notorious 200' platform, now the coastal fringe sloping up to 75 m OD. The hinterland was of hills and ridges dominated by erosion-resisting Old Red Sandstone and Upper Carboniferous Pennant sandstones. Intervening slacks in softer mudrocks and thin sandstones formed drainage basins and low ground that would eventually become bays.

LANDSCAPES CUT IN TIME

Global developments

Era	Epoch	Age
	Quaternary	0 Ma
		2.58 Ma
Neogene	Pliocene	5.33 Ma
	Miocene	23.03 Ma
Paleogene	Oligocene	33.9 Ma
	Eocene	56.0 Ma
	Paleocene	66.0 Ma
	Cretaceous	

Climate scale: POLAR ICE CAPS — TEMPERATE — SUB-TROPICAL — TROPICAL — Paleocene-Eocene Thermal Maximum

Global sea level: +200 +100 0 m — Long term, Short term — Increasing Antarctic ice

Regional and local developments

Glacial-interglacial sea-level fluctuations

Subsidence ← Marine planation reaches 75 m OD ?

Uplift and dynamic support from mantle plume diminish as the North Atlantic opens and Britain moves away after 54 Ma

Subaerial peneplanation:

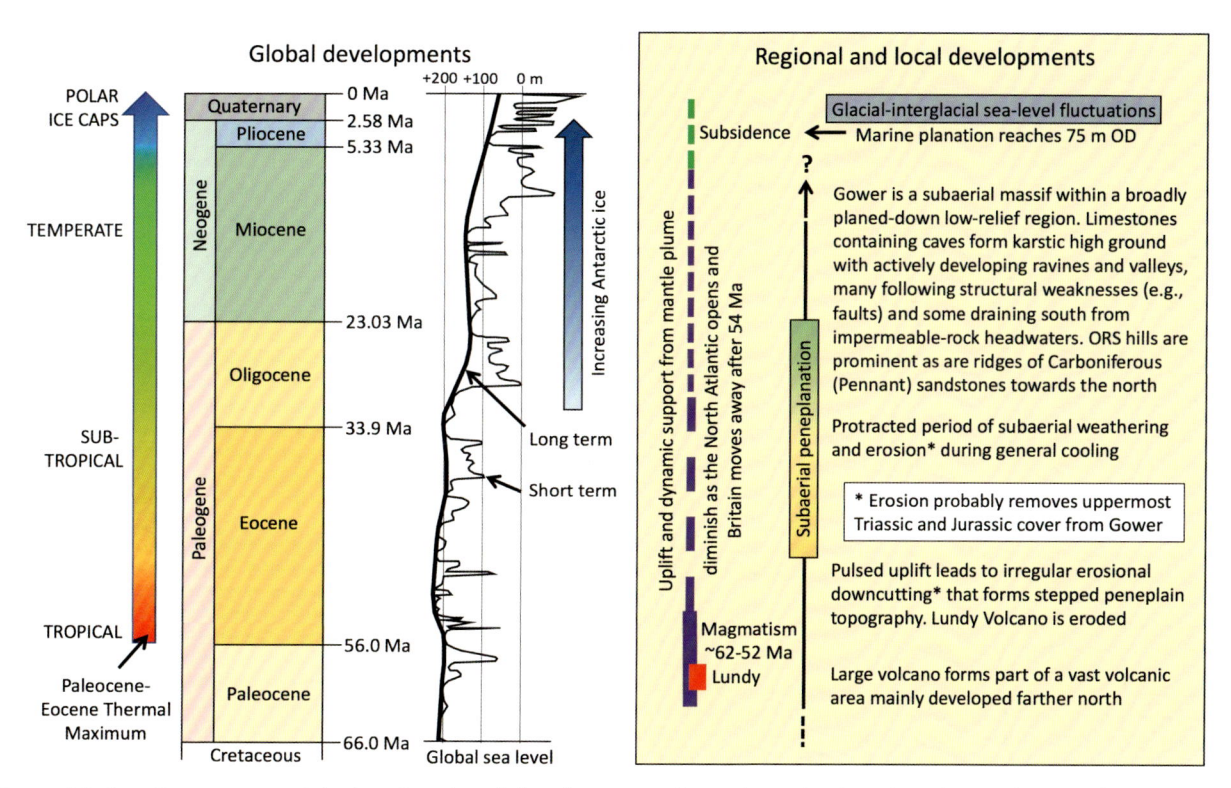

Gower is a subaerial massif within a broadly planed-down low-relief region. Limestones containing caves form karstic high ground with actively developing ravines and valleys, many following structural weaknesses (e.g., faults) and some draining south from impermeable-rock headwaters. ORS hills are prominent as are ridges of Carboniferous (Pennant) sandstones towards the north

Protracted period of subaerial weathering and erosion* during general cooling

* Erosion probably removes uppermost Triassic and Jurassic cover from Gower

Pulsed uplift leads to irregular erosional downcutting* that forms stepped peneplain topography. Lundy Volcano is eroded

Magmatism ~62-52 Ma Lundy — Large volcano forms part of a vast volcanic area mainly developed farther north

Figure 39. Pre-Quaternary global and regional developments (see Fig. 21) related to the evolution of Gower landscape. The question mark in the right-hand panel reflects uncertainty regarding the Early to Mid-Pliocene manner and timing of the marine advance onto the land surface.

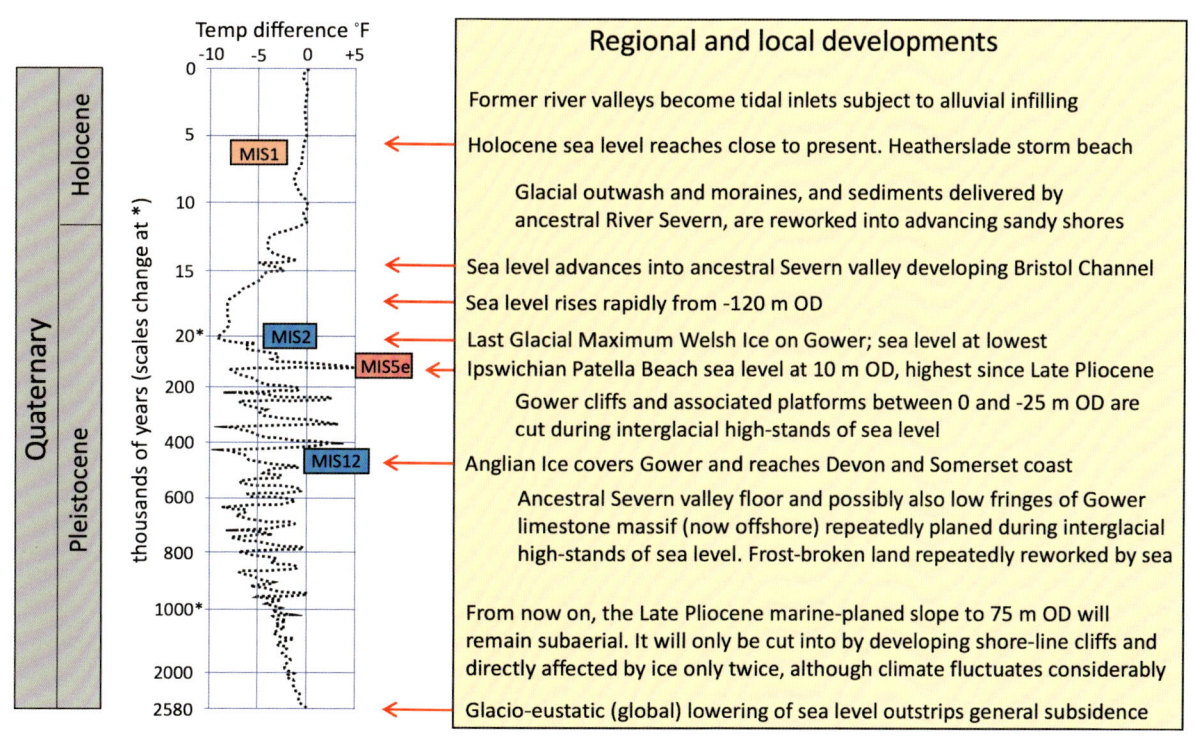

Regional and local developments (Quaternary)

Temp difference °F: -10, -5, 0, +5

Holocene — MIS1 (5 ka):
- Former river valleys become tidal inlets subject to alluvial infilling
- Holocene sea level reaches close to present. Heatherslade storm beach
- Glacial outwash and moraines, and sediments delivered by ancestral River Severn, are reworked into advancing sandy shores
- Sea level advances into ancestral Severn valley developing Bristol Channel
- Sea level rises rapidly from -120 m OD

Pleistocene:
- MIS2 (20 ka) — Last Glacial Maximum Welsh Ice on Gower; sea level at lowest
- MIS5e — Ipswichian Patella Beach sea level at 10 m OD, highest since Late Pliocene
- Gower cliffs and associated platforms between 0 and -25 m OD are cut during interglacial high-stands of sea level
- MIS12 — Anglian Ice covers Gower and reaches Devon and Somerset coast
- Ancestral Severn valley floor and possibly also low fringes of Gower limestone massif (now offshore) repeatedly planed during interglacial high-stands of sea level. Frost-broken land repeatedly reworked by sea
- From now on, the Late Pliocene marine-planed slope to 75 m OD will remain subaerial. It will only be cut into by developing shore-line cliffs and directly affected by ice only twice, although climate fluctuates considerably
- 2580 — Glacio-eustatic (global) lowering of sea level outstrips general subsidence

Figure 40. Quaternary global and regional developments related to Gower landscape evolution. The temperature variation curve is redrawn from Lisiecki and Raymo (2005), Hansen et al. (2013) and Jouzel et al. (2017).

Perhaps the most striking image showing the true karst surface of the limestones is of the near offshore (Fig. 41). Here the 'late' cover of glacial material that obscures most limestone karst on Gower has been removed by the rising sea to reveal superb and striking etching of the folded and fractured limestones. This karst surface developed over many millions of years and it graphically reveals the tight folding and accompanying faulting that affected the limestones, mainly in latest Carboniferous to earliest Permian times, with further faulting and fault dilation later during the Triassic Period (Woodcock et al. 2014).

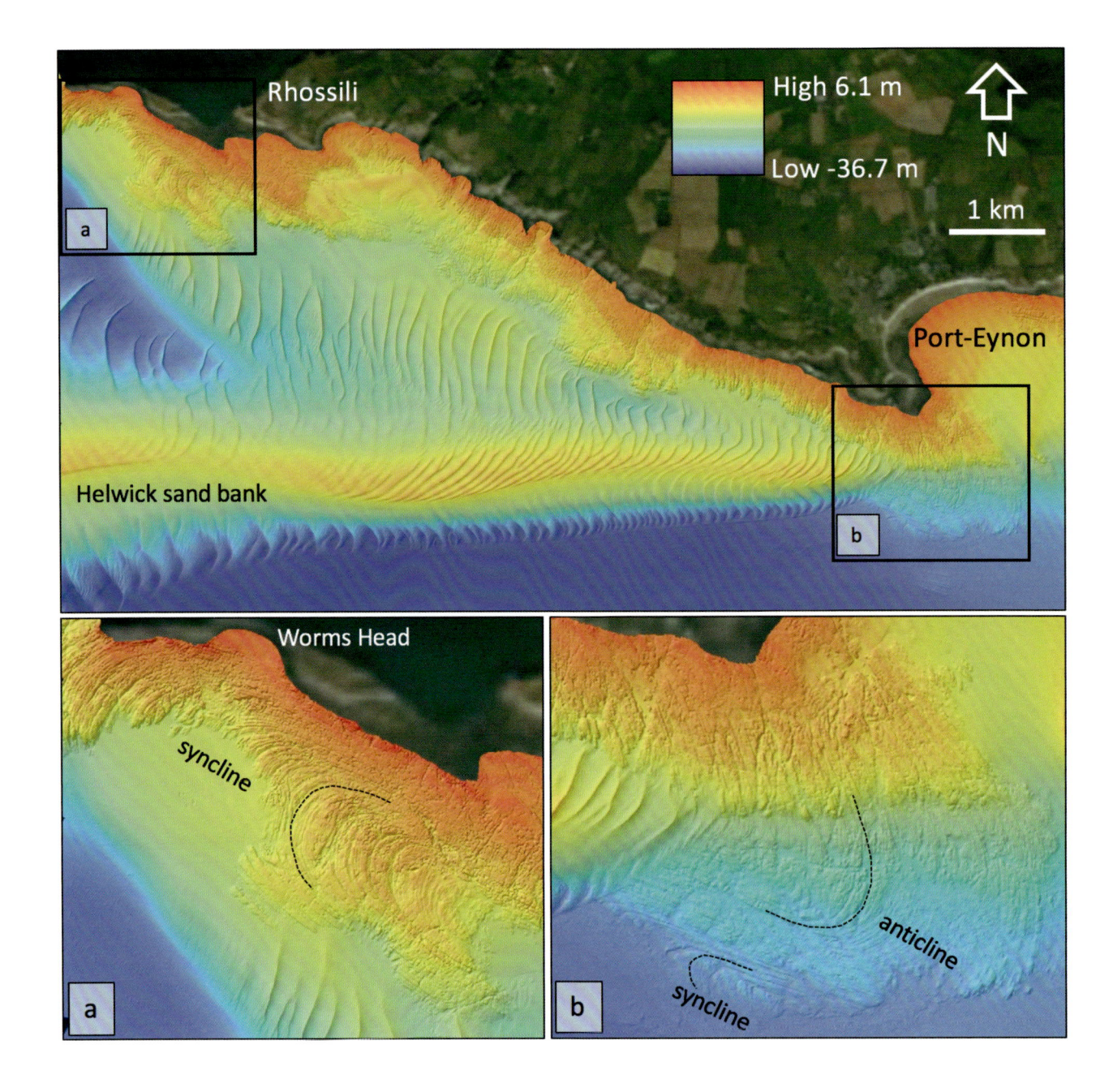

Figure 41. (Page opposite). Detail of the near-offshore karst surfaces in deeply eroded limestones showing folds (anticline and syncline) and fractures that formed mainly in Late Carboniferous to Early Permian times and were first eroded then. The images graphically reveal the deep erosion that has cut through the folds over some 290 million years. The surfaces now continue south into the floor of the Bristol Channel, but before the Late Pliocene relative rise of sea level 3 or 4 million years ago, as well as during glacial intervals of the Quaternary times, this karstic terrane formed a subaerial massif standing proud above the ancestral River Severn Valley floor, from a level now some 35 m below sea level. Panel (b) clearly shows a sloped platform and drowned cliff extending west to east from the covering tip of the Helwick sand bank (see such cliffs marked in Fig. 11). **Depths are relative to the level of the lowest astronomical tide: LAT.** *(Base image multibeam bathymetry, at 2 m x 2 m resolution, from The Admiralty Marine Data Portal, kindly processed and provided by Gareth Carter).*

Whereas the Late Pliocene age of the marine planation of the massif has long been generally accepted, it remains unclear, to this author anyway, what might have occurred as the sea encroached to that level while the land subsided; hence the question mark in Figure 39. Although the present coastal fringe above the cliffs appears as a continuous slope, i.e., not stepped, implying steady advance of the Late Pliocene coastal planation, it seems conceivable that part of our offshore rocky apron could have been cut into by the sea early during that advance, possibly forming the submerged cliffs at about -15 to -25 m OD.

After the Late Pliocene marine planation to 75 m OD, 'all' that remained was to cut the modern cliffs during Quaternary relatively mild episodes (interglacials) and ultimately to drape the land with glacial and periglacial debris along with frost-shattered rocks in screes and thin soils. Perhaps it is fair to say that the post-Pliocene landscape evolution constitutes for us a beautiful and intriguing 'late dressing' of the main surfaces that had previously evolved over considerable deep time.

Harry on Spaniard Rocks.

Landscapes Notes

[1] Geological time is divided formally into Periods. We live in the Quaternary Period known for its glacial and interglacial (milder) episodes, which started happening some 2.6 million years ago. The last glacial episode ended about 11,650 years ago, which is taken to define the end of the Pleistocene and start of the current Holocene epoch. Use of the term 'Ice Age' can be confusing; many would say the Pleistocene epoch was the Ice Age, but in fact our current Holocene time could be just another of the several interglacial episodes that have occurred in Pleistocene time, so that the age of ice may not be over yet. It was warmer here than now in a previous Pleistocene interglacial episode, some 123,000 years ago, as we shall see. I use 'Ice Age' here broadly for the ongoing Quaternary Period as we have no way of telling yet whether there will be another glacial episode in the geological long term. The existence of substantial northern polar ice, including that on Greenland, has characterised the Quaternary Period and without it sea level would be many metres higher than now, so, I say, long live the 'Ice Age'!

[2] Most of the movement ultimately relates to convective circulation in which hot Earth materials rise due to their buoyancy relative to adjacent colder materials. The heat partly comes from decay of small quantities of radioactive elements deep in the Earth and also from friction in the gravitational reorganization (convection), with some remaining from the time of planet formation. The movement is slow; the fastest at the surface of the planet is horizontally at about 15 cm per year, for ocean-floor crust, while most is far slower.

[3] Finding precisely what the impacts of global warming will be on Earth's coastal communities is difficult and dependent on computer modelling that has to deal with uncertain Earth properties and interactions. Happily, for understanding Gower now, we do not need to worry about this.

[4] Earth's ocean-floor crust is created by volcanic activity along 'oceanic spreading ridges'. Newly created ocean-floor crust is hot and buoyant, and it subsides as it spreads away from the ridge and cools. Ultimately, relatively cold ocean-floor crust sinks back into the underlying mantle at 'subduction zones'. Ocean-floor creation at spreading ridges is not uniform in space or steady in geological time, so that the average age and hence average depth of oceans varies through time.

[5] That is, height above the Ordnance Datum, which is mean sea level recorded at Newlyn, Cornwall, between 1915 and 1921.

[6] An unconformity is a break in a succession of rocks that represents an interval, possibly a long time in geological terms, during which there may have been uplift and erosion, commonly (not always) relating to folding or tilting in mountain building. Thus, an unconformity typically has rocks of contrasting types above and below, with the lower ones widely deformed and with strata at an angle to those above. Where the unconformity surface has been cut by the sea, e.g., marine planation, that surface will tend to be planar. Where is has been cut subaerially, e.g., by rivers, it can be extremely irregular, representing ancient topography. The unconformity beneath the Triassic rocks on Gower is of the latter type.

[7] Historically, fine red ochreous sediment, or reddle, was excavated from several fault fissures on south Gower, including that forming Red Chamber (SS426867) southeast of Mewslade Bay and an occurrence in Port-Eynon ("deposit of painte mineral" exported from the Sedgers Bank in late 17th century; Tucker 1951), and, of course, the famous hominid remains at Goat Hole (Paviland; SS436858) were deliberately stained with the ochre, hence the original epithet 'Red Lady'.

[8] In Chapter 4 it becomes clear that caves on Gower initiated beneath a limestone land surface significantly above the present surface, so on the one hand we know that the sub-Triassic unconformity must have

existed 'not far' above the present surface, while the caves require there to be some missing thickness of limestone beneath any Triassic surface. Heaven alone knows the ultimate resolution; we must imagine an irregular Triassic-age eroded limestone landscape 'not far' above present, but 'far enough'.

[9] https://en.wikipedia.org/wiki/Thecodontosaurus; https://en.wikipedia.org/wiki/Bristol_Dinosaur_Project

[10] European Marine Observation and Data Network initiated by the European Commission.

[11] 1.5-1.9 mm/year between 1901 and 2010; 2.8-3.6 mm/year between 1993 and 2010 (IPCC, 2014: Climate Change 2014: Synthesis Report. Contribution of Working Groups I, II and III to the Fifth Assessment Report of the Intergovernmental Panel on Climate Change [Core Writing Team, R.K. Pachauri and L.A. Meyer (eds.)]. IPCC, Geneva, Switzerland, 151 pp.). At the time of finishing this book, in August 2021, the Sixth Assessment Report was published highly confident that the rise during 2006-2018 was 3.2-4.2 mm/year.

[12] Limestone is not dissolved by 'ordinary' seawater and it is speculated that the considerable development of bubbles and turbulent mixing in breaking waves admits uptake of atmospheric CO_2, which produces a weak carbonic acid that can very slowly dissolve the rock, especially where the flow at the contact is rapid.

[13] Banner refers to the apparent 'tying' of these tidally reworked banks to coastal promontories according to the direction of strong currents. There are several fine examples in the Bristol Channel.

[14] Planet Earth is now understood to be continuously evolving as materials rise buoyantly from great depths, cool, and sink again, while the more rigid crust on which we live moves about the surface as large 'tectonic' plates. Earthquakes and volcanoes are obvious manifestations of this dynamic system and fit into a 'model' – a more or less complete explanation – of how it is all supposed to work.

[15] Granite is formed from slowly cooled silica-rich magma and in some volcanoes this cooling can occur at the relatively shallow depth of about 1 km; widely, though, it occurs deeper below the volcano summit.

[16] Mantle plumes are large-scale physical systems, the source(s) of which are not universally agreed upon. Certainly, there are various styles and origins of upwellings of mantle, generally called plumes. The present 'essay' is no place to evaluate the pros and cons of the plume model that is applied here; it has enjoyed wide acceptance and explains a great deal, but, strictly, this does not mean it is a fully understood reality. Nevertheless, the physical consequences we need here are real, irrespective of the ultimate origin.

[17] Remnants of Triassic sediments on south Gower typically dip some few degrees southwards, as does the offshore counterpart (Fig. 10). Part of this tilt may be attributable to effects of the Iceland Plume.

[18] Technically 'lithosphere', comprising the relatively cool and hence more rigid crust and uppermost mantle.

[19] This 'absolute' is perfectly suitable for our purpose regarding Gower, although at any particular instant in time absolute sea level globally varies in altitude relative to the Earth's centre, by up to 180 m, primarily due to density-related regional variations of gravity and the non-spherical shape of the Earth.

[20] In Cornwall, Late Pliocene marine sediments, the St Erth Beds, record sea level at 35-45 m OD while earlier Oligocene and Miocene deposits record deep subaerial weathering at higher levels (Walsh et al. 1987; Roe et al. 1999). In Pembrokeshire, at Trefgarn gorge (SM958240) 65 m OD (Fig. 12), a subaerial weathering profile has admixed marine sediment near its top and is cut by river deposits (Battiau-Queney 1984). This shows the low levels to which prior subaerial peneplanation occurred and that the marine incursion reaching there barely removed the weathered material before the reversion to

subaerial conditions due to Pleistocene sea-level fall. Similarly, at Flimston in Pembrokeshire (Fig. 12; SR92689528), a pocket of Oligocene lake-bed clays (pipe clay) remains at 50 m OD within the marine platform cut across the Carboniferous limestone.

[21] There are also numerous ice-melt collapse depressions, kettle holes, not to be confused with sink holes (Section 3.13).

[22] The reader is directed to Wikipedia (https://www.wikipedia.org) for a potted account of 'karst'. The characteristic pavements are poorly exposed on Gower because they are widely overlain by various soils and glacial debris, mostly weathered till, and blown sand. Nevertheless, the cliff-top edges commonly reveal well developed karstic surfaces that formed before any visits by ice.

[23] Inland it may not have been seawater that re-flooded passages, if rainwater draining from a higher surface had sufficient head (pressure) to sustain submarine springs.

[24] The local adage is "When you can see Lundy it's going to rain, and when you can't it's raining already".

[25] Glacials and interglacials refer to the cycles associated with Earth-orbital forcing of climate, while stadials and interstadials are defined by the actual temperature record found in seafloor sediments and referred to the Marine Isotope Stages (MIS).

[26] Easier said than understood, these "main coastal cliffs" are rather complicated. The box below, showing Fall Bay (SS414874), highlights the problem; there are many cliffs, and it is not always clear when they formed (the indicated dates are approximate).

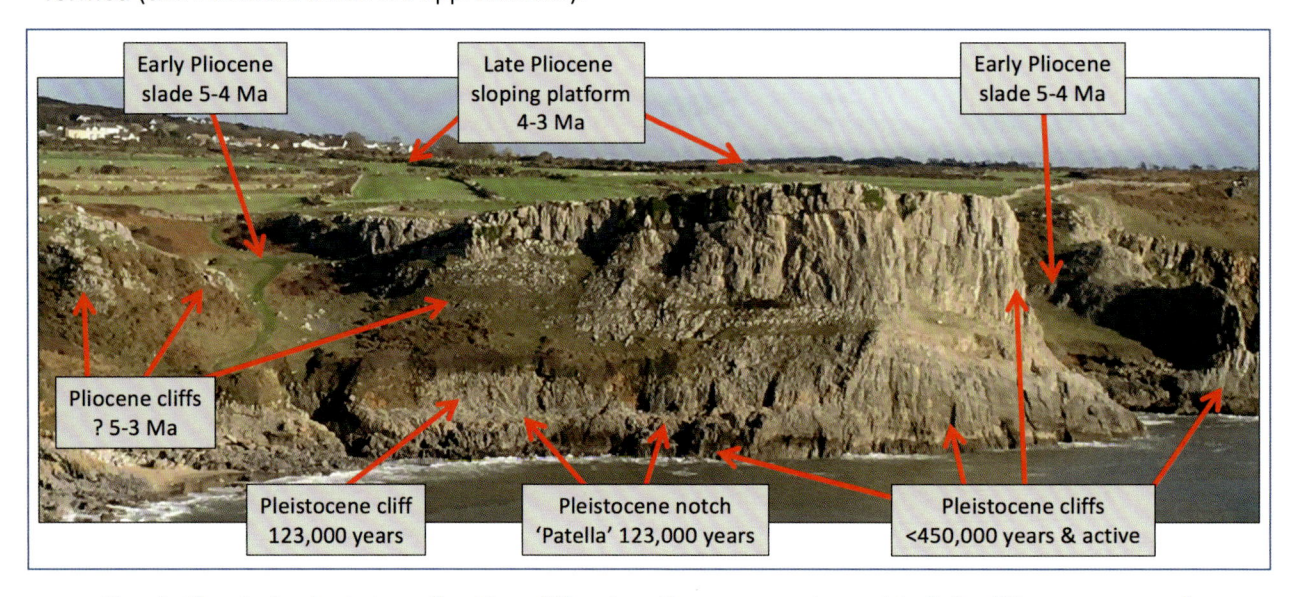

Clearly the slades had steep flanking cliffs when they were cut roughly 5-4 million years ago, long before the sea arrived. We know that the platform was cut across the slades (4-3 Ma), but it is not clear what that marine incursion did to the pre-existing cliffs, hence those closest to what would have been exposed coast are tentatively identified as simply Pliocene. We know that the coast was 'visited' by waves several times during Pleistocene interglacials, probably after the Anglian glaciation around 450,000 years ago (see text). The Patella notch and platform at around 10 m OD allow us to distinguish some cliffs as cut at 123,000 years ago. Thus, what we loosely refer to as "main coastal cliffs" were far from simply created. The right-hand side of the large upper cliff, Lewes Castle, had a massive pinnacle we as youths would often climb and perch on to belay; long since fallen off it is a scary reminder that the wasting of the cliffs is active.

[27] The boulders, of cemented scree and head, could be post-Anglian and pre-Ipswichian.

[28] One must not confuse these with the pre-Patella 'Inner Beach' excavated in Minchin Hole (SS555868). This comprises 2 m of stratified sands attributable to an earlier high stand (Sutcliffe et al. 1987), but its associated interglacial age is unknown.

[29] A similar fossil record exists in Bacon Hole, 525 m to the east (SS560868).

[30] Reported by George (1932), but the original exposures are now obscured by slumped and loose slope materials and vegetation.

[31] This layer is very sparsely fossiliferous but nevertheless yielded this Patella shell and nearby a single small robust periwinkle, probably Melarhaphe neritoides (below). This cave site and Patella deposit is best accessed by scrambling up from the beach platform during low spring tides.

[32] Limited study of the wind-blown sand deposits at Devil's Kitchen (Fig. 33c) and on the northeast coast of Oxwich Point (SS510856) indicates wind from the south or southeast.

[33] Although amusing to imagine, it would of course have been impossible for hippos, rhinos, elephants and mammoths to negotiate the steep Gower cliffs to enter the caves. It is well established that carnivores such as hyenas, wolves and eventually humans brought the animal parts into their shelters.

[34] Carbonate deposition from carbonate-rich stream waters is commonly enhanced by removal of dissolved CO_2 via the metabolism of bacteria, algae and plants, or by degassing in agitated flows, and it cannot occur during total freezing conditions (permafrost). It is likely to have been enhanced in interglacial or post-glacial times and is discussed further in Chapter 3.

[35] The main driver of 'ice ages' is the global configuration of Earth's continents and sea-water circulation in the oceans around them. The presence of continental Antarctica at the south pole with its massive accumulations of ice became the main driver of global sea-level change and cooling conditions since 33 million years ago and then, since 2.58 million years ago, the additional development of grounded ice sheets at northerly latitudes led to the Quaternary 'Ice Age'.

Hollie, Hayley and Horton.

3 GLACIATION OF GOWER

3.1 Introduction

Ice advanced onto Gower twice during late Pleistocene time (Fig. 40), first in the Anglian glaciation about 450,000 years ago and again in the Devensian Last Glacial Maximum some 23,000 years ago. The Anglian ice crossed over Gower from the west-northwest and reached the north coasts of Somerset and Devon, at least. On Gower it delivered sparse but distinct erratics plucked from local outcrops and from Pembrokeshire and farther north. Various igneous (once molten) rocks, such as rhyolites, dolerites and gabbros distinguish Anglian erratics, having been brought by the ice from Skomer, Ramsey Island and the coast northeast from St Davids Head into the Preseli Hills (Fig. 12). Rocks from farther afield, some from Scotland, came to South Wales via the Irish Sea Ice stream[1].

Anglian till[2] is generally thin but widespread on south Gower beyond the Devensian limit (Fig. 42 and see below). It is locally exposed beneath Devensian till, and along the south coast its erratics can be found amidst reworked Devensian debris, commonly with distinctive red Triassic rock fragments. Its soils are a rich brown, commonly sandy, and can drain moderately well. Anglian till occurs high on the ridges of Cefn Bryn and Rhossili Down. Boulder fields there, beneath outcrops of Upper Old Red Sandstone (UORS) quartz-pebble conglomerates, probably represent degraded tors that for some time existed as nunataks[3] protruding from the Anglian ice. UORS erratic boulders commonly rest quite fully exposed on the Anglian till and typically show deep weathering around protruding pebbles, reflecting almost half a million years of exposure to the elements, including the Devensian freeze. The Devensian ice did not reach as high on Gower as the Anglian ice did.

Gower Devensian debris in many coastal locations rests on so-called Patella Beach deposits, which record the Ipswichian interglacial episode around 123,000 years ago (Section 2.14). The Late Devensian glacial maximum, known as the Last Glacial Maximum (LGM), was protracted (26,500-22,000 ka). The maximum ice advance out of central Wales onto Gower is considered to have occurred about 23,000 years ago, according to exposure-age dating of Arthur's Stone on Cefn Bryn (see below; Phillips et al. 1994).

This chapter focusses on the Last Glacial Maximum (LGM) and describes aspects of the melting away (recession) of that Welsh Ice. New limits of the ice advance and recession are described here for the first time, as inferred from mapping of the glacial deposits, especially the Devensian till and its contained erratics, but also some reworked representatives. LiDAR-terrain and Google Earth images proved particularly useful for mapping. A significant new finding is that ice crossed over eastern Gower, from the north, and continued south beyond what are now coastal cliffs, possibly extending 3-4 km from the present shoreline.

Before painting-in the findings, it seems worth first explaining the principles and practice of this study of the glaciation of Gower. Here the glacial deposit limits are taken to define 'lobes' where ice evidently reached. The Paviland Moraine[4] in western Gower was already known and described (Shakesby et al. 2017). An additional remnant of that feature is identified, and an ice limit, marked by ice-melt features and deposited debris, is drawn to identify a 'Paviland Lobe'. The ice limit connected to this lobe has been mapped eastwards along the north flank of Cefn Bryn to define an 'Ilston Lobe' and this appears truncated by the newly discovered 'Pwll-du Lobe', which trends southwards across Gower and probably beyond.

The 'lobes' are deposit-defined ice limits; the ice-advance directions and upland origins suggest that a continuous ice sheet extended onto Gower within which there were topographically directed ice streams that formed the distinct lobes. It has long been recognised that Devensian glacial deposits at Broughton and in Rhossili Bay are characterised by their content of 'Coalfield erratics', comprising abundant Pennant sandstones rather characteristically with cobbles and small boulders of shale and abundant ironstone nodules. This content differs from the Paviland Lobe so that a separate ice stream is inferred crossing north-westernmost Gower.

Regarding the directions of ice flow, we know that the Welsh Ice came from the north and northeast and that, broadly, northern peninsular Gower constituted a barrier lying transverse to the main flow. Evidently much of the thick ice was deflected southwest down the valley that would become the Loughor Estuary. Nevertheless, ice covered all of north Gower at some time; it deposited till and fluvioglacial debris on all of the pronounced ridges of Pennant sandstones that extend from near Penclawdd (SS536954) in the west to Townhill (SS640938) and Kilvey Hill (SS671940) above Swansea in the east (British Geological Survey 2011). These barriers reach 100 m to 190 m OD, but the base of the ice would locally have been below Ordnance Datum during the glaciation.

In this study we will distinguish where and via which route, in general, ice surmounted peninsular Gower, as well as discussing in detail some elements of its melting back, i.e., its recession. It is worth remembering that the ice that surmounted Gower was near to its ultimate limit, on the verge of break up and melting away, although there is evidence of ridge truncation to form steep spurs in north Gower showing that some deeply erosional sculpting did occur.

We will consider evidence of protracted flow of ice to maximum reaches as well as what appears to have been brief encroachment of thin and wasting ice. It is taken for granted and only briefly mentioned that LGM ice occupied Swansea Bay, having originated from flows mainly down the Swansea and Neath valleys. The contact between the Pwll-du ice that crossed Gower from the north and the ice in Swansea Bay is inferred from the terrain; the erratic contents of these two seem indistinguishable.

Figure 42. (Page opposite). Anglian till and surfaces on Gower.

*(a) Inverse-coarse-tail-graded basal Devensian till resting on typical erratic-poor Anglian till, northeast of Llethrid (SS533915). The "inverse- etc." means that the large stones, which are embedded in typically yellowish and clay-rich fine-grained matrix, get larger upwards[5], because they arrived and were deposited here later than the lower small ones. The contact (**dotted line**) is not sharp because the underlying Anglian till has been partly incorporated above by shear-related mixing as the Devensian ice and debris scaped over the Anglian till.*

(b) Reddenhill Farm field (SS537890) during a ploughing match, with characteristic rich-brown soil. Erratics, sizeable examples of which typically were cleared from the fields over past centuries, are rather scarce in Anglian till, in marked contrast to the Devensian occurrences.

(c) Anglian erratic of local Upper Old Red Sandstone quartz-pebble conglomerate (Reddenhill SS535893). The rock is closely similar to the UORS on Gower and dissimilar to the north-crop counterpart (see below).

(d) Surface of Anglian coarse-grained diorite erratic (>40 cm) near to (c), likely to be from northern Britain and brought south by Irish Sea Ice (Reddenhill SS536894).

(e) Surface of Anglian medium to coarse-grained dolerite erratic (>60 cm), sparse but characteristic of Anglian till on Gower and most probably derived from north Pembrokeshire. The white mineral is plagioclase, the black is pyroxene, and the green-brown matrix is their alteration product (61 m OD above Caswell SS595878; coin is 17 mm).

(f) Surface of local UORS quartz-pebble conglomerate showing the deep weathering of the matrix that, over a period of some 450,000 years, causes the pebbles (here including deep-red jasper) to protrude substantially. Such deep weathering does not prevail on Devensian-age surfaces (Rhossili Down SS419897).

GLACIATION

Finally, in this preamble, it is important to know that what remains as a glacial or periglacial[6] deposit is not likely to record all processes that affected that site. This has proved important in this study where ice has advanced over an area but left little or no primary record of it, e.g., as till. What does remain is the product of the recession during which any till may be reworked in debris flows, fluvioglacial outwash and/or slope-gravity mass wasting, as in slumps and landslides. The Devensian episode of ice advance and retreat was protracted and certainly unsteady; successive recessional moraines[4] prove this point in the retreat. Similarly, there is evidence that ice margins and drainage from the ice must have been quite complicated and changing through time. Ideally one would want to see a stage-by-stage 'movie' of developments. The maps produced in this study are hopelessly inadequate to represent the evolving glaciation, so they need to be viewed as a physical record of material evidence that relates to a long-lasting succession of processes and conditions.

3.2 Cryptic limits

Although the Gower peninsula is internationally renowned for its 'marking' of the limit of the Welsh Ice that encroached here from the north during the LGM (Fig. 4), surprisingly little has been established previously concerning the actual ice-reach limits. Key coastal exposures at Broughton and Rhossili bays, near Horton, near Southgate and as far east as Rotherslade have been written about, and argued about, a lot! Numerous confusing and rather bitty academic papers reflect conflicting interpretations and changing of minds, although some more recent studies give relevant useful detail on glacial sediment fabrics and compositions. Adding to the confusion, chemical analyses meant to determine the age of formation of fossils, and hence of their host deposits (amino acid dating[7]), ultimately proved misleading in the way they were interpreted.

The best catch-up review to inform the innocent reader is the field guide: 'The Quaternary of Gower' edited by Richard Shakesby and John Hiemstra (2015). In it, figure 2.1 shows six widely differing proposed Last Glacial Maximum (LGM) ice limits on Gower, published from 1929 to 1988. Knowing Gower, it is not clear why such different limits were drawn inland; the childhood picture creation 'join the dots to make a doggie' springs to mind. Nobody, it seems, actually mapped the limits continuously inland. Refreshingly, a recent detailed study by Shakesby et al. (2017) stands out in defining part of the extent of the LGM, by characterizing the Paviland Moraine in western Gower (Fig. 23). Otherwise there has seemed to be little development of 'the big picture'.

The exploration described below has been one of the most enjoyable elements of this study. It led mostly 'off the beaten track' and, apart from a few bramble crashes, proved not difficult to accomplish. Continuous traces marking the limits of the Devensian ice on central and eastern Gower have been mapped. The Paviland Moraine and its extrapolated till upper limits (fill-lines[8]) in the west are now traced eastwards along the north flank Cefn Bryn, including Arthur's Stone, and across Green Cwm and Ilston Cwm. Perhaps most interestingly, this limit terminates at the margin of another lobe that marks ice that crossed Gower by an eastern route. This ice certainly crossed the present shoreline, between Southgate and Mumbles Head, such that it covered Pwll-du Head, Pwll-du, Caswell, Langland and Rotherslade. It has not previously been recognised and is referred to here as the 'Pwll-du Lobe'. Its eastern limit is beyond Rotherslade and so includes the site of the (rather scruffy) coastal exposure that figures in all six previous LGM limits. Seabed features recently imaged are *tentatively* interpreted to suggest that the Pwll-du ice may have extended 3-4 km beyond the present coast.

Lastly, several previously undescribed features of the general recession of the Welsh Ice off Gower are described, before the 'big picture' is painted to show a new model of the last glaciation of Gower. Before setting out on this mission, however, it is probably important to know how and why the work proved "not difficult to accomplish".

3.3 Black Mountain Rosetta Stone

Ancient Egyptian recorded history became clearer to decipher when a decree in hieroglyphic text, issued in 196 BC, was found alongside an Ancient Greek version of the same thing. Three versions of the decree were carved on a stone named after Rosetta (also Rashid), the town on the Nile Delta where it was discovered in 1799. The Rosetta Stone thus became a most important key to unlock an ancient history.

The rock succession well exposed on the west flank of Black Mountain, north of Brynamman (SN713142), has similarly yielded a strikingly clear key for deciphering the glaciation of Gower (Fig. 43). The recognition of this northerly source of distinctive erratics is certainly not new. The two Geological Survey Memoirs concerning Swansea and West Gower (Strahan 1907a, 1907b) contain numerous observations of the tills and gravels, alluding to what they knew to be origins in the north crop of the Coalfield, rounding of "far-travelled" erratics, and more local incorporation of blocks of Pennant sandstone. Here, the tacit understanding of those field geologists is fully explained, and the evidence illustrated, rather than taken for granted.

Our 'Rosetta Stone' is the presence in the north-crop of the coalfield – best seen on Black Mountain – of extremely distinctive tough, white grits and sandstones, viz. the Upper Old Red Sandstone Grey Grits and the Marros Formation Basal Grit and Twrch Sandstones. These rocks do not occur in the geological succession on Gower. Original variations of sedimentary environments in Marros times had quartz-rich sand beaches and deltas form on the limestones in the north, while muds with some thin sand layers were deposited on the limestones towards the south, as seen on Gower. Ganister fragments on Gower can only have come from the north crop.

The key to understanding the maximum extent of the last glaciation – the Last Glacial Maximum of the Welsh Ice on Gower – is in mapping the *distribution limits* of erratics delivered to Gower from the north crop in the vicinity of Black Mountain. As the ice from central Wales (Fig. 4) surmounted the Black Mountain escarpment, which is the north crop, it plucked and eroded the bedrock there so fragments, some huge (e.g., Arthur's Stone, see below), were transported south to Gower.

Importantly, while there are similar rocks of the Marros Formation in Carmarthenshire, west of Gower, they cannot have been transported from that direction because of the essentially northeast-to-southwest flow of the Devensian ice, which is widely marked by mound alignments and scratch striations. Also, a large proportion of our distinctive quartzite erratics on Gower are supremely well rounded (Fig. 44) and this rounding can only have occurred in the energetic rivers that flowed from the north ahead of, or beneath, the southwards advancing ice, mainly down the course of the present River Loughor. Debris around the advancing ice would have been incorporated in the glacial outwash to form braided river plains of rock fragments that were rounded during their transport.

North-crop limestones that must also have been eroded by the ice are absent or at least scarce amongst the far-travelled Gower erratics, probably because they were obliterated by attrition in transport. Limestone dissolution is likely to have been minimal in the cold barren landscape where biotic CO_2 production would have been limited and the waters thus not especially acidic. Limestone fragments in Devensian till exist only locally where the ice crossed Gower limestones. Of course, along with the distinctive white quartzites, robust fragments of the LORS Brownstones and UORS quartz-pebble conglomerates were also transported (Fig. 44), so the erratic populations of rocks are mixed, but there is no doubt about the Black Mountain, north-crop, origin.

Figure 43. Black Mountain geology (vicinity of SN7319).

*(**a** and **b**) Top of Afon Clydach waterfall. Lower Old Red Sandstone (LORS) Brownstones Formation. The lower slopes to the north are softer LORS rocks while here are tough pink-purple gritty sandstones similar to those exposed on the north flank of Cefn Bryn (Fig. 122).*

*(**c** and **d**) Upper Old Red Sandstone (UORS) quartz-pebble conglomerates similar to those exposed on Cefn Bryn, Rhossili Down and Llanmadoc Hill, but subtly different. This is the rock of Arthur's Stone glacial erratic on Gower. Relative to Gower counterparts these conglomerates are rich in weathered brown shale fragments (**arrow in c**) and relatively poor in the deep red jasper pieces characteristic of Gower (Fig. 42). (continued...)*

Figure 43 (c and d continued). These quartz-pebble conglomerates contain abundant well rounded and near-spherical shapes, commonly with a 'ghostly' translucent grey colour, in contrast to the lower sphericity and salmon pink shades of Gower counterparts.

*(**e** and **f**) Upper Old Red Sandstone Grey Grits. These tough white quartz grits and sandstones with few impurities do not occur in the rock succession on Gower.*

*(**g**, **h**, **i**, and **j**) Twrch Quartz Sandstone quarries in the Carboniferous Marros Formation on top of Carboniferous limestones (around SN7218). Inset in (g) is the Marros basal quartz-pebble grit, including a red jasper fragment that suggests derivation by reworking of the UORS conglomerates (SN874160). The quartz sandstones are very pure and tough, originally mainly deposited in shorelines of a river delta; they include classic 'seat-earth' ganister, which can underlie coals and is distinctive for included fossil plant roots (**arrow in i** and grey carbonaceous streaks in j). The lack of minerals other than quartz renders these rocks chemically inert so that they were used to line furnaces.*

Figure 44. Devensian north-crop erratics on Gower commonly show extreme rounding that must have occurred in energetic rivers ahead of the Welsh Ice advancing from the north, most probably down what became the valley of the River Loughor. The 20 cm cobble on the right is from till at the west end of Cefn Bryn (SS471909) and is of tough Marros quartzite; the purple one is LORS Brownstone (sandstone) and below it is UORS quartz-pebble conglomerate, both also from till on Cefn Bryn.

Added to the distinctive Devensian north-crop erratics is a variable abundance of variously blue, green, brown or yellow sandstone fragments. These are less well rounded than the quartzitic rocks and commonly rather slabby in shape, reflecting a tendency to split along depositional bedding. They are from the Upper Carboniferous Pennant Sandstone Formation, which forms the uppermost (coal-bearing) rock unit in the region and is the most widely used building stone in South Wales. Ice from the north must cross Pennant sandstones so their occurrence as erratics is no mystery, but they are especially abundant in the newly recognised Pwll-du Lobe. In contrast to the other ice that reached Gower farther west, the Pwll-du ice must have cut through the prominent ridges of Pennant sandstones that are barriers in the northeast (see below).

3.4 Rotherslade revisited

As pointed out by John Hiemstra and Richard Shakesby in the Quaternary of Gower field guide of 2015, Rotherslade has been included in all versions of the drawn southern limit of the Last Glacial Maximum ice on Gower, because the oft-visited coastal exposure there (SS612872) was deemed to be of *in situ* Devensian till. Hiemstra et al. (2009), however, reported a detailed study showing that actually none of it is *in situ* primary till and they suggested instead that it represents material reworked from ice of the Swansea and Neath valleys (Tawe – Nedd Glacier) that had reached but not overtopped Mumbles Hill (SS624875).

Fieldwork for this study finds complete agreement regarding the reworking of the Rotherslade material, but, rather than it coming from the east, the mapping shows that it is reworked material of the newly discovered Pwll-du Lobe and its ice stream, which came from the north. The recently widened and locally re-routed coastal footpath, particularly between Rams Tor and Limeslade Bay, provides excellent additional exposures (Fig. 45). Most of the slopes in this vicinity are steep and it is hardly surprising that the deposits manifestly are slope-modified mixtures that reflect origins in transport and deposition from ice, from debris flow, and from aqueous reworking. The populations of larger fragments clearly are north-crop erratics, dominated by quartzites with conglomerates and Brownstones, and with abundant slabby chunks 1-60 cm in diameter of Pennant sandstones that are angular or sub-rounded. These latter erratics show considerably less rounding compared with the north-crop fragments and are interpreted as plucked during passage through the north Gower barrier of Pennant sandstone ridges.

The coastal Devensian deposits between Rotherslade and Limeslade (Fig. 45) are variable laterally but tend near the base to show mixture of some erratics with pre-existing limestone breccias that had formed in the frost before arrival of the ice. The overlying erratic-rich layers locally contain abundant cm-scale, well-rounded pebbles of tough rocks that reflect fluvioglacial reworking; the pebbly deposits are similar to those of the outer edge of the Paviland Moraine.

Evidently no original ice-deposited till remains; everything here is a mixture that reflects reworking. Towards the small-pebble-rich top, angular limestone fragments occur suggesting that naked limestone occurred nearby when reworking was active.

*Figure 45. Reworked Devensian till and fluvioglacial outwash between Rams Tor (SS619869) and Limeslade Bay (SS625871), with a-d in local ascending succession (scale is 10 cm). (**a**) Close to limestone bedrock, angular frost-shattered (thermoclastic) limestone fragments are mixed with sparse rounded Devensian erratics (**arrows**). (**b**) Characteristic assemblage of erratics of the Pwll-du Lobe with relatively large and well-rounded north-crop quartzite cobbles associated with abundant angular and sub-angular fragments of Pennant sandstones (**arrows**). The sandy matrix contains abundant angular to rounded fragments of Pennant sandstone mixed with small rounded pebbles of quartzite and Brownstones. (**c**) The deposits are locally rich in well-rounded pebbles that represent the full spectrum of Pwll-du Lobe erratic rock types. This is closely similar to the outer-fringe outwash deposits of the Paviland Moraine inland of Thurba Head to Red Chamber (SS424868). (**d**) Near to the top of the sequence angular limestone fragments (**arrows**) are mixed with pebble-rich material.*

3.5 New map of Last Glacial Maximum limits on Gower

Figure 46 summarises the new findings of the Devensian Last Glacial Maximum on Gower. The limit of what is here defined as the Ilston Lobe has been traced continuously from the northwest end of Cefn Bryn, where it joins the Paviland Lobe, to the vicinity of Ilston, where the newly discovered Pwll-du Lobe limit cuts north-to-south to reach the coast at Southgate. The Pwll-du Lobe is defined by a distinctive assemblage of abundant large erratics (Fig. 47), mapped occurrences of which are marked on the map as purple or red dots (red dots indicate subject of a photograph). The assemblage typically has north-crop quartzites and ganister, with quartz-pebble conglomerates and Brownstones, along with less-well rounded slabby pieces of Pennant sandstone. All of these rock types form boulders up to 60 cm in diameter, locally up to 80 cm (e.g., Brandy Cove, SS585873) and in places 1 m (e.g., near Moorlakes Wood, SS560907). Many occurrences are in field boundaries reflecting centuries of field clearance and numerous examples are incorporated in church walls (e.g., Pennard SS565887 and Bishopston SS577893). Limestone fragments occur in the till towards the southwest where manifestly they were derived locally (e.g., near Widegate, SS565883). Numerous small limestone quarries show that the Devensian till widely is very thin in the Pwll-du Lobe. Walls of associated limekilns commonly include north-crop erratics (e.g., near Kilvrough, SS563894). No erratics have been found on Mumbles Head, which is taken as defining an eastern limit to the Pwll-du Lobe (Fig. 46).

Figure 46. (Opposite). New map of the Last Glacial Maximum (LGM) limits on Gower. (The shaded-relief LiDAR image was processed using base data © Natural Resources Wales).

The Paviland Lobe margin is continuous with the Ilston Lobe margin, which traces along the north flank of Cefn Bryn, across Green Cwm to Ilston Cwm. These ice maximum limits probably were broadly contemporaneous.

The Pwll-du Lobe cuts across and is distinctly different from the eastern reach of the Ilston Lobe, being marked by substantially larger quartzitic north-crop erratics (NCEs) with abundant relatively angular Pennant blocks and limestone (only towards the southwest), in addition to 'normal' north-crop LORS Brownstones and UORS quartz-pebble conglomerate. The purple and red dots mark the main exposures that distinguish the Pwll-du Lobe.

The most obvious Anglian erratics within the Pwll-du Lobe are dolerite, but Gower-type UORS quartz-pebble conglomerate boulders are also Anglian. Coastal exposures commonly show a mixture of north-crop erratics with a few Anglian erratics and some Triassic sandstone fragments. These coastal deposits show evidence of aqueous and debris-flow transport, with normal steep-slope solifluction, and they represent reworking of the post-LGM landscape during ice recession.

The Whitford recessional moraine with its characteristic Coalfield erratics is shown in Figure 143. Dunvant Gap is the pronounced cutting through the sandstone ridges of northeast Gower. Unlike other breaks in the ridge topography, this most prominent breach is not obviously controlled by any faults (British Geological Survey 2011).

The recessional moraines are discussed in Sections 3.14 and 3.15.

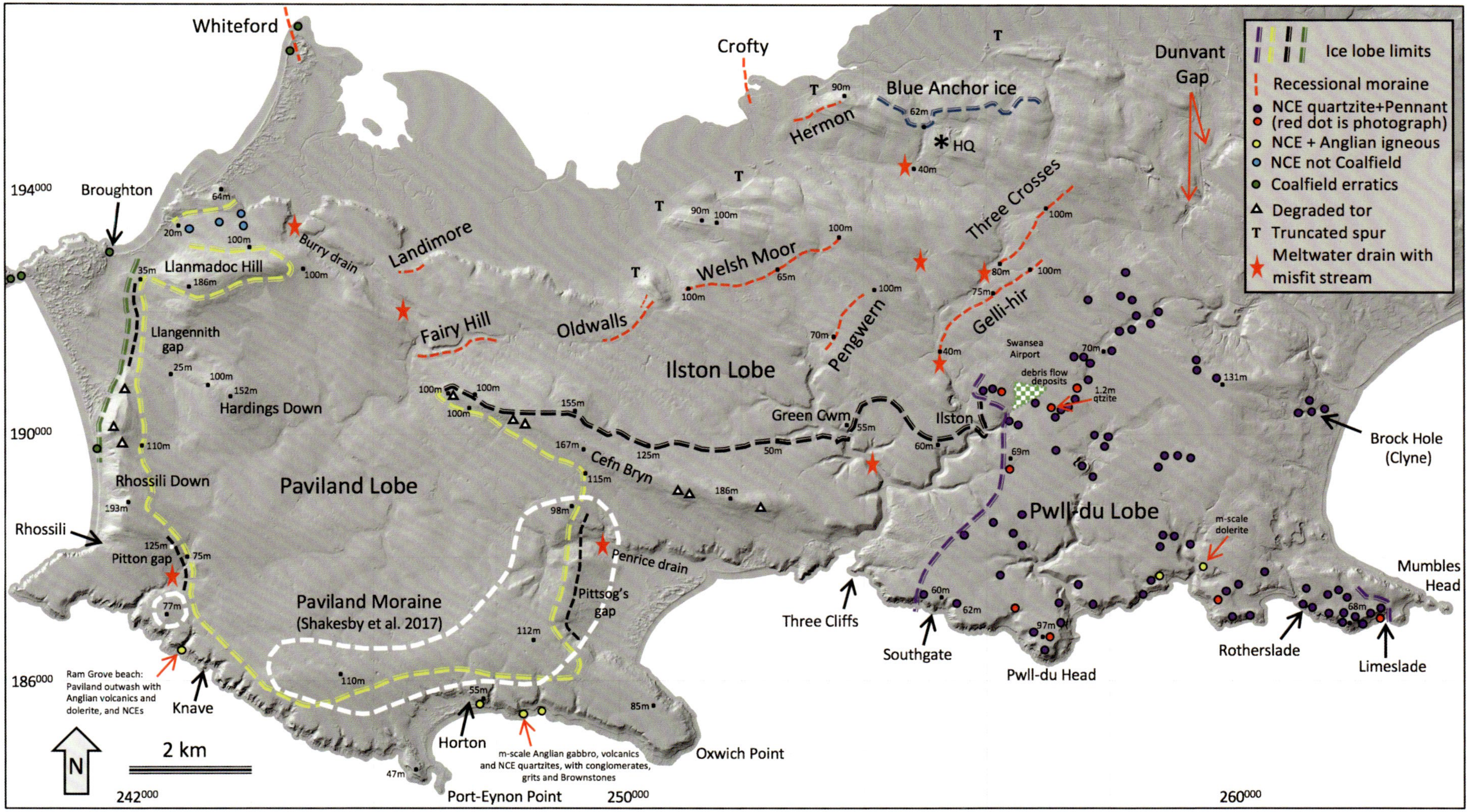

Figure 46. New map of the Last Glacial Maximum (LGM) limits on Gower.

ALL OUR OWN WATER

80

Figure 47. (Page opposite). Devensian north-crop and Pennant erratics in the Pwll-du Lobe on Gower; the generally large size of the quartzites and the abundance of Pennant sandstone blocks distinguish this glacial till lobe from the others to the west, especially the eastern end of the Ilston Lobe.

(a) Langland golf course quartzite cobbles and boulder (SS599872).

(b) Pwll-du Head ganister with carbonaceous plant remains. This specimen is one of many erratics around the summit of Pwll-du Head (97 m OD) where there are other distinctive examples that must register passage of ice and cannot be due to any gravity-driven reworking (unlike Fig. 45).

*(c) Hunts Farm wall with incorporated Devensian erratics of quartzite, **q**, quartz-pebble conglomerate, **c**, and Pennant sandstone, **p** (SS564871).*

(d) Facetted cobble of Brownstones Formation sandstone on Pwll-du Head, formed by abrasion on rock substrate producing a flat face on the one side while embedded in moving ice.

(e) Prominent 1.2 m diameter quartzite erratic near Swansea Airport (SS570905).

(f) Quartzites and slabby Pennant (top right) in Kilvrough limekiln wall (SS563894); piles of similar erratics occur nearby in field clearances.

(g) Quartz-pebble conglomerate; one of a super-abundance of large erratics near Swansea Airport (SS560907).

(h) Cattle-shed wall entirely composed of the diverse erratics of the Pwll-du Lobe (quartzite, Pennant sandstone, Brownstones and conglomerate), at Swn-y-Coed, east of Fairwood Common (SS575908).

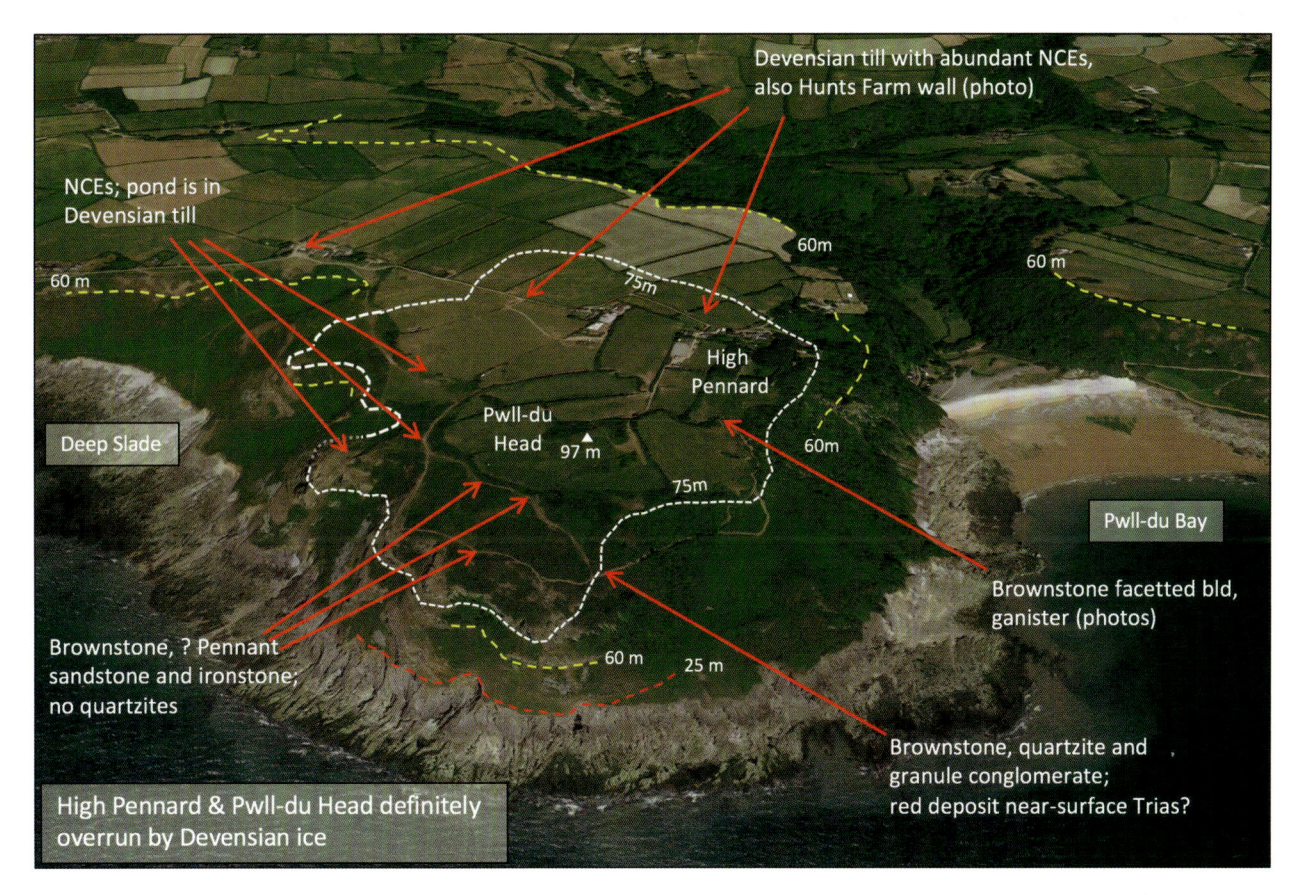

Figure 48. Typical record of fieldwork, here locating erratics on Pwll-du Head, using an oblique Google Earth image (see also Fig. 23). This is a key locality for proving original existence here of Devensian ice. This and similar coastal occurrences between Southgate and Mumbles Head require that Devensian ice continued south across the present coastline. NCE is abbreviation of North-Crop Erratic.

3.6 Pwll-du Lobe: taking the plunge

With the limits of the Pwll-du Lobe defined on land, and then by inference the likelihood of original continuity of ice flowing widely across the cliff line of the time, it was logical to seek evidence for its existence beyond the present coast. Figure 49 shows the seabed offshore from the on-land lobe margins and there are some tantalising features there. What follows here is *necessarily speculative*, although the existence of relatively elevated seafloor features that are 'in the right place' and not obviously related to anything else is intriguing.

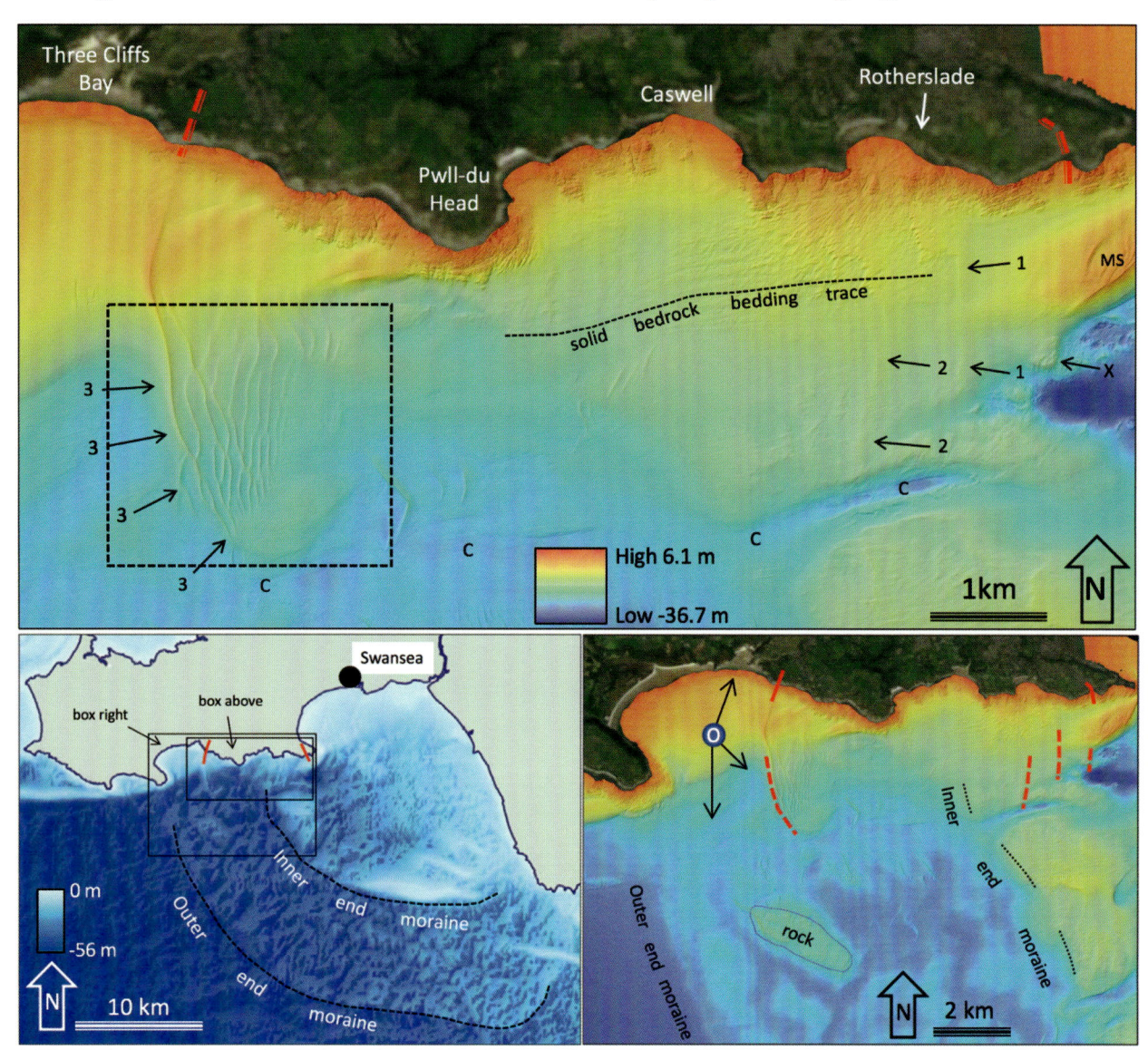

Figure 49. Seabed features where the Pwll-du ice is known to have continued beyond the present coast. The mapped lateral limits of the ice on the coast are marked in red.

*(Top) In the east **X** marks an irregular mound partly buried by the Mixon Shoal sand bank (**MS**). **Arrows 1** and **2** indicate two sets of elongate mounds that are transverse to underlying rock bedding. To the west, in the **box**, is an elevated area of seabed on which sand waves are able to form. This area terminates abruptly to the west where the seabed falls away to deeper levels, indicated by arrows (**labelled 3**). (continued...)*

*Figure 49 continued. Features X and 1-3 are tentatively interpreted as moraine remnants of the Pwll-du ice lobe, which must have flowed over cliffs that at the time were some 80-120 m high (60-90 m OD plus -20 to -30 m below present sea level) and then extended 3 to 4 km farther south. **C** marks a channel. Depths are relative to the level of the lowest astronomical tide: LAT. (Base image multibeam bathymetry, at 2 m x 2 m resolution, from The Admiralty Marine Data Portal, kindly processed and provided by Gareth Carter).*

*(**Bottom left**) End moraines of the Tawe – Nedd ice (from Gibbard et al. 2017) shown in relation to the other images.*

*(**Bottom right**) Seabed features identified in the top panel and accentuated here as **red dashed lines** do not obviously relate in either orientation or continuity to the main end-moraine features of the Tawe – Nedd ice (Bottom left), which would favour their origin being related to the Pwll-du ice lobe. However, their relative age is unclear and ice on Gower at its maximum extent probably coexisted with ice in Swansea Bay. Mumbles Hill may have remained protruding from the ice as a nunatak. The identified channel clearly cuts the Tawe – Nedd moraines and probably was formed by meltwaters during retreat of that ice. It is tempting to suggest that the subtle channel in deep water off Three Cliffs Bay, labelled **O**, could be related to the copious meltwater that must have drained from Gower ice through both Green Cwm and Ilston Cwm, via Parkmill and Pennard valley to join meltwater from what is now Swansea Bay. A similar discharge must have exited via Bishopston Valley and Pwll-du Bay, but the evidence here does not warrant further speculation.*

Regarding the discovery in this study that Devensian ice crossed over eastern Gower, it should be acknowledged that Danny and Bethan McCarroll, in Shakesby and Hiemstra (2015, p. 129), surmised that Devensian ice probably continued south beyond the present shore. They offered no direct evidence and even considered ice covering the whole of the Gower peninsula, but they did point out the problem of having ice conveniently stop just at the cliff top so as to have its till reworked over the edge. Other authors had needed to suggest that the ice delivered material to somewhere on the platform behind the cliff tops from where it was reworked over them. It is true, however, that the south Gower *coastal* localities of Devensian erratic-bearing deposits show no evidence of deposition directly from the base of a glacier; just as at and near the classic locality at Rotherslade (see above), all the coastal deposits show down-slope and aqueous or debris-flow reworking.

This study shows that large erratics, commonly up to 60 cm in diameter, occur all along the eastern coastal fringe (Fig. 46), including on the high points such as Pwll-du Head, and that the ice must have crossed southwards to what is now offshore. The simple interpretation of the manifest reworking is that it occurred during recession of the ice, when it was absolutely inevitable that material left on the slopes would be reworked.

3.7 Ilston Lobe

Figure 50 shows the Ilston Lobe southern limit mapped between the Paviland and Pwll-du lobes. No other limits of the lobe are defined. It is probable, however, that the Paviland ice and the Ilston ice coexisted at their farthest limits, while the Pwll-du Lobe cuts across the Ilston deposits and, on the argument of differing erratic content, seems to represent a separate ice stream. The important issue of main pathways of the ice is dealt with in concluding this chapter, when the available evidence has all been laid out. Although no clear distinction between Paviland and Ilston erratic populations is known, Ilston limits are rather variable and indistinct towards the east and the ice there may have been relatively thin, perhaps on the order of only 10-20 m thick and thus susceptible to breaking around low topography. Ilston Lobe accumulations of erratics are irregularly scattered from the vicinity of Green Cwm eastwards. The 'truncation' by the Pwll-du Lobe deposits is marked by a dramatic coarsening of the north-crop erratics as well as the additional content of Pennant sandstone blocks.

From the western end of Cefn Bryn to the vicinity of Arthur's Stone (highlighted in green in

Fig. 50), and beyond there to just west of Green Cwm, the Ilston Lobe limit is marked by moderate concentrations of north-crop erratics, up-slope of which they are absent and give way to sparse erratics of deeply weathered Anglian UORS conglomerate. The limit traces continuously via intermittent exposures and a few ditches reveal yellow clay-rich till. Although the limit southwest of Arthur's Stone is a minor ridge of moraine, there are no other obvious mounds or continuous topographic breaks. Holy Well springs (see Section 5.13) are located just south of the limit where permeable weathered Anglian till on the slope above gives way to relatively impermeable Devensian till below. This permeability contrast probably initiated emergence of the springs, which have now eroded back by destabilising the slope above them.

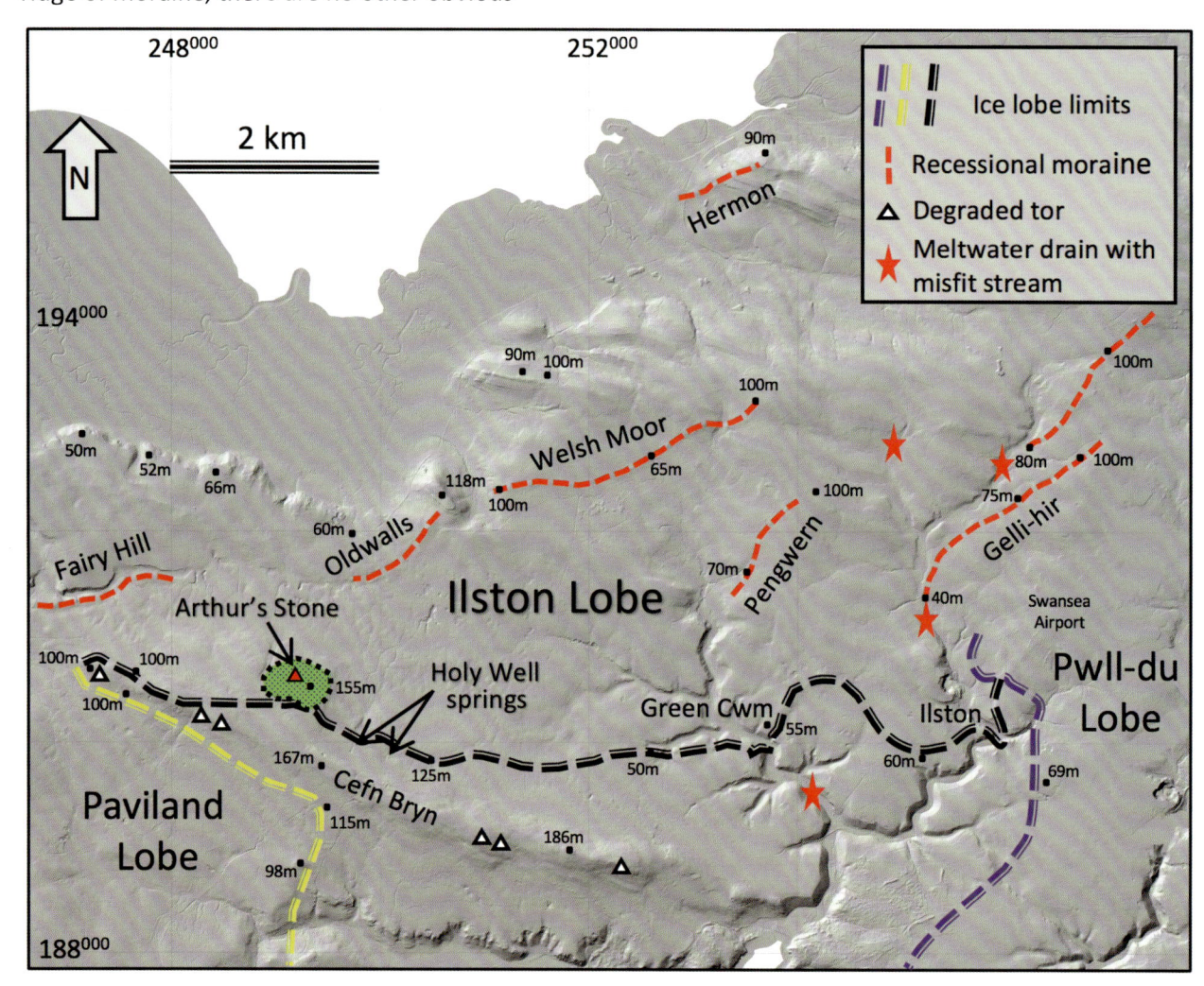

Figure 50. Ilston Lobe Devensian ice limit connected to the Paviland Lobe limit and cut across by the Pwll-du Lobe limit northeast of Ilston. The Pwll-du Lobe limit north of Ilston is lost in the extensive 'made ground' of the original wartime airfield that is now Swansea Airport. The area highlighted in green is where there is a super-concentration of large north-crop erratics, including Arthur's Stone at 147 m OD. This concentration is at the highest part of this lobe limit, probably representing maximal sustained throughput of ice via the relatively low section of cliffs immediately to the north (see below). (The shaded-relief LiDAR image was processed using base data © Natural Resources Wales).

Figure 51 shows an example of the detailed mapping across Green Cwm; the limit shown on the map of Figure 50 appears coarse at that scale, but the boundary generally is known to within a few metres. Mapping here, e.g., noted as the 'Church Problem' (Fig. 51), found the boundary to vary from considerable accumulations of Devensian north-crop erratics, up to metre-scale, to erosional cutting into and mixing with Anglian substrate. Isolated boulders of north-crop origin in places are on Anglian till tens of metres beyond the evidence of the Devensian ice basal contact. They are presumed to reflect melt out from thin ephemeral ice or to have fallen from near the top of the ice front.

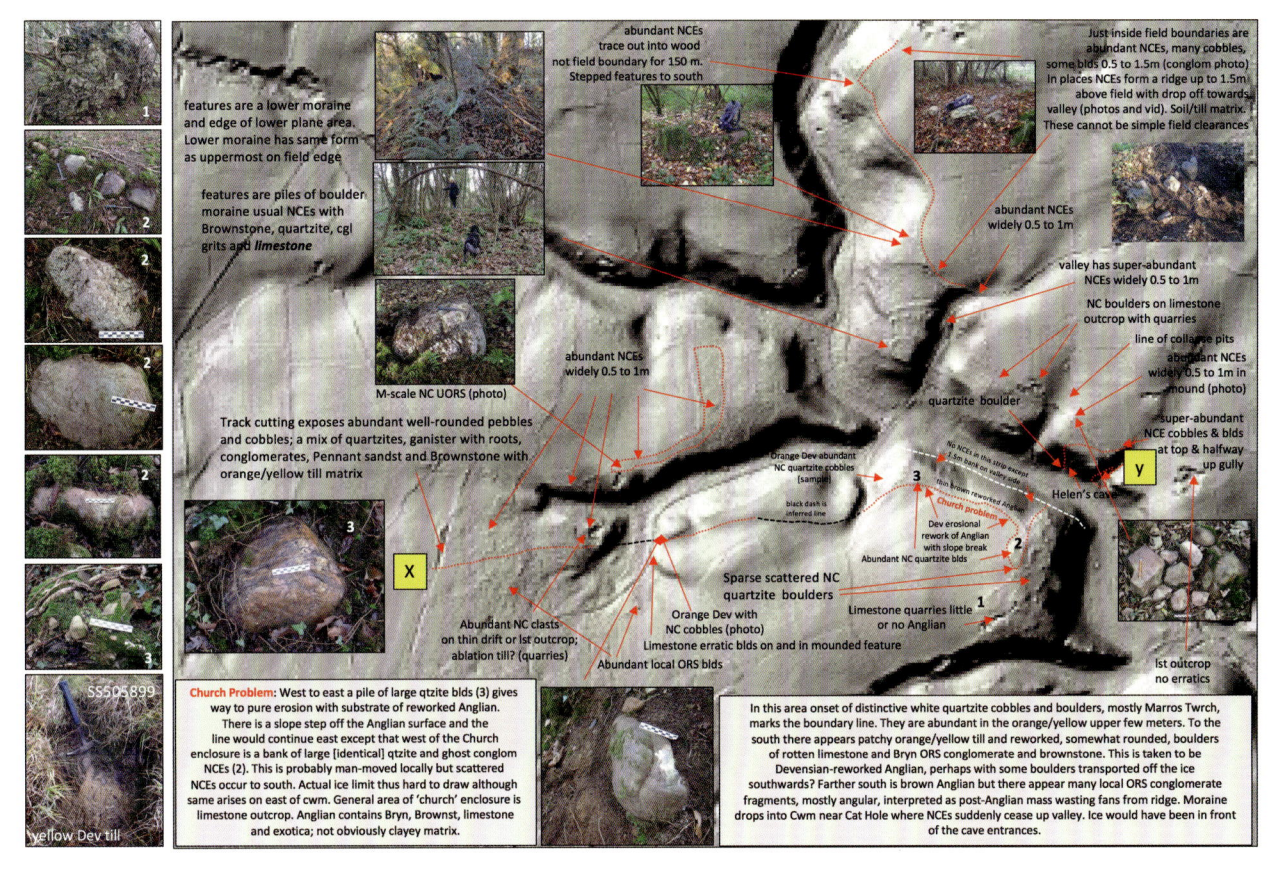

*Figure 51. Sample field record of mapping of the Ilston Lobe limit towards and across Green Cwm (see Fig. 50). The valley sides are mostly mature woodland, so mapping onto images produced by LiDAR, which 'sees through' vegetation, was especially helpful. The boundary mapped here is about 1.4 km from **x to y**. Location y is close to Cathole Rock and Cave (SS537900), opposite which, to the southwest, is Church Hill, ostensibly named for remains of a chapel (SS535898), although there is no evidence for this and abundant Romano-British building materials and artefacts have been found there[9]. In recording this fieldwork, salient evidence is precisely located and backed up by photos. The 'thumbnail' photo images are of original file size so that they can be fully enlarged. (The shaded-relief LiDAR image was processed using base data © Natural Resources Wales).*

Centuries of field clearance of the abundant erratic boulders, especially on the northeast side of Green Cwm, has tended to build some features difficult to distinguish from those due to ice front collapse and melt out. On the east flank within the cwm there are terraces formed of north-crop debris (e.g., SS532904, apparently remodelled by humans around SS533902), which are taken to reflect lateral accumulations from the ice during its recession. The modern valley floor is strewn with large, up to metre scale, north-crop erratics, well exposed where the dry-valley floods have incised a channel (e.g., SS531904) and also by the entrance to the reconstructed Parc le Breos chambered tomb (SS537898). Recessional features are discussed in detail in later sections.

East of Green Cwm to Ilston the ice limit is tortuous and large (>60 cm) boulders are scarce. The ice evidently bridged Ilston Cwm, shedding erratics and presumably copious water into the valley, but it was thin. Outcrops of limestone on the east side of Ilston Cwm are deeply and distinctively 'karstified', fluted and forming pinnacles, which is here interpreted as reflecting continuous exposure since Anglian times some 450,000 years ago (above Cannisland steam and near Courthouse Farm; SS560901). Such ancient karst is rare inland on Gower[10].

3.8 Ice, ritual and ancient preference

Arthur's Stone (Maen Ceti) is one on the best-known Neolithic (before about 4,500 years ago) tombs in South Wales, prominent on the skyline of Cefn Bryn when viewed from the north. It is steeped in legend, myth and mystical powers that are perhaps more interesting than it simply being a 25+10 tonne split erratic of quartz-pebble conglomerate from the north crop (Fig. 52). It has an 'exposure age' of ~23,000 years (Phillips et al. 1994), meaning that its surface has been exposed to the Sun (solar radiation) for that duration and that, as an erratic some 300 m from the southern limit of Devensian ice here, it records the time of the Last Glacial Maximum.

Arthur's Stone rests level with the ground that is outside and beyond what is manifestly a reworked and excavated depression around it. People who know about these things suggest that the stone is unmoved and that the tomb-chamber 'supports' were placed there in space that was dug underneath. Originally the whole tomb would have been buried or partly buried within a broad mound.

Now, not wishing to arouse those persons who prefer mysterious powers to physics, and hence with some trepidation, it is pointed out that the area surrounding Arthur's Stone, highlighted in green in Figure 50, is characterised by a concentration of exceptionally large (metre-scale) north-crop erratics. Many of these appear to be burial sites, marked by the large stones with small ones collected in addition. Sir John Gardner-Wilkinson, an Egyptologist who moved into Reynoldston, mapped the area and referred to it as a "large cemetery" [11] (Wilkinson 1870).

While some may invoke ley lines and paranormal forces that the ancient peoples knew, or felt, this simple geologist finds the real interest here to be the fact that this localized super-concentration of very big erratics at around 145-155 m OD, including the largest north-crop erratic exposed on Gower, is at almost the highest reach of the Devensian ice on the peninsula[12]. There is no question that this area, some 800 x 500 m, has the highest abundance, topographically and numerically, of really big north-crop erratics on Gower, commonly around 60 cm in diameter but with many of metre-scale, and of course Arthur's Stone itself.

Figure 52. (Page opposite). Arthur's Stone (Maen Ceti) on Cefn Bryn (SS491905).
*(**Top**) Neolithic burial chamber with the large fragment on the right broken and fallen from the main capstone. The break probably exploited an existing fracture that ultimately failed due to stress applied by point-loading and/or freeze-thaw. An interesting time to be here must be when (it is written) local maidens bring cakes and crawl around the stones three times to improve their chances of finding a reliable partner. This inevitably would be much more effective than visiting the nearby King Arthur Hotel.*
*(**Bottom left**) The north-crop derived capstone rock has characteristic fairly abundant weathered clay-ironstone (brown mudstone pointed out; see Owen 1964), some small black chert pieces, and few of the dark-red jasper fragments that characterize the local conglomerate (Bottom right).*
*(**Bottom right**) Deeply weathered outcrop surface with characteristic red jasper, exposed since Anglian times (Rhossili Down SS419897).*

The high ground around Arthur's Stone yields both intriguing glaciology and fascinating insights into the preferences of men and women in ancient times. The Bronze Age Round Cairn (Fig. 53) is a deliberate, but seemingly indiscriminate, massive collection of readily portable north-crop erratics. It constitutes a large sample of the ice-delivered debris, populated by abundant quartzites, including ganister, with UORS quartz-pebble conglomerates and LORS Brownstone Formation sandstones; many of the fragments have the typical highly rounded form (Fig. 53b).

Mainly towards the south-southeast of Arthur's Stone, but dotted all around, are the numerous burial sites that Gardner-Wilkinson considered a "large cemetery". Most of these have large north-crop erratics and many appear to be local collections of them. One burial site (Fig. 53c-e), which appears 'disturbed', is notable for its remaining arrangement of similarly sized quartzite stones including many well-rounded examples. These manifestly are a deliberate collection made to adorn this burial site. Compared with the mixed collection of erratics in the Bronze Age Round Cairn, this site records distinct human preference for and selection of the most attractive white and rounded pieces.

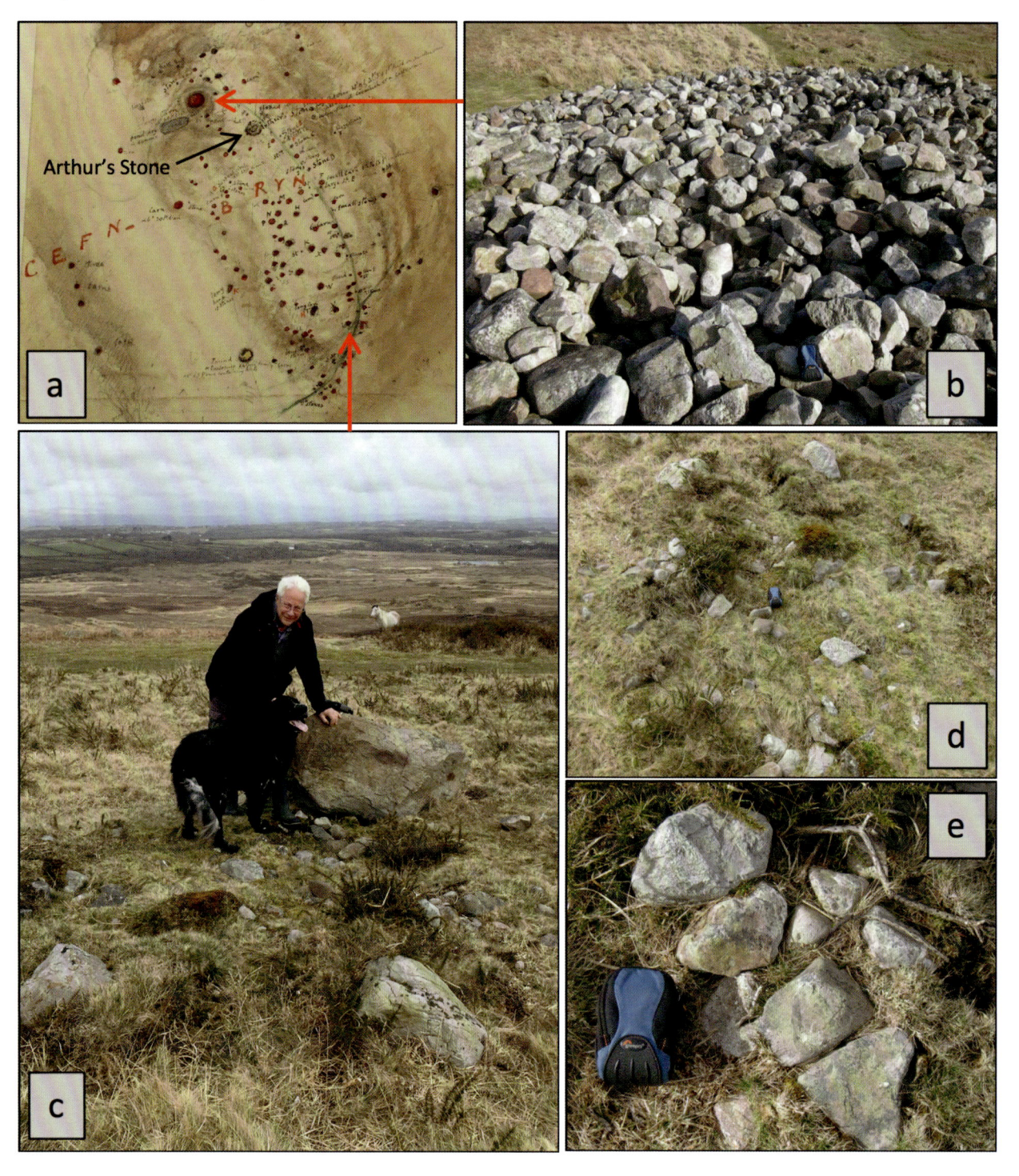

*Figure 53. (Page opposite). Arrangements of north-crop erratics on Cefn Bryn. (**a**) Part of a field map made by Sir John Gardner-Wilkinson showing as red dots burial sites in what he referred to as a "large cemetery"[11]. (**b**) Close to Arthur's Stone is a Bronze Age Round Cairn, which comprises a massive collection of erratics representative of this locality. (c, d and e) One of the small cairns (remains)[13]. The site (**c**) extends towards the camera from a supposed original capstone (and the dog). (**d**) Site viewed in the opposite direction showing an oval outline and (**e**) a representative clutch of north-crop quartzite obviously selected and carried here to ornament the burial or ritual site; such polished and rounded stones are not representative of what is generally available, as reflected by the diverse rocks collected in the Round Cairn (b). (Gardner-Wilkinson's map is in the Bodleian Library, Oxford, and this part is reproduced with permission; © National Trust).*

3.9 Forces of nature

While the monumental site that includes the 'cemetery' on Cefn Bryn is indeed atmospheric for its conjured visions of various early and more recent rituals, there has to be a physical explanation for coincidence of prominently large erratic debris at (almost) the highest reach of the Devensian ice on Gower.

Rather obviously (Fig. 50), this high reach on Cefn Bryn closely faces to the north the least protected, lowest section of north Gower, where ice from the north would not have had to surmount the high sandstone ridges of Cilifor Top (118 m OD) and Llanelen (90 and 100 m OD), or the main Pennant sandstone barriers farther northeast. Although ice did surmount all of these barriers and apparently truncated the spurs they formed, the ice stream that came through with least blocking is likely to have advanced farthest. And, with Cefn Bryn presenting a barrier, that stream is likely to have advanced somewhat higher too, before its sustained collapse and melting at that limit.

Figure 54 shows a highly simplified conceptual view of the effects of the topographic barriers of north Gower. The arrows are not glaciers but are intended to indicate the relative rates and directions of flow of internal parts of the advancing ice sheet – i.e., ice streams; arrow 1 indicates the most impeded flow and arrow 3 the least impeded. In reality, 1 through to 3 would be a continuum within the advancing ice. Thus, the aerial view would be of a complete cover of ice extending to the Ilston Lobe limit while the arrows suggest that the thicker parts of the sheet would be successively less impeded by the topography towards the west.

It is important to remember that the actual barrier presented by north Gower cliffs would have been higher when the ice filled what is now the estuary, although the depth of the valley remains largely unknown. What is not shown here but will be discussed in the conclusion of this chapter, is that the southwards deflection of ice onto Gower, i.e., the curvature of the three arrows, is driven by other ice advancing more directly southwards bearing coalfield erratics and blocking the westwards flow(s) down the main valley.

All else being equal, ice flows farthest where it flows fastest. If flow is constant to a collapse and melt-out limit, then transported debris will continuously arrive at that limit (despite also being deposited as 'lodgement' till from the bottom of the ice during flow). Here it is suggested that the concentration of large erratics high on the barrier presented by Cefn Bryn is an indication that north-to-south flow was (1) relatively strong towards this point (arrow 2 in Fig.54) and (2) sustained for sufficient time for freighted erratics to accumulate in abundance there, near the melt-out terminus of this 'ice conveyor'. The relative strength of this part of the ice flow is attributed to its passage through the low part of north Gower, after severe restriction farther northeast. It is emphasised that other ice would simultaneously have surmounted the barriers and spread, perhaps less vigorously, towards the lower level reaches of the Ilston Lobe, towards Green Cwm and Ilston in the east. Obviously, from Figure 54, the extension of this hypothesis would be that the Paviland Lobe reflects an even greater mass flux of ice that passed between the west end of

Cefn Bryn and the east flank of Llanmadoc Hill, but more on that later.

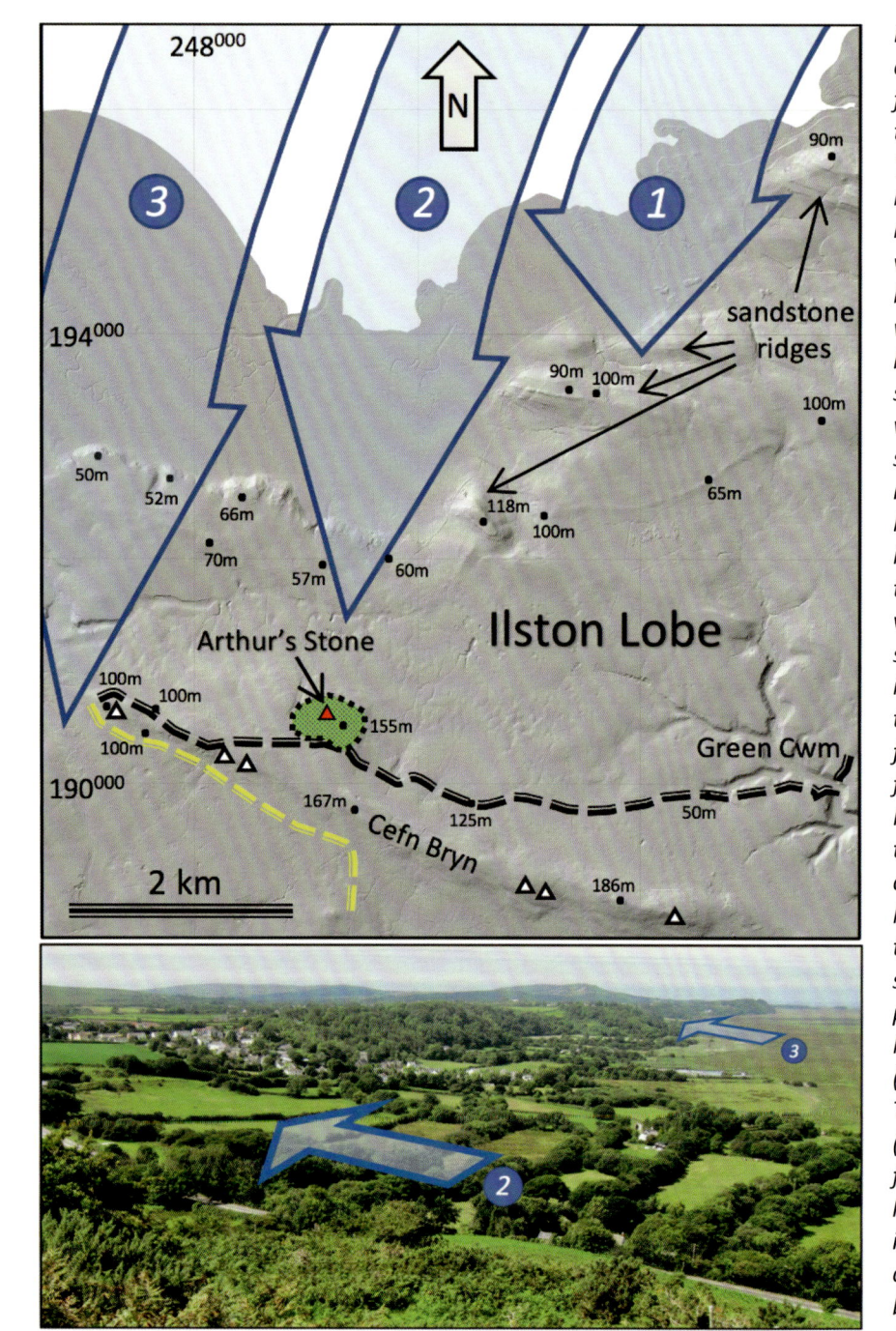

*Figure 54. (**Top**) Diagram depicting relative effects of ice-flow impedance by barrier topography on north Gower. The 'flows' 1-3 are not glaciers but represent ice streams that in reality are parts of a continuous variation of flow velocity. The barriers presented by the cliffs would have been significantly higher than the spot heights shown when the present estuary was fully occupied by ice. The super-concentration of large north-crop erratics at the highest reach of the Ilston Lobe ice, highlighted in **green**, is taken to reflect the relative vigour of ice advance in notional stream 2 in comparison to the impeded stream 1. Stream 3 in this scheme would be that feeding through the low section farther west to form the Paviland Lobe. An aerial view of the ice would show a continuous cover south to the Ilston Lobe limit, with a radial spread towards the southeast. (The shaded-relief LiDAR image was processed using base data © Natural Resources Wales). (**Bottom**) View west from Cilifor Top, near spot height 118 m (above), over Llanrhidian in the foreground showing the till-lined low pass exploited by notional ice stream 2, with 3 in the distance. The barrier would have been greater when the ice filled the original valley on the right.*

3.10 Paviland Lobe

Paviland Moraine was known, and its age debated, for some time before Shakesby et al. (2017) established its LGM origin. From the internal architecture of part of it, they revealed a low-profile succession of debris-flow and fluvioglacial outwash deposits sourced from ice to the northeast. Figure 55 shows the moraine as previously recognised together with a maximum ice limit inferred in this study as the Paviland Lobe.

The Paviland Lobe limit does not simply follow the high points on the outlined moraine, although clearly the ice at some stage had to occur behind and above deposits originating from its front. From the south flank of Cefn Bryn, just below 115 m OD (Fig. 55), around the southeast and south as far as Pitton, the limit is drawn to include the numerous kettle holes recognised in the shaded-relief LiDAR image (see below). This is based on the inference that large (>10 m diameter) calved ice blocks, which formed the kettles, would not be transported far on the outer moraine depositional slopes; catastrophic floods (jökulhlaups) would not have developed except perhaps very locally at one or more drains on the ice margin. Defined in this way the glacial deposits are up to at least 35 m thick, e.g., along the moraine ridge in the south near Hills (110 m OD; SS453861) and towards the northeast around Kittle Top (98 m OD; SS492888).

Shakesby et al. (2017) found debris up to pebble grade near Slade, but the moraine surface and numerous exposures throughout the Paviland Lobe show typical occurrence of north-crop cobbles and boulders in till (Fig. 56). The southern limit around Horton (Fig. 55) may well be conservative in reach, because metre-scale erratics reworked by debris flows widely occur down to the present coast and indicate till coarser than seen elsewhere in this lobe (Fig. 57). Erratics there include Anglian boulders, with gabbros and silicic volcanic rocks most probably from north Pembrokeshire (matching St David's Head and Skomer rocks), presumably swept from the substrate as the Devensian ice advanced broadly southwards. There is, however, no found evidence that Paviland ice crossed the cliffs now at the shore, as did happen for the Pwll-du Lobe; the sea floor is masked by modern sands beyond the rocky apron that extends down to about -25 m OD.

Widely across the east flank of Rhossili Down and on Llanmadoc Hill, the Paviland Lobe ice limit is drawn at the high limit ('fill-line'; see below) of Devensian till, generally marked by changes of slope, drainage and vegetation, and hence often land use. This line is also fairly clear for some 2 km southeast along Cefn Bryn but then trends south around a distinct cluster of kettle holes north of the mapped outline of the Paviland Moraine (west of Penrice Home Farm, SS494889). There are, however, several substantial uncertainties regarding the limit. Evidently the ice came from the northeast, between the end of Cefn Bryn and Llanmadoc Hill, but it would not have been blocked or supported at three places where the ground surface dropped away, designated as Llangennith gap, Pitton gap and Pittsog's gap (Fig. 55).

In the vicinity of Llangennith gap it is conceivable that Paviland ice came in contact with ice from the north, which is inferred from a distinctly different assemblage of erratics, referred to as 'Coalfield lithologies' and dominated by Pennant sandstones. This study has no evidence for what might have occurred in the Llangennith gap, except to say that smoothly continuous till extends down close to the present coastline, at around 25 m OD near Broughton Bay (see below). Conceivably the ice simply wasted away by piecemeal collapse and melting as a small hanging glacier, which is likely at least to have occurred during the post-maximum general recession. There is no evidence for other ice at the Pitton gap and for some unknown time meltwater must have flowed out south-westwards from there through what is now Mewslade Bay (SS421871), conceivably from another hanging glacier. In the bay, abundant boulders of UORS quartz-pebble conglomerate litter the upper beach, along with a few exotic igneous rocks, suggesting an Anglian derivation. Similarly, at Pittsog's gap the underlying softer Marros mudstones there might have been eroded readily beneath and in front of the ice, ultimately forming the Penrice drain. The edge of the glacial debris is under the Oxwich marsh.

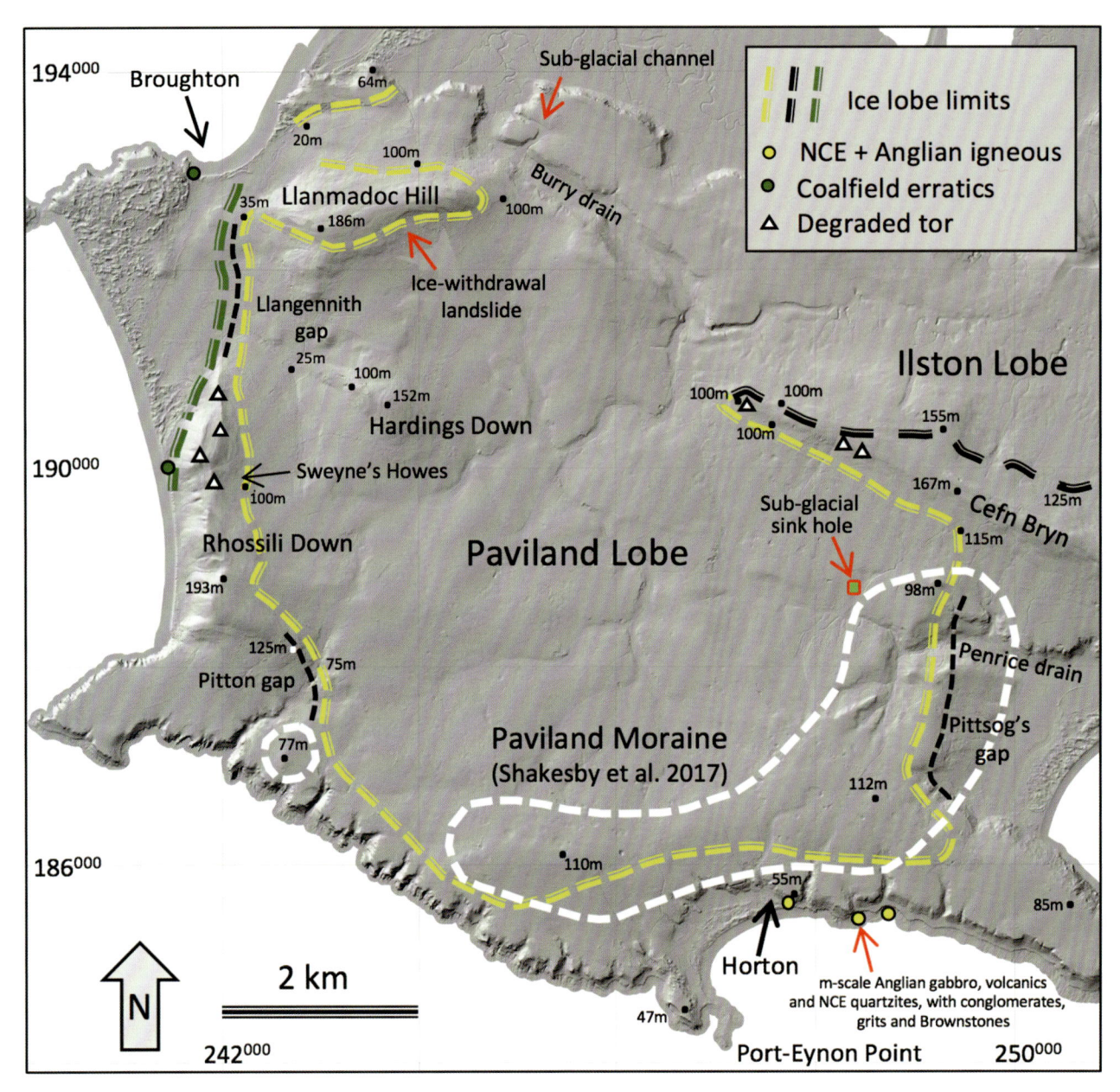

*Figure 55. Paviland Lobe Devensian ice limit inferred from the Paviland Moraine, from the distribution of kettle holes, and from the till fill-line against high ground. The inferred limit (**yellow line**) is the supposed maximum extent of ice, but there is considerable uncertainty about its position in relatively low ground, at the Llangennith, Pitton and Pittsog's gaps (**black dashed lines**), and also at Horton, where unusually large erratics litter the shore; in any case the limit probably would have fluctuated. Also shown within the Paviland Lobe are the locations of The Bulwark ice-withdrawal landslide and the Little Reynoldston pre-Devensian karst sink hole (SS483888), which are described in the text. The Ilston Lobe limit is well defined where north-crop erratics are scattered up to 155 m OD (Arthur's Stone Devensian erratic is nearby at 146 m OD). The limit of the ice bearing coalfield erratics in the vicinity of Llangennith is not so well defined. That the Coalfield Lobe traversed a one-time sea-bed substrate to the north is shown by marine gastropod mollusc shells found in the glacial deposits below Rhossili Down (Prestwich 1892, p. 292). (The shaded-relief LiDAR image was processed using base data © Natural Resources Wales).*

Figure 56. View to west along the topographic crest (110 m OD) of the Paviland Moraine; the ice came from the right. Worms Head is barely visible in the middle distance. Inset is the view to east from the position located by the arrow, showing abundant pebbles, cobbles and small boulders of north-crop erratics (SS451861).

The elevated ground attributed to the Paviland Moraine in the vicinity of Horton is close to the modern cliff tops. Inevitably during maximum ice advance some meltwater and debris would have been channelled southwards through the cliff line via the several slades (Fig. 57). We know now that the slades initially formed as more extensive valleys that were planed off before Quaternary times (Section 2.10), and it appears that Devensian ice meltwater and debris did very little to modify their profiles here. Thermoclastic limestone breccia that formed before arrival of the ice widely lines the floors of the slades, showing that little erosion occurred there before some incursions of debris flows bearing large erratics. The Penrice drain in the east must have taken some meltwater and debris, but otherwise sub-glacial drainage probably occurred principally northwards via Burry drain (Fig. 55), the head reaches of which are mostly within 1 km of the inferred maximum ice limit.

*Figure 57. Glacial erratics of the Paviland Lobe at and east of Horton. (**a**) North crop quartzite in a garden wall at Horton (SS477855). (**b**) Shore below Western Slade (SS484855) where in past decades coastal erosion has been sufficient to both displace the coastal path inland and make new collapsing exposures of reworked Devensian till. The debris fans here comprise Devensian thermoclastic limestone breccias with layers and lenses of glacial debris. On the axes of the dry valley slades here (Section 2.10), reworked till with boulders 60 cm to 1 m in diameter of both Anglian and Devensian origin occur, feeding the erratics onto the beach. The **red arrow** marks the location of (c), with (d-f) nearby. (**c**) All brownish, non-grey rocks are north-crop quartzite, grit and conglomerate erratics, commonly >60 cm in diameter. (**d**) Metre-scale north-crop quartzite, **q**, and Anglian volcanic conglomerate, **v**. (**e**) Rhyolitic volcanic conglomerate (v in panel d) originally delivered to Gower by Anglian ice, most probably from Pembrokeshire (conceivably Skomer volcanics). (**f**) Metre-scale Devensian north-crop quartzite, **q**, and LORS Brownstones Formation sandstone, **b**.*

3.11 Little Reynoldston subglacial sink hole

When ice advanced from the north onto Gower limestones it spread across perforated karst ground that was widely smeared with old (weathered) Anglian till that had been vegetated. It seems likely that at least initially some basal ice-contact water would have drained away, and it is tempting to suggest, as others have (e.g., Shakesby et al. 2017), that advance of the ice over sink-hole-perforated limestone would have influenced its mobility. At this near-terminus and low-altitude location of the Welsh Ice, the glacier would have been 'warm based' and so to an extent lubricated by meltwater along basal contacts, as opposed to frozen to the ground as occurs with 'cold-based' ice. This study contributes little to this idea except to say that the till widely deposited from the Paviland ice is at least locally sufficiently rich in clay to be essentially impermeable, as evident from the numerous small ponds. Thus, while encroachment of the ice might initially have found some free basal drainage, further advance seemingly sealed some of the ground.

South of Little Reynoldston, a small quarry (Fig. 58) exploited the exposed limestone lip of a sink hole. On the north wall of the hole, Devensian erratics are cemented in a calcareous coarse-grained sandy matrix, and this fill would originally have been free draining. Paviland ice must have advanced over the hole, which would have drained subglacial water away from this vicinity. The erratics have not yielded definitive north-crop quartzite, but the rounding of the tough sandstones and conglomerates indicates fluvioglacial modification, which suggests the deposit is Devensian and not Anglian. The time of carbonate cementation is unclear. An active sink 70 m to the east was deliberately circumvented by a large drainage channel that was cut by farm-estate workers into the nearby Devensian till to distribute water from Cefn Bryn to the fields below. This early water management, probably by the Lucas estate at Stout Hall, highlights the variable thickness of the Devensian till and its generally low permeability.

Figure 58. Limestone quarry with partly preserved debris-filled pre-Devensian sink hole, south of Little Reynoldston, viewed towards east (SS483887). Boulders and cobbles in pebble-to-sand-grade matrix are cemented with ferruginous calcite against the steep wall of the sink hole, 2-3 m below the quarry top. The large rounded fragments are erratics of UORS conglomerate, brown sandstone, vein quartz (from the conglomerate) and limestone. The rounding and mixture indicate Devensian glacial derivation. (Notebook is 15 cm). Quarries such as this typically started where the limestone was exposed above surrounding till to form an escarpment that facilitated excavation.

Sink holes are common features of Gower limestones, but many are at least partly obscured by till and, as discussed below, some Devensian till has abundant depressions related to melt-out of ice blocks, which can be confusing. The best place to see clearly the incidence of true sink holes is where the Devensian ice did not go, e.g., along the south flank of Cefn Bryn between Perris Wood (SS502887) and Nicholaston (SS520884). Aligned multiple depressions in till are evidence of fault control of sink-hole locations, e.g., the 0.5 km line of a dozen or so holes that crosses the main roads near Cillibion (SS514913). Somewhat disturbingly, buried or partly buried sink holes also routinely distinguish themselves by their ongoing sinking activity, as reflected in the regular need for road repairs. There were four non-trivial instances of road collapse in the period of late 2019 through 2020.

3.12 The Bulwark landslide

It is common during deglaciation for steep slopes originally flanked and possibly eroded by ice to collapse in slumps, in which the debris is deformed, and slides where the debris moves *en masse* on a basal detachment. Typically, in nature, there are elements of both styles. As the ice surface lowers during recession upper parts of slopes become unsupported while water saturation and elevated pore-water pressure remain in the slope below; this 'paraglacial[14] development' is the main condition leading to slope failure. Such a collapse is inferred to have occurred on the south flank of Llanmadoc Hill, where continued glacial-ice flow during recession apparently transported landslide material westwards (Fig. 59).

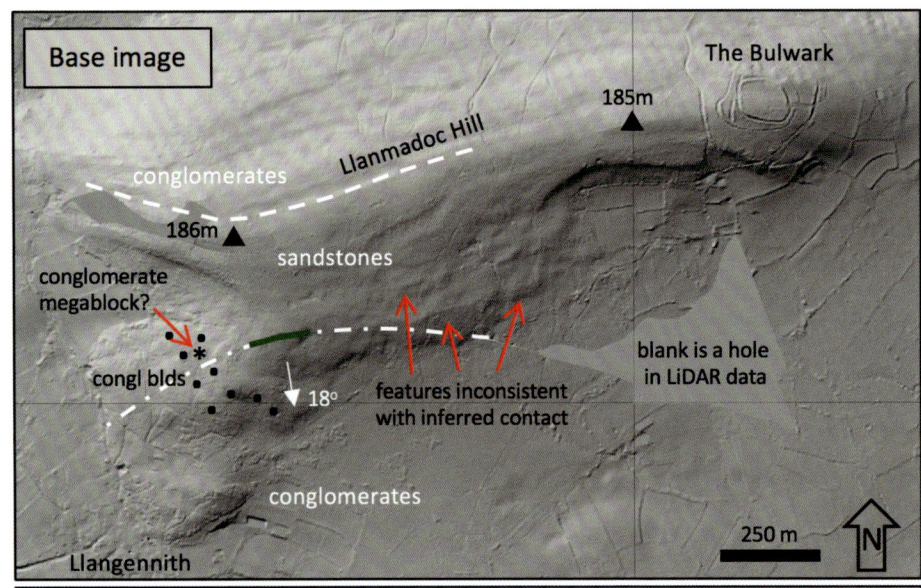

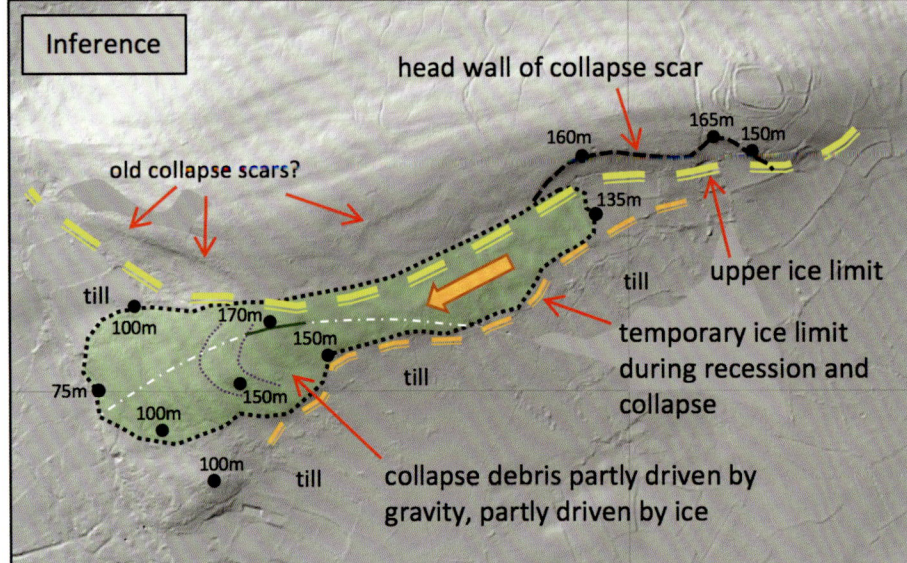

Figure 59. Inferred ice-withdrawal landslide on the south flank of Llanmadoc Hill, viewed with shaded-relief LiDAR data.

*(**Top**) The Bulwark Iron Age hillfort on the eastern shoulder of the hill lies above a distinct steep 'bite' that cuts into sandstones of the Lower Old Red Sandstone (LORS) Brownstones Formation (geology in **white** from British Geological Survey 2002). The contact with Upper Old Red Sandstone (UORS) conglomerates to the north is plain to see and to trace on the ground. The same boundary to the south is not obvious and the contact, shown here as a **white dot-dash line**, probably has been inferred during geological mapping and extrapolated from a <u>locally</u> distinct change in loose debris: conglomerate fragments to the south, sandstone to the north, marked by the **short green line**. (continued...)*

*Figure 59 continued. Whereas the contact and parallel stripes due to the rock layers are distinct along the ridge and to the north, the ground to the south is irregularly mounded. Here ridges are not consistent with the interpreted geology but are parallel to the inferred landslide deposit in near-source areas, and with indistinct west-facing bulbous forms farther west (**fine dotted lines**). Large conglomerate boulders are abundant in a zone across the inferred contact and one occurrence, large enough to be solid outcrop (on the 'wrong' side of the inferred contact), probably is a loose mega-block; it occurs with sandstone blocks. No solid rock dipping south at 18° was found.*

*(**Bottom**) It is inferred that the south flank of Llanmadoc Hill records a gravitational collapse that left a headwall escarpment near The Bulwark and that, instead of the landslide simply moving freely away from the headwall scar, it collapsed against ice that constrained it to flow westwards and probably also transported debris that way. Extending the inference, it seems likely that the initial collapse resulted from lowering of ice that earlier had erosionally over-steepened the rock slope, which must have included UORS conglomerate. The debris deposit is 1.3 km long and 0.3 km wide in the lobate snout; its northern edge rises westwards to slightly higher (170 m) than the collapse scar (165 m), suggesting ice support and transport. The northwest end of the lobate snout rises sharply from the till slope there and appears superimposed upon it (see Fig. 60). (The shaded-relief LiDAR image was processed using base data © Natural Resources Wales).*

*Figure 60. Devensian till in the Paviland Lobe. The **white dashed line** is the 'fill-line' of the upper limit of the till. (**Top and middle**) Till wrapping the base of Llanmadoc Hill down to low levels of around 25 m OD, viewed from the same place near Delvid (SS421926). (**Top**) View east-southeast towards the lobate termination of The Bulwark landslide, which rests on till continuous to the foreground. This shows that at its maximum extent the Welsh Ice of the Paviland Lobe extended down close to the modern shoreline and that the landslide occurred afterwards, during recession, onto the freshly exposed till slope. Soliflucted head and boulders on the left probably reflect destabilisation of the slopes above owing to a spring line that extends at the slope foot from Delvid to Lagadranta. (**Middle**) View northeast showing the tongue of Paviland-type till that extends down into the coastal blown sands and marsh behind Broughton Bay (see Fig. 55). (**Bottom**) Till fill-line across the east flank of Rhossili Down, viewed from Sweyne's Howes Neolithic burial chamber remains (near and middle-right). The fill-line is distinctly marked by changes of slope, drainage and vegetation.*

3.13 In recession: kettles and kames

There are hundreds of shallow depressions in the glacial debris of the Paviland Lobe. They are typically circular and range widely in diameter from a few metres to hundreds of metres; many support ponds. The depressions are called kettles, or kettle holes, formed by melting away of ice blocks that became buried, or partly buried, by glacial sediment. The holes tend to be circular irrespective of the ice-block shape, because the glacial debris that is left unsupported in the melting fails on arcuate fractures; these successively step outwards as the debris collapses inwards and disintegrates. Eventually the collapse forms a smooth-floored pond (Fig. 61). In the Paviland Lobe the kettles register recession of the ice front where collapsed, or 'calved', ice blocks successively were buried farther away from the original ice limit. Short alignments of large kettle holes in the upper parts of the Paviland Moraine suggest simultaneous calving of long strips of the ice front in recession (Fig. 62).

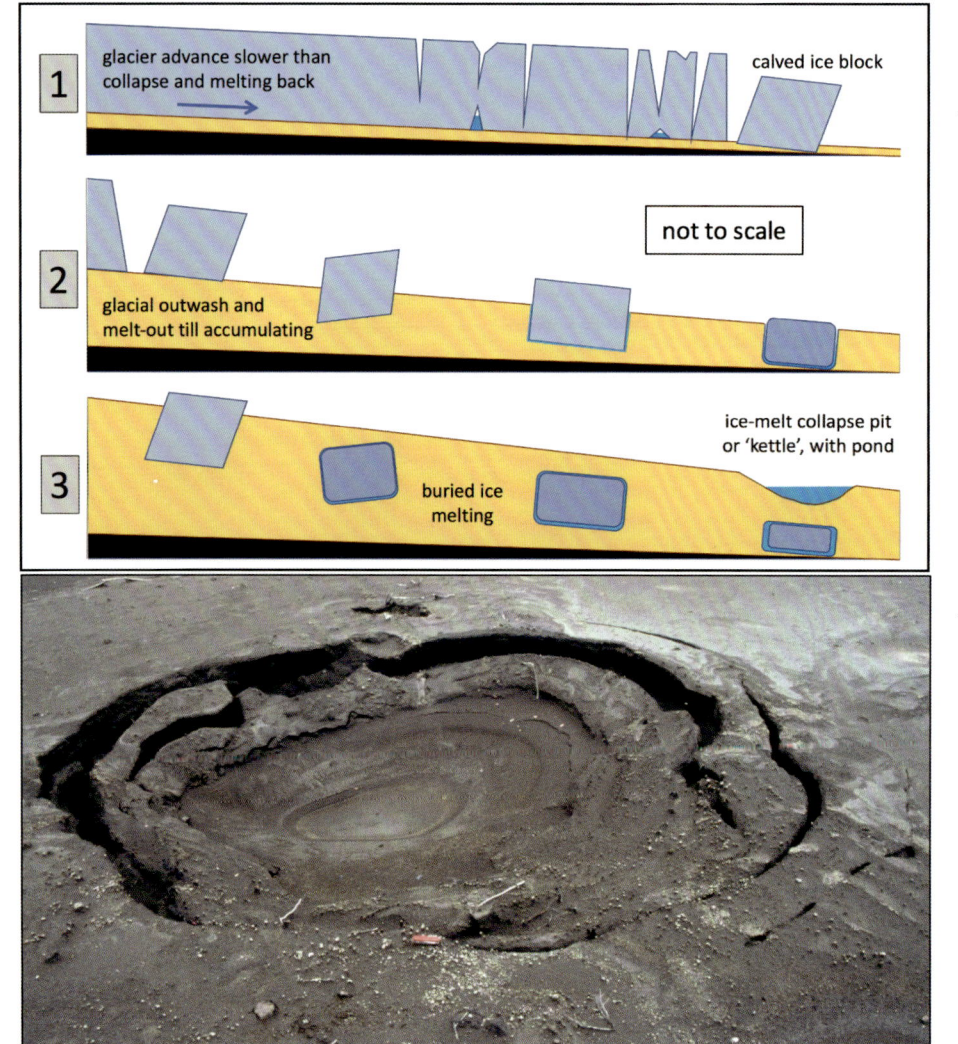

Figure 61. Kettle formation.

*(**Top**) Simplified stages of kettle formation during glacial recession. Although the glacier may still advance from its upland source, its front recedes by collapse and melting.*

*(**Bottom**) Ice-melt collapse in progress, 3.5 m wide, showing outwards stepping increments of collapse of unsupported debris on arcuate fractures (photo courtesy of Mike Branney).*

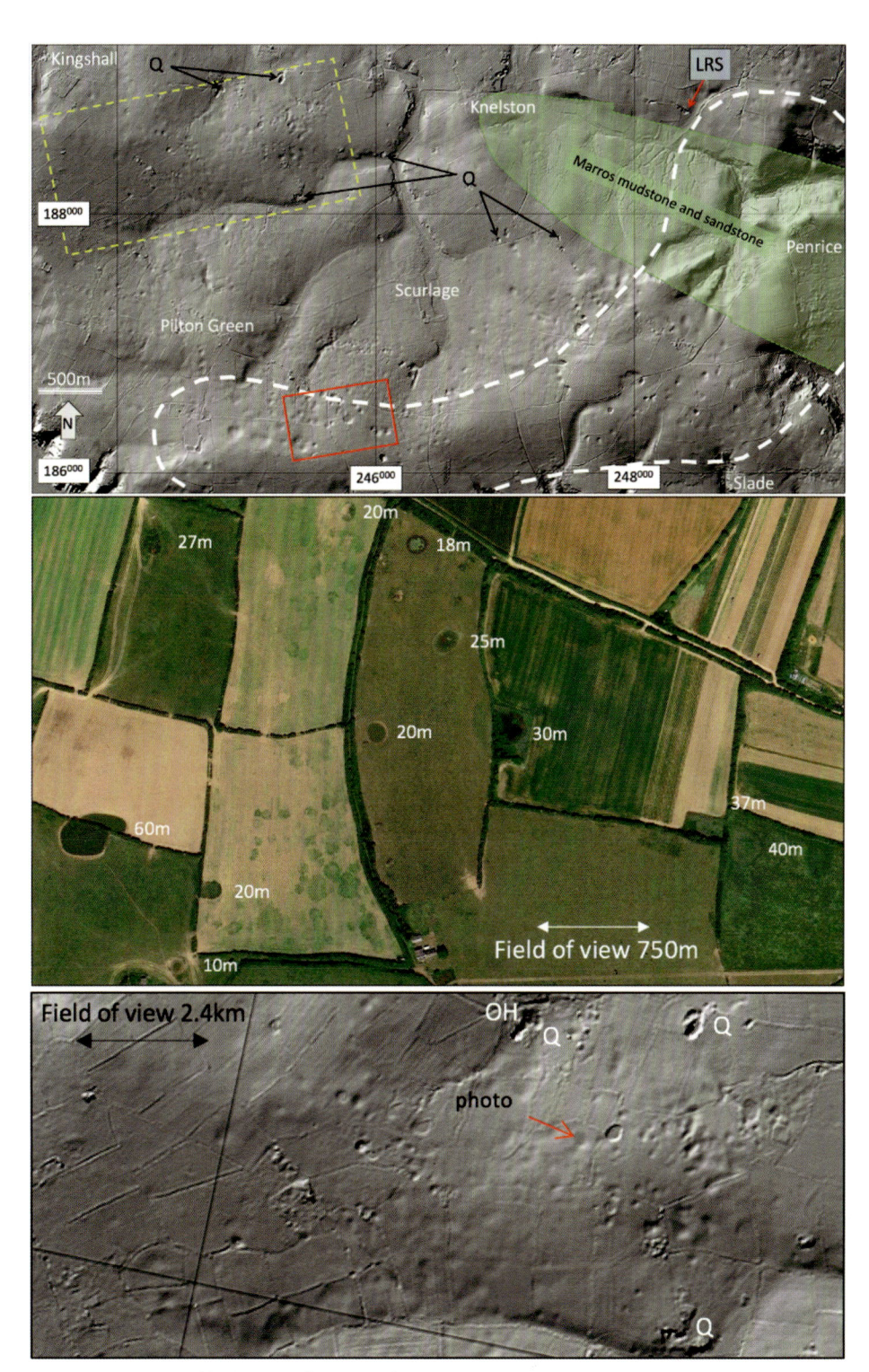

Figure 62. Kettles and kames of the Paviland Lobe near Scurlage (SS464876).

(**Top**) *Shaded LiDAR view of part of the Paviland Moraine (**white dashed outline**) and the drainage basin behind it. The **red box** locates the middle image and the **yellow box** the lower image. Quarries (**Q**) generally appear sharply defined and irregular; kettles are no more than a few metres deep, generally circular in outline, 10-200 m in diameter and softer edged; they considerably outnumber quarries. There are >50 kettles in the ~1.25 km^2 north and northwest of Slade (bottom right). **LRS** marks Little Reynoldston subglacial sink hole.*

(**Middle**) *Google Earth image of kettles, with their diameters, some forming ponds reflecting clay in the substrate. The 60 m-diameter kettle (left) is the easternmost of a row of three large ones (top image) that probably reflect simultaneous calving of ice blocks from the receding ice front. The building at the middle lower edge of the image is Littlehills (SS457861) situated on the crest of the moraine.*

(**Bottom**) *Shaded LiDAR view of Old Henllys (**OH**) kettle and kame field.*

*Irregular low-profile kames (mounds) amidst the kettles reflect ice-melt stranding of debris that widely is a few tens of metres thick; the quarries (**Q**) are at around 50 m OD and the accumulated debris rises towards top left to 105 m OD. Surface drainage was active as the ice melted, causing some kettle margins to slump into it (bottom left of image). The **red arrow** marks the position of the kettle shown in the upper image in Figure 63. (The shaded-relief LiDAR images were processed using base data © Natural Resources Wales).*

*Figure 63. (**Top**) Shallow kettle hole 40 m in diameter with desiccated clay pond floor 20 m wide showing pale green (SS450886).*

*(**Bottom**) View towards south of a pond, 27 m in diameter, in a kettle hole near Penrice Home Farm (SS493890; access courtesy of Thomas Methuen-Campbell). The conical mound to the left beyond the pond is Kittle Top (SS492888), which is an erosional remnant of this northern end of the Paviland Moraine (Fig. 55), and the distant skyline is the summit of the moraine above the south Gower coast. The pond is at 94 m OD, the summit of Kittle Top is at 98 m and the skyline high points are 112 m and 110 m. The floor of the 'basin' behind (to the right or west of) the moraine has original low parts at 45-50 m OD, so the Paviland ice must have been at the very least some 70-80 m thick when it receded from the moraine crest shedding ice blocks into the accumulating debris.*

3.14 Recessional moraines and drains

The Welsh Ice began to retreat as climatic conditions changed from those of the glacial maximum here, after 23,000 years ago. With warming and less snowfall, the ice at source thinned and at its extremities began to recede. During recession ice still flows from its upland source, but the increasing terminal collapse and melting progressively reduce its reach. The ice is widely acknowledged to have gone completely from Wales by 16,000 years ago (Clarke et al. 2012) and in the thousands of years of the recession there inevitably were fluctuations of the ice flow, possibly including some temporary re-advance. Periods of relative stability affecting the ice caused its reach to be temporarily maintained, so that freighted debris was dumped along a static front.

Recessional moraines are evidence of temporary halting of glacial retreat and Gower displays several, some perhaps previously unnoticed (Figs 64 and 65). From Figure 64 the order of recession is, broadly, from east towards northwest, but there may be missing elements. It is not clear whether the Pengwern and Three Crosses features are contemporary, although Welsh Moor, Old Walls and Fairy Hill have long been identified as part of the same temporary limit. During recession the meltwater that was released, along with normal precipitation runoff, would adjust courses to drain away, typically carrying suspended fine sediment. Gower reveals that some of this 'fluvioglacial outwash' was substantial and locally a moderately powerful agent of erosion.

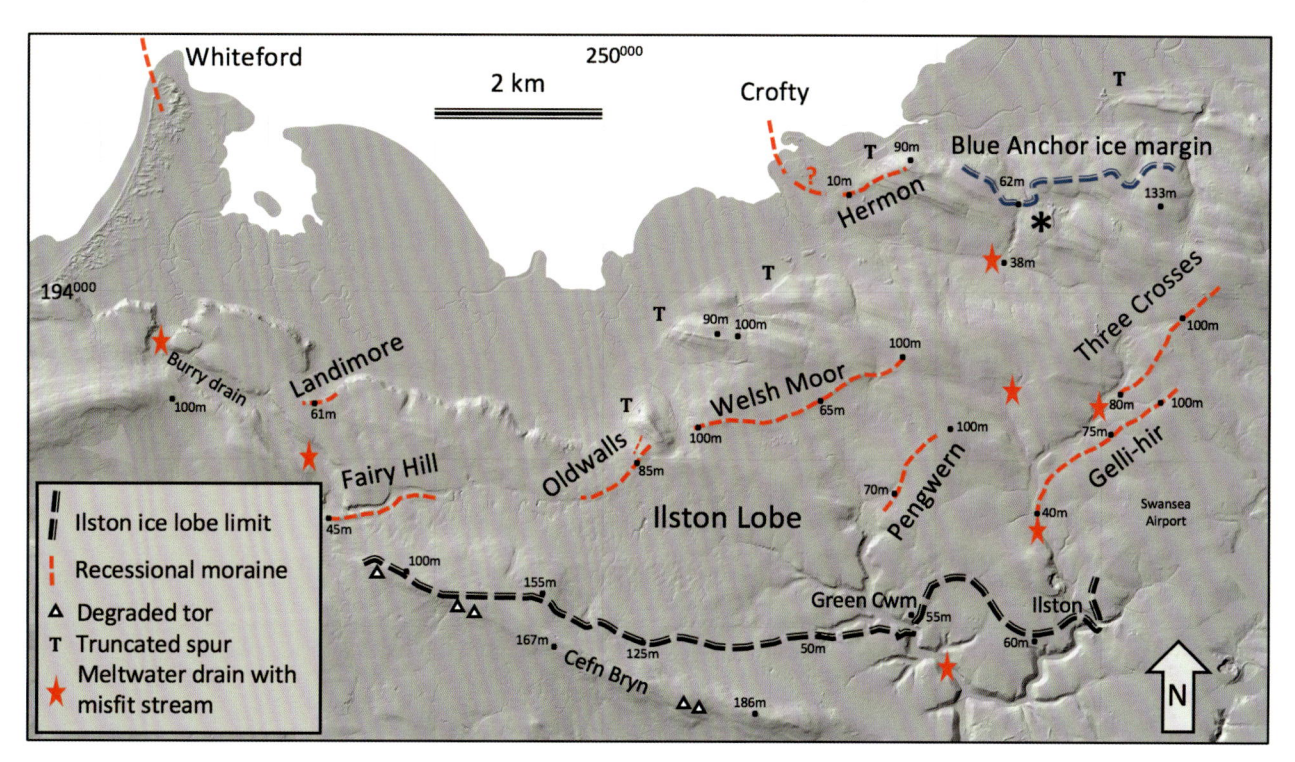

Figure 64. Post-Last Glacial Maximum moraines formed during the collapse and melt-out of the Ilston ice lobe and, ultimately, of ice occupying the valley of the present Loughor Estuary, as marked by Whiteford and Crofty moraines and the Blue Anchor ice margin (Section 3.15). (The shaded-relief LiDAR image was processed using base data © Natural Resources Wales).

Apart from one minor morainic ridge southwest of Arthur's Stone, very little of the Ilston Lobe ice limit is marked by any pile-up of moraine in the sense of 'terminal moraine', in contrast to Paviland Moraine. The limit is mostly marked by densely scattered erratics with little topographical expression, although the terrain in the vicinity of Green Cwm, with its complex of steep cliffs, throws the limit into some considerable irregularity (Fig. 51). The Ilston ice probably was relatively thin near Green Cwm, perhaps only a few tens of metres thick at most, and in this distant spreading reach it would have been sluggish and, in recession, prone to forming a featureless sheet of till (Fig. 65, scenario 2).

The Gelli-hir moraine is extensively (not everywhere) marked by a steep topographic drop of at least several metres on the northwest side, with boulders strewn along the high point; it corresponds to scenario 3 in Figure 65. As the ice melted away from the Gelli-hir moraine, the topographic step locally became modified by meltwater that flowed generally south-westwards along and close to it. Figure 66 shows the southwestern part of the Gelli-hir moraine where it lies along part of the drainage that extends from Three Crosses to Three Cliffs, via Ilston Cwm (Fig. 46). The diagram shows the drainage upstream from Cartersford (SS551913) to the bridge below Gelli-hir Wood (SS559926), a distance of 1.5 km in which the present stream drops 26 m as the moraine crest drops 34 m.

ALL OUR OWN WATER

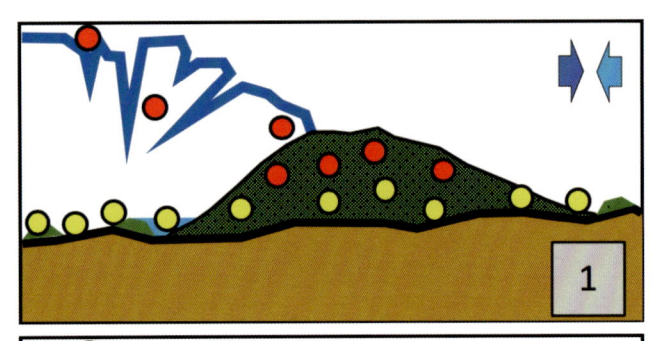

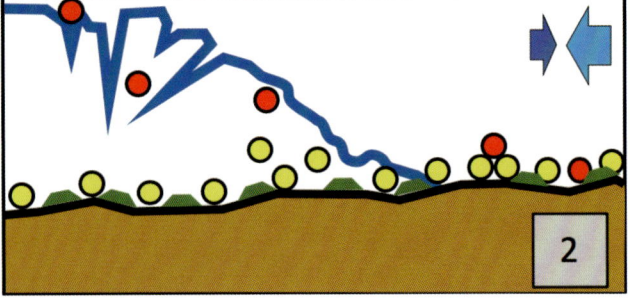

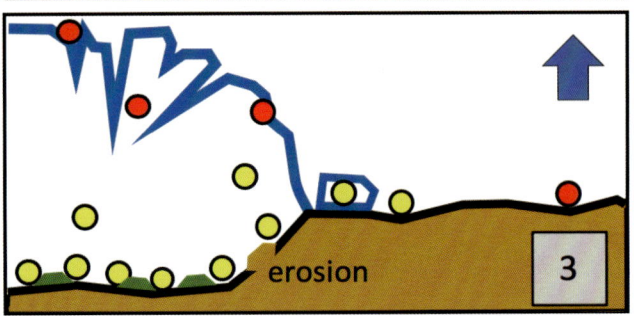

Figure 65. Schematic illustration of the three main types of Devensian recessional ice contact found on Gower. Red and yellow circles represent erratic boulders (i.e., ≥25 cm diameter) initially transported at different levels in the ice, green is finer grained till, dark blue arrows represent ice flow and pale blue arrows represent the melt-back trajectory. The brown substrate is erodible till. This can be Anglian in age at the ice limits and locally within them, but widely is Devensian beneath the recessional ice, having been deposited during the advance.

(1) Ice advance and melt-back are temporarily balanced so that the ice limit is static; a terminal or recessional moraine is built. The Pengwern and Welsh Moor recessional moraines are continuous long mounds of this type rich in pebbles, cobbles and boulders.

(2) Melt-back exceeds ice advance, forming a discontinuous thin layer of till and stranded boulders. Parts behind the Ilston Lobe limit to the west and east of Green Cwm are of this type. Extensive featureless areas between the recessional moraines in the Ilston Lobe are of this patchy scatter wherein large erratics are sparse.

(3) Ice movement is practically parallel to the contact and erosive into substrate; some erratics are scattered from melted fallen ice blocks or individually tumbled from the ice margin. Such a lateral contact occurs locally above Green Cwm (Fig. 51, the 'Church Problem'), just behind the limit, and also extensively along the Gelli-hir recessional moraine (Fig. 64). It may arise only where the ice is confined against an original valley side. In the course of melt-back a topographic step is left capped with erratics.

Figure 66. (Page opposite). Shaded LiDAR image of Gelli-hir moraine and associated fluvioglacial drainage some 1.5 km upstream from Cartersford bridge (SS551913). Several generations of drainage are evident, but times of their initiation are not all clear (see text). It is certain, however, that the present stream is a meandering misfit within the fluvioglacial outwash plain surface, which averages about 100 m wide. The bedrock base here slopes generally towards southwest, trending off the higher ground of Lower Coal Measures down towards the softer Marros Formation mudstones. Thus, drainage to southwest is inevitable and the gorge cut through solid rock (mid-left) could have formed before the arrival here of Devensian ice, or even beneath it. (The shaded-relief LiDAR image was processed using base data © Natural Resources Wales).

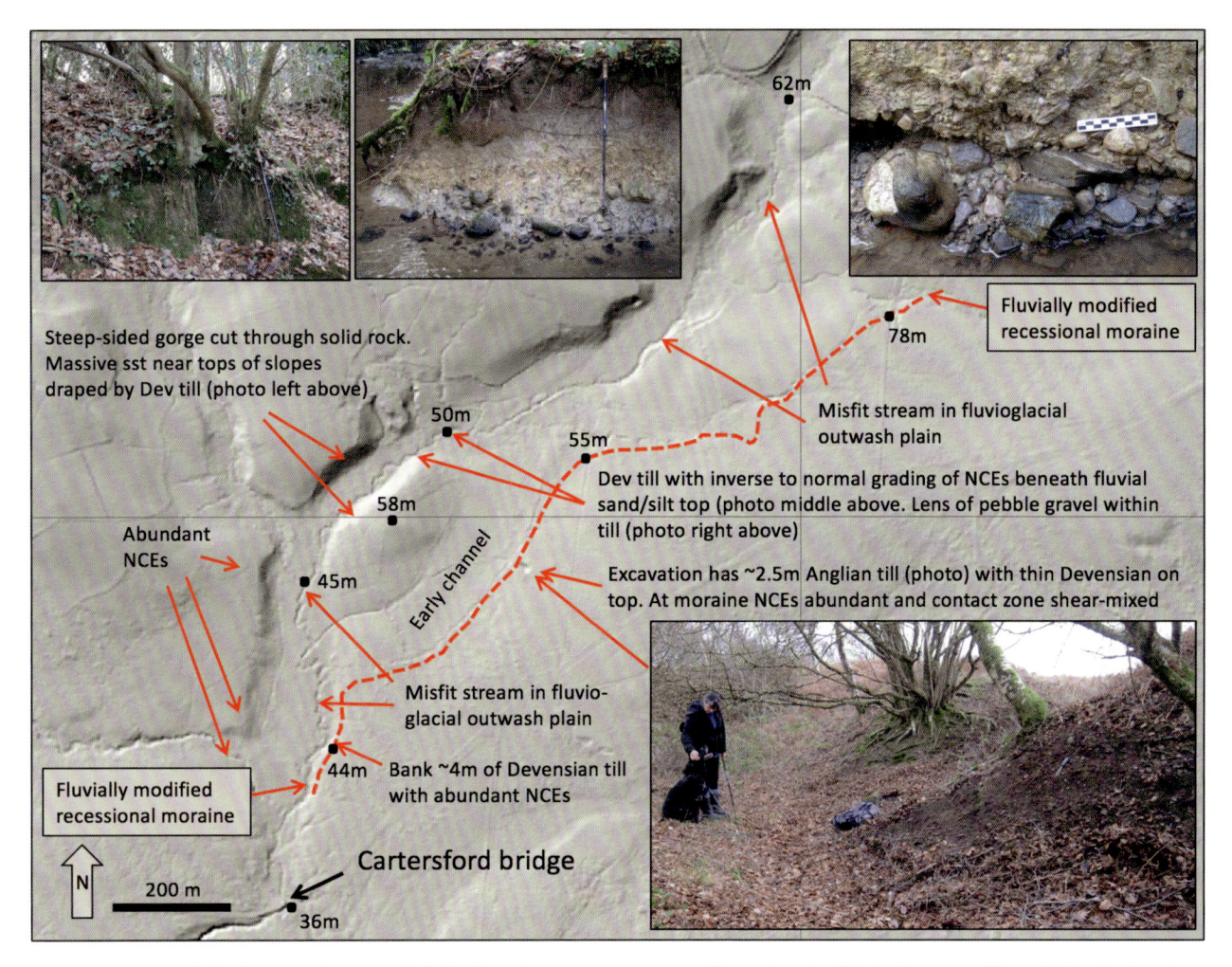

[Figure 66: Annotated LiDAR/relief map near Cartersford bridge with photo insets. Labels on map:]

- Steep-sided gorge cut through solid rock. Massive sst near tops of slopes draped by Dev till (photo left above)
- Fluvially modified recessional moraine
- Misfit stream in fluvioglacial outwash plain
- Dev till with inverse to normal grading of NCEs beneath fluvial sand/silt top (photo middle above). Lens of pebble gravel within till (photo right above)
- Excavation has ~2.5m Anglian till (photo) with thin Devensian on top. At moraine NCEs abundant and contact zone shear-mixed
- Abundant NCEs
- Early channel
- Misfit stream in fluvio-glacial outwash plain
- Bank ~4m of Devensian till with abundant NCEs
- Fluvially modified recessional moraine
- Cartersford bridge
- Elevations: 62m, 78m, 50m, 55m, 58m, 45m, 44m, 36m
- N, 200 m scale

It is not straightforward to establish the times of development of the drainages shown in Figure 66. Given that the solid-rock base generally falls away south-westwards, drainage in that direction is inevitable. Who knows what the Anglian landscape that formed 450,000 years ago would have been like by the time of the freezing cold that came with the Late Devensian glaciation? We can safely assume that the Devensian ice broadly approached from the north and northwest and more certainly that it receded towards northwest. At this near-maximum reach of the Welsh Ice, total freezing would not have persisted, if it occurred at all, so there would have been fluvioglacial outwash ahead of the advancing ice, followed by active drainage beneath the ice when at its full extent, and then outwash again during the recession.

That the Gelli-hir feature (Fig. 66) is indeed a recessional moraine and not merely an ancient riverbank is established by the marked concentration of north-crop erratic boulders (NCEs) all along the mapped line[15]. Southeast of that line Devensian (north-crop) erratics are scarce in Fairwood Common and a few exposures show that some of the till thickness there is Anglian in origin. Northwest of the line the floor of the modern valley reveals both bedrock and Devensian till (Fig. 66). Somewhat problematically, the steep-sided gorge cut through solid rock (mid-left in Fig. 66) seems likely to predate the Devensian recession. It appears that as the ice withdrew from the Gelli-hir moraine a broad shallow channel (the 'Early channel') formed locally along it but was abandoned as the later fluvioglacial outwash cut more deeply via a western route. It seems highly improbable that this later channel would have cut through the solid rock to form the gorge at that time. Why would it? Why not deepen the 'Early channel'? Hence, it appears that the main drainage formed earlier, before the recession, and that this was re-established by erosion of till during it. The valley floor exposes typically yellow, graded Devensian till at the entrance to the gorge (Fig. 66) and the original drainage could well have been established long before the Late Devensian incursion of the Welsh Ice.

Far clearer than the early drainage development, the final outwash evidently partially filled the valley to form a wide flat surface of sand and silt. This locally rests on till although elsewhere it rests on fluvioglacial gravels that in places are cut into and undermine the till (Fig. 66). Simplest of all, the modern stream clearly is a meandering misfit.

The stream section north of Cartersford bridge (Fig. 66) is part of the drainage that passes through Ilston Cwm and joins the Green Cwm drainage in Parkmill, before exiting into Three Cliffs Bay (Fig. 46). Upstream of Parkmill, which is the limit of high tides and modern estuarine sedimentation, the cwms are distinctly flat floored with pebble, cobble and boulder gravels of fluvioglacial origin, within which the modern streams are misfit and tend to meander. The impact of fluvioglacial sedimentation in the pre-Devensian caves of Green Cwm (Llethrid and Tooth caves) is described in Chapter 4.

The Burry drain (Fig. 55), which initiated as a subglacial channel taking water from the Paviland ice lobe northwards, evidently was extended northwards during the recession, exposed as the ice receded. It too is widely characterised by a flat floor of fluvioglacial outwash gravels with a meandering modern misfit stream, although the final 2 km to the high tide limit below Llanmadoc and Cheriton (SS448932) is physically complicated. The subglacial channel north of Cheriton church (SS450931) (Fig. 55) was cut at an uncertain time, while a borehole on the plateau northwest of Landimore moraine (SS461931) (Fig. 64) unexpectedly intersected some 50 m of sands and clays that must fill a hidden channel.

3.15 Blue Anchor ice margin

The Blue Anchor ice margin is a hybrid of subtle 'moraine' in the east, where the grain of the solid rock becomes hidden beneath thicker till, and then, in the village of Blue Anchor at 62 m OD, a narrow saddle at the head of a steep south-directed drain (Fig. 67). Farther west the margin is inferred to have continued above the particularly steep slope that drops to the coast at Penclawdd, to link westwards with the Hermon moraine. The common altitudes and spatial configuration of the Blue Anchor and Hermon moraines suggest that they represent the same ice limit so that the ice must have formed a hanging glacier exposed above and supplying water into the Bryn-hir drain. Hermon moraine, which plunges steeply down the truncated spur to the west, seems likely to have been continuous with the Crofty moraine (Fig. 64), in which case the glacier snout then would have been over 100 m high.

*Figure 67. (Page opposite). **(Top)** Strongly shaded LiDAR image of the vicinity of Penclawdd, north Gower (SS544957). The Blue Anchor 'moraine' is the slight feature at the edge of thicker till to the north; it cuts diagonally across and buries the grain of the underlying Pennant sandstone strata. Bryn-hir ice-margin drain extends to 62 m OD at a narrow saddle north of which the slope immediately drops steeply away towards Penclawdd. There is no head-water land catchment for the Bryn-hir drain so that glacial ice must have supplied the eroding water. The steep valley north of the 62 m OD saddle would then have been sub-glacially eroded. (The shaded-relief LiDAR image was processed using base data © Natural Resources Wales). **(Bottom left)** Wall almost entirely of north-crop quartzite and conglomerate erratics abundant in the Blue Anchor moraine (photo located by **white star**; SS557952). **(Bottom right)** Reflecting an industrial heritage, a wall in Penclawdd made mostly of locally available smelting slag is adorned with contrasting north-crop quartzite erratic boulders.*

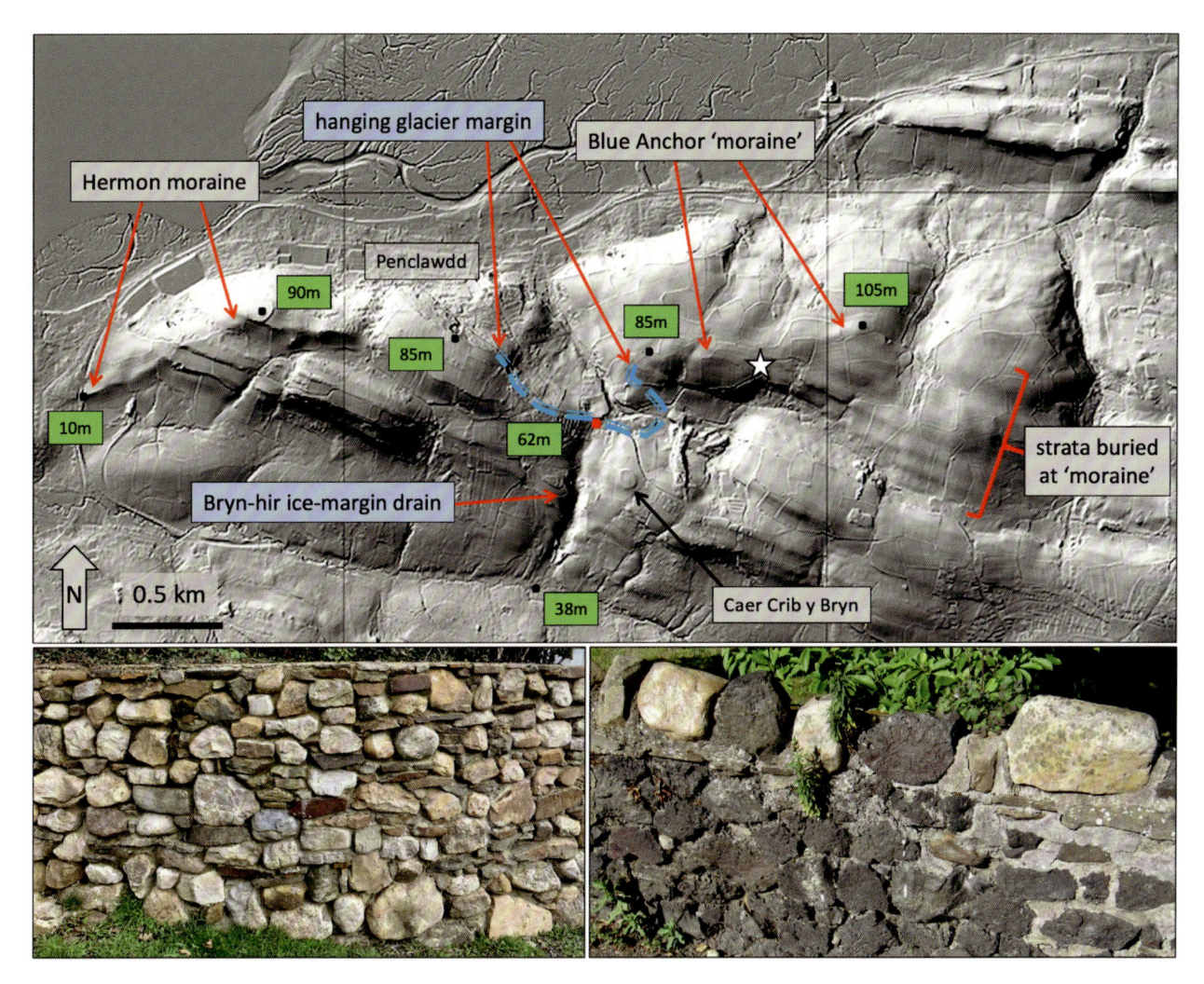

3.16 Accommodating talus

It is now worth imagining the views there would have been of the cliff-edged massif that would become Gower, both after the Anglian glaciation and when the Late Devensian ice had receded. While periglacial freezing conditions persisted and the sea was far away, there would have been substantial frost shattering of the limestone cliffs, which at the time would have overlooked barren outwash plains soon to become tundra. As with any cliffs susceptible to freeze-thaw fragmentation, talus (scree) aprons gradually built up around them. Talus aprons are widespread on and beneath Gower cliffs and would originally have extended farther down before post-glacial raised sea levels reached them and waves eroded their toes.

Figure 68 shows the two key features of the periglacial talus aprons remaining today. First, because talus piles build at the 'angle of repose'[16], they have uniform slopes around the limestone cliffs that were shattered to form them. Second, and most obvious when they have been partly eroded, the uppermost layer is strongly cemented with calcite (calcium carbonate) such that it commonly forms overhangs beneath which are various caverns. Many of these formed rock shelters that were used by animals and humans.

Some of the talus-roofed caverns became important 'bone caves'. On the one hand remains include creatures that enjoyed the earlier mild post-glacial conditions, such as the straight-tusked elephant, hippopotamus, soft-nosed rhinoceros, hyena, and red deer, and, on the other hand, creatures tolerant of the cold are represented, such as the cave bear, woolly rhinoceros and mammoth, and, evidently, humans. As it is hard to imagine a hippo getting into Ravenscliff Cave (SS546873), or for that matter a woolly rhino climbing into Leather's Hole (SS529876), one needs to assume that hunters and scavengers carried only parts of the big animals into the shelters.

Figure 68. Periglacial calcite-cemented talus and head deposits undermined and partly cleaned of uncemented debris owing to post-glacial sea level rise and wave erosion.

   *(**Top**) Cliff pinnacle at Mewslade Bay (SS41878722) flanked by cavernous hollows formed beneath overhanging cemented talus and head deposits. Vestiges of the cemented layers extend seawards and before marine erosion the talus would have reached much farther down. The pinnacle is reminiscent of the Protestant church Hallgrímskirkja in Reykjavik, Iceland (inset).*

   *(**Bottom**) View west across Deep Slade (Hunts bay) towards Oxwich Bay highlighting (**red arrows**) cemented talus and head that forms overhangs and protected hollows where marine erosion has cut the foot of the periglacial talus slopes that originally reached farther down. (continued...)*

Figure 68 continued. **BH** *points out the location of the famous bone cave and human occupation shelter known as Bacon Hole (SS560868), which, facing southwest, is hidden in this view. The* **yellow arrow** *shows the shelter used by the author as a refuge preferred during summer school holidays, when the walk to winter quarters was too far and the hunting and gathering of sandwiches were easy going.*

Cementation of the talus clearly is most pronounced in the surface layer(s). Although the angular limestone fragments form an open framework, in places with infiltrated soil, they are strongly bound at point contacts by calcite precipitated there. Although biogenic release of CO_2 is likely initially to have been limited in the cold periglacial desert, climate 'improvement' with advance of tundra and eventually temperate vegetation and rainfall must have yielded dissolved CO_2 (as it does today) so that dissolution of limestone by carbonic acid occurred, soon followed by its precipitation as calcite during degassing and drying of wetted surfaces. Calcite 'flow stone', stalactites and stalagmites formed where percolated water degassed and evaporated within the caverns. The cementation continues today.

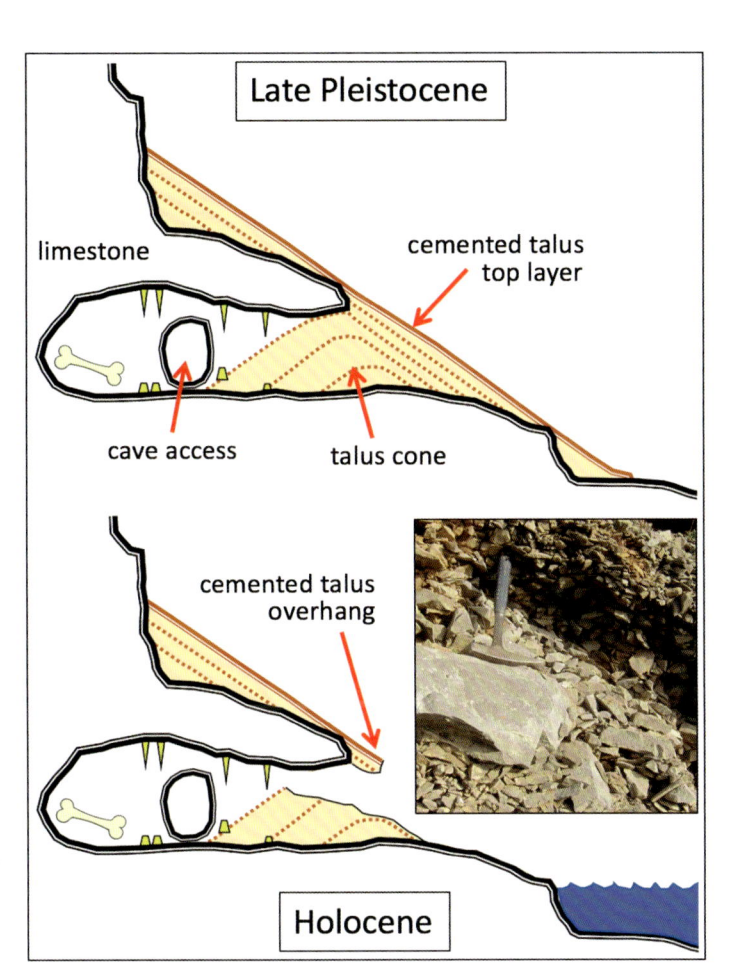

Figure 69. **(Top)** *Illustration of how periglacial talus partly buries Gower limestone cliffs, here hiding a cave by progressive building of a talus cone; the separate access is schematic but is not uncommon for Gower bone caves. Remains from an earlier relatively mild interglacial time could be beneath the talus cone.*

(Bottom) *Sea level rise and consequent wave erosion has exposed the cave rendering its record of occupation accessible and prone to invasion by successive waves of archaeologists. Minchin Hole by Southgate (SS555868) is a fine example.*

Figure 70 shows that in some cases periglacial limestone talus was quarried instead of intact rock. Here, at North Hill Tor, it appears that an original bone chamber entirely within the talus was destroyed by the quarrying. Most Gower bone caves were excavated and material removed from them before the advent of systematic recording, so that important information was liable to be lost. We could hope that one or more pristine examples remain to be discovered behind limestone talus on Gower.

*Figure 70. North Hill Tor limestone talus quarry (SS452938) now and when active in 1862 (Royal Photographic Society via Getty Images). Lines **A to C** and **D to E** mark the top surface of the periglacial talus that originally mantled this limestone outcrop. The limestone talus was quarried rather than the solid rock and comparison of the two views shows the considerable amount that was removed. Typical cemented talus overhangs are evident at the quarry. The site is of interest for its excavation, by Colonel ER Wood in 1869, of a cave that contained Pleistocene mammal bones including hyena, cave bear and woolly rhinoceros, along with Upper Palaeolithic flint tools. It is recorded that the cave was destroyed by quarrying, and, in the absence of notable solid-rock cave remains, it is suggested here that it was within the talus, above the location marked F. (Drone photograph courtesy of Andy Freem).*

3.17 The big picture

Figure 71 summarises inferred pathways of the LGM ice to peninsular Gower 23,000 years ago. Black Mountain is shown highlighting the extensive outcrops there of the 'Rosetta Stone' rocks, particularly the quartzites and ganister that distinguish Gower ice (as well as ice in the valley of the River Tawe). Exposures of those same strata farther west are topographically less prominent and far more limited, so that the distinctive erratics were far fewer in the Coalfield ice. The divide between the Gower ice and the Tawe – Nedd ice is speculatively drawn; exposures above Swansea Bay, e.g., at Brock Hole gardens by Clyne Castle (SS615904) are of quartzite-laden Devensian till indistinguishable from that on Gower and along the Swansea valley, so this study makes no distinction on the basis of material. However, till and bench-like recessional features in fluvioglacial gravel that step down towards east in Swansea Bay strongly suggest placement of the lobe limit of the Tawe – Nedd ice as marked. This contrasts with west Gower where Gower ice left smooth till slopes down to near the present coast (Fig. 60). The Tawe – Nedd glacial debris contains blocks of Pennant sandstone as found in deposits in the Pwll-du Lobe, both parent ice streams having surmounted and cut through gaps in the Pennant sandstone ridges, such as the Dunvant Gap.

Figure 71. (Page opposite). Ice pathways at the Last Glacial Maximum, showing the main source exposures of the Black Mountain 'Rosetta' rock types. Three major ice streams are suggested, based on their dominant erratic types; the 'streams' are not individual glaciers and seen from above they would make up a continuous flowing ice sheet. Arrow directions in the north and northwest follow ice-moulded features. (continued...)

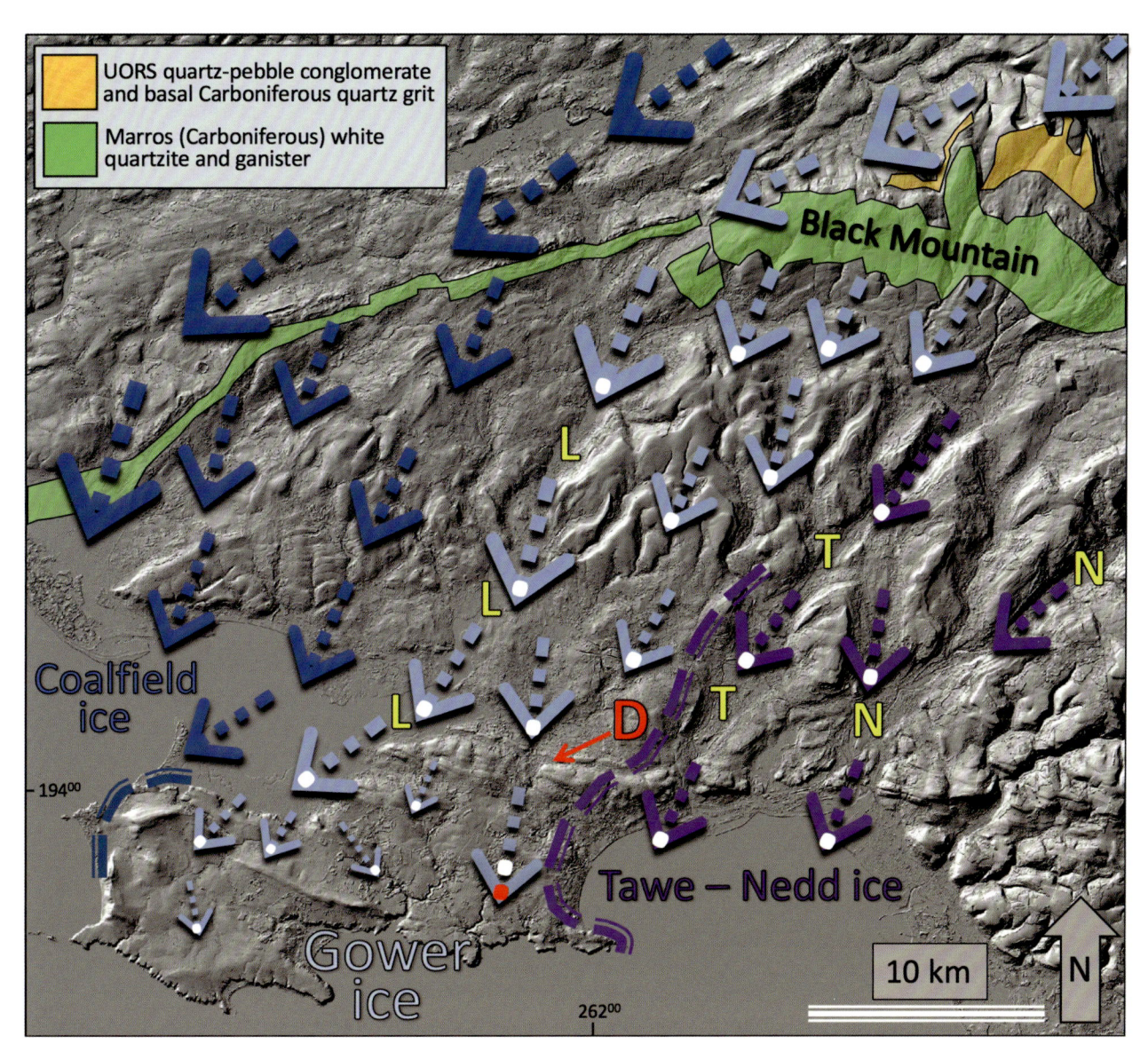

Figure 71 continued.

Coalfield ice is dominated by sandstone erratics, with Pennant sandstone prominent along with shale and small ironstone nodules (and some coal), but it also carries conglomerates, grits and quartzites consistent with the limited extent of their upstream source exposures.

Gower ice is distinctive for its quartzites in particular, represented by the **white dots**, derived from extensive high-ground exposures on Black Mountain. Fluvioglacial rounding is characteristic of a significant proportion of these erratics and is interpreted as due to Gower ice passing down its own major outwash drainage along what is now the Loughor river (**L**), and down the drainages to the east towards the Dunvant Gap (**D**). The Gower ice that formed the Pwll-du Lobe is shown with its stream passing through and over the Dunvant Gap, where it picked up its additional freight of Pennant sandstone blocks, indicated by the **red dot**. Erratic populations along the Tawe valley (**T**) and on land in Swansea Bay are indistinguishable from those of the Gower ice.

Tawe – Nedd ice (Neath valley, **N**) is known from its moraines formed offshore (Fig. 49) and its wrap around the Mumbles Head nunatak, at the south-eastern extremity of peninsular Gower. (Base image provided by BGS © UKRI. Derived in part from NextMap Britain elevation data provided under licence to BGS from Intermap Technologies. Geology from British Geological Survey 1975, 1977.)

While this study has used the erratic content of the Gower Devensian ice to establish new understanding of the plan-view extent at the Last Glacial Maximum, 23,000 years ago, the thicknesses of the ice are not so obvious. Glacigenic deposits on top of the dissected ridge north of Swansea Bay (Figs 3 and 71), which rises to 190 m OD, indicate this order of ice thickness at least northeast of the Dunvant Gap, where solid rock is a few metres above OD before disappearing under alluvium. On Gower peninsula the ice surmounted the ridges west of the Dunvant Gap, which reach almost 150 m OD in Three Crosses (SS574943), but the minimum thickness argument then fails farther west as the bedrock bottom of the Loughor Estuary falls away to an unknown maximum depth, locally at least to -32 m OD. Boreholes show that along the present course of the Loughor drainage solid rock lies at about -10 m OD at both Pontarddulais (SN259203) and Loughor Bridge (SS562980) and at a similar depth beneath the Crofty moraine (SS521954), off the main valley axis (unpublished report). North of the estuary, east of Llanelli and around Loughor, coal pits penetrated up to 32 m of alluvium before hitting bedrock (Strahan 1907).

Evidently all of north Gower was surmounted by ice at the full maximum. Various features described suggest that at its southern on-land limits the ice reduced to several tens of metres thick as it wasted away. One may assume that north of the Arthur's Stone high point in the Ilston Lobe (155 m OD) the ice was perhaps 100 m thick, and similar in the middle of the Paviland Lobe relative to the high on Llanmadoc Hill (170 m OD). The greater extent and till volumes of the Paviland Lobe suggest its ice stream was faster (less restricted) than the stream that reached onto Cefn Bryn at 155 m and dumped the Arthur's Stone erratic assemblage (Section 3.8). Although Gower ice reached around the north flank of Llanmadoc Hill (Fig. 46), it appears that farther north there was Coalfield ice that must have deflected the Gower streams southwards.

The scale and penetration south of the Pwll-du ice is unknown, but it appears that there were two main streams split on either side of the main Pennant sandstone ridges above Penclawdd and Gowerton, one, the Pwll-du ice, going south via the Dunvant Gap and the other southwest down the Loughor valley and forced onto Gower (Fig. 71).

It appears that at the glacial maximum Llanmadoc Hill, Rhossili Down and Cefn Bryn stood proud of the ice, locally with small tors of tough conglomerate, while Oxwich Point, part of the south flank of Cefn Bryn, and cliffs across to Southgate also remained exposed (Fig. 46). The Paviland Moraine outer face would have been visible, as would the various drains and slades on its outer edges. It would be a very long time before Oxwich sand dunes and the marshes behind them obscured the eastern toe of the Paviland Moraine. Viewed eastwards from Oxwich Point the Pwll-du ice should have been clear to see, appearing beyond Three Cliffs Bay and descending over the ancient cliffs east of Southgate, perhaps extending some kilometres southwards to join the Tawe – Nedd ice.

Recessional moraines on peninsular Gower suggest a staged retreat, which may have taken a few thousands of years. Quite graphically, the interpretation of the Blue Anchor ice in the recession (Fig. 67) suggests a glacier snout steeply rising from the Loughor valley to over 100 m high, so there was still some time to go then before complete decay of the Welsh Ice some 16,000 years ago.

As the ice left the Gower limestone cliffs, freeze-thaw again formed aprons of talus around them, soon to become vegetated tundra. And then the wonderful animals came back followed by our relatives hunting them. One may wonder if those hunters were aware or found the evidence that their own ancestors had already visited before the Late Devensian glaciation.

Glaciation Notes

[1] Throughout this chapter we consider ice *sheets*, which are laterally extensive and continuous when viewed from above. Ice *streams* are those parts of the sheets that flow relatively fast, commonly down underlying valley topography and extending farthest.

[2] Till is the superficial debris widely deposited from glacial ice, generally from the base of a moving ice sheet or by melting away of stagnant debris-laden ice. It typically consists of a mixture of large stones set in a fine-grained matrix, forming a rather smooth layer up to a few metres thick. Mostly in this work, till that forms the ground surface (not beneath other superficial material) is to an extent 'weathered' or reworked and forms various soil types according to place, hillslope or field, drainage and vegetation. Nevertheless, the primary deposit remains recognisable according to its poor sorting, content of erratics and commonly a clay component.

[3] Nunataks are like islands of rock in a sea of ice.

[4] A moraine is generally a mounded and elongate feature consisting of mixed glacial debris. Moraines typically form by accumulation at the margins of moving ice, e.g., along sides against slopes (lateral moraines) or at the farthest reach (terminal moraine). Ice will persistently deliver debris to a 'fixed' termination where melting-away balances the ice advance, and this occurs during recession too when ice reaches a certain limit for some duration before further melting back; this forms recessional moraines of which there are several good examples on Gower.

[5] This coarsening upwards used to be interpreted to result from shear-related migration of large fragments up and away from the basal contact, which would require the entire thickness to be in motion and to become sorted as in the so-called 'muesli effect' where, when cereal is dilated by shaking, the nuts end up on the top and fine powder percolates towards the bottom.

[6] Periglacial conditions occur in the vicinity of glacial ice where freezing and thawing occur.

[7] Amino acid dating utilises known rates of the molecular degradation that causes ratios of amino acid types to change with time.

[8] Fill-line here refers to the mapped limit of till, which, used advisedly, is a record of ice extent. In practice a fill-line may be a thinning irregular 'feather-edge' of a till deposit according to the nature of the ice presence there. In places on Gower where the ice barely reached as it broke down and melted the fill-line is indistinct, whereas on a long-lived lateral margin it can form a sharp break of slope (e.g., Fig. 60). Isolated glacially delivered boulders that can be several tens of metres beyond a fill-line suggest that the ice had significant thickness there, so that the boulders probably fell and rolled away.

[9] The site was first extensively studied by Philippa Grove of Parc le Breos Farm (Grove 2008) and excavated in 2008, details of which were published in Evans et al. (2010) available at: http://www.ggat.org.uk/cadw/church_hill/Church%20Hill%20excvation%20and%20survey%20final%20report.pdf

[10] It is mostly seen in profile near cliff tops where the Devensian ice did not venture (e.g., above Little Tor; SS526879), and vestiges also occur widely beneath drift at or just above the upper limit of the pre-Devensian *Patella* 10-15 m OD notch or superstorm surge deposit (Section 2.14), e.g., around Oxwich Point (Fig. 35) and at the back of Bantam Bay on Pwll-du Head (SS573864).

[11] The several Neolithic chambered tombs on Gower are not associated with such clusters of small burial sites. There are several Bronze Age sites in the *immediate* vicinity, so, without archaeological information

to the contrary, it seems reasonable to *guess* that the "large cemetery" is of the Bronze Age (about 4,500 - 2,800 years ago). The timing of the Bronze Age varies according to the spread of the culture in which copper and tin smelting and the forging of tools, weapons and ornaments (or the trading of these) succeeded the late Neolithic primary use of stone and bone. This age range applies to Britain whereas Bronze Age cultures existed far earlier in the Near East, e.g., in Egypt.

[12] Hardings Down reaches 152 m but seemingly lacks erratics on top and ice may have reached 170 m on Llanmadoc Hill before the slope failure there.

[13] The author is no archaeologist; nothing was moved. The site mentioned here was fully exposed and, like others, probably was previously 'disturbed' by treasure hunters or amateur historians.

[14] Paraglacial processes occur after and as a consequence of previous glaciation; slope instability and landslides are typical.

[15] For a while during mapping, this feature was mistaken for a terminal moraine, because of the substantial concentration of north-crop erratic boulders there, with, at that time, nothing of note found farther east in Fairwood Common. It seemed initially that the line would join the Ilston Lobe limit near Green Cwm (Fig. 64), until the boundary 'excursion' into Ilston was discovered.

[16] Any loose granular material, such as salt, sand, gravel and scree, can be expected to pile up to a slope angle that is characteristic of the (dry) inter-grain friction properties. Any attempt to add material at the top of the slope will lead to avalanching. The stable maximum slope angle is 'the angle of repose'. For normal dry sand the angle is 34° and for angular fragments it can be steeper.

Broughton Bay from Llanmadoc Hill. View towards northwest across Carmarthen Bay whence cometh the Anglian ice bearing erratics from Pembrokeshire, and some cockles (Figure 100).

4 CAVE DEVELOPMENTS

Caves referred to in this chapter are potentially hazardous, some requiring specialist equipment and skills. Access to some caves is seasonally or permanently restricted according to fauna or archaeology and some are subject to landowners' specific permissions. Description in this chapter must not be taken to imply a right of public access or any measure of safety to explore.

4.1 Plenty of time

Caves have formed in Gower limestones for more than 200 million years. The folding and faulting of the rocks that would form peninsular Gower started in latest Carboniferous times, roughly 300 million years ago, and persisted for some 10 million years into earliest Permian times. As the rocks were heaved up, while folds and thrust faults developed, they were at the same time being eroded down so that the ground surface exposed all the deformed limestones and other layers broadly as they are today, but at a higher level (Section 1.3). The erosion persisted in Triassic deserts where fault-fissures opened and became filled with collapsed debris and water-lain red sediments, reflecting flash floods (Section 2.4).

At the close of Triassic times, 201 million years ago and heralding the marine conditions of the Jurassic Period, sea level rise supported water tables beneath the land surface allowing phreatic dissolution of limestone and hence the earliest cave development. Evidence for this on Gower is in the pothole formed at the end of Worms Head (Fig. 10), with further evidence on Oxwich Point (Section 6.5). Such early cave development is not peculiar to Gower; it occurred widely elsewhere in Carboniferous limestones of South Glamorgan and the Mendip Hills, where potholes and caverns contain Triassic sediments, floras and faunas, notably early reptiles (Section 2.4).

Cave development occurred again in what became a subaerial limestone massif when the cover of Jurassic and any Cretaceous rocks was eroded off according to Paleogene and Neogene uplift, conceivably from some 62 million years ago (Fig. 39). It could continue until the marine inundation and planation of Late Pliocene times, some 3 or 4 million years ago. The perforation of Great Tor (Section 2.11) shows that caves in the evolving limestone massif that would become Gower preceded the erosion that formed the slades (dry valleys), placing that cave development *before* Early to Middle Pliocene time, which was 4 to 5 million years ago (5-4 Ma). Further, through Quaternary times, from 2.68 million years ago until now, those limestones were mostly above sea level and so then could also support cave development. In this interval, however, they were sometimes frozen, twice inundated by ice and once briefly, but significantly, encroached by the Ipswichian Patella seas that covered them up to 10 m OD (Section 2.14). So, there were vast lengths of time during which caves could and did form in Gower limestones.

4.2 Little did we know

Looking back, as a teenager, cave exploration on Gower was rather frustratingly limited at a time when many caves were being dramatically extended nearby around the north crop of the limestones, especially in the Swansea Valley. The new discoveries there often found fine open-draining (vadose) stream passages upstream of the prominent exits where cavers naturally looked for access (e.g., Dan yr Ogof and Ogof Ffynnon Ddu). On Gower we pushed into myriad tubes and slots hoping that the few accessible metre-scale passages might continue for exploration downstream. However, as upper passages proved too small, we would be forced downwards and then thwarted by the water table where we could see rifts or shafts flooded with clear, imperceptibly-slow-moving water (e.g., in Llethrid and Tooth Cave). Similarly, with the special exception of Ogof Ffynnon Wyntog (see below), the springs and resurgences proved impenetrable, even in drought and to cave divers. In flood conditions, the water tends to well up from conduits that are rooted below the water table and seem too small for the discharge.

When one appreciates that there must have been enough time for caves to form naturally with stream discharges at unconstricted exits, it appears that Gower caves may differ fundamentally from counterparts in the north crop that are well above sea level. It is perhaps helpful to view the Gower situation by imagining how it would be if the Swansea Valley was flooded by the sea, so that the lower stream exit passages there became stagnant and blocked, forcing flood waters to surface at higher levels via misfit small (under-sized) passages.

Here is the proposed explanation. In the chapters on the landscape and glaciation it is shown that by far most of the limestone karst and cave development of Gower occurred when the sea was nowhere near, through periods amounting to, at least, some 60 million years. The absolute base level of drainage was latterly fixed by the ancestral River Severn, 20-30 km south of Gower and reaching down to 50 m below present sea level (-50 m OD). The effective base level for drainage from the limestones, however, would have been at the southwards-dipping contact with the low-permeability Triassic rocks, which generally cover the limestones downwards from around 25 m below sea level (-25 m OD).[1] In other words, when the sea was nowhere near, which was for most of a very long time, caves could have their waters exit to surface some 25 m lower than present sea level and hence at a considerable distance from the modern shore, e.g., about 3 km south of the mean sea level position (0 m OD) at Three Cliffs Bay.

Evidence from the known substantial caves of Gower (Llethrid – Tooth Cave system and Ogof Wyntog – Ogof Ffynnon Wyntog; see below) and the coincidence at or just above 10 m OD of many barely penetrable significant springs and resurgences (e.g., see Fig. 104: Wellhead [14], Trinity Well Ilston [20], Pennard Castle [22], Bishopston Valley, Caswell Valley, Leason [1] and Llanrhidian [3]), indicate that there are below-sea-level parts that must have become partly blocked when sea level rose above their original outlets, as it did on several occasions (Chapter 2).

The coincidence of springs and resurgences at around 10 m OD is taken to result from the Ipswichian (MIS 5e) high stand of the sea, which reached that level around 123,000 years ago. Water would have stagnated in the caves below 10 m OD, for several thousand years, so that passages gradually became filled with sediment. And, of course, such blockage would have resumed with the relatively recent Holocene rise of sea level near to its present position some 7,000 to 5,500 years ago (Fig. 13), as valleys and estuaries became filled with alluvium.

So, it is reasonable to infer that there are deep flooded parts of Gower cave systems that remain relatively stagnant. The deepest parts probably were originally phreatic passages, but others, shallower than about 20 m below modern sea level, were probably vadose before becoming choked. Owing to the absence of any significant topographic height of Gower limestones, there is unlikely ever to have been sufficient head of water (pressure) to keep passages clear for springs to remain significantly active below sea level.

Perhaps it is a peculiar feature of Gower karst that the seaside setting coupled with Quaternary interglacial conditions have caused many resurgent waters to arise via misfit conduits above inaccessible 'lost' depths and relatively stagnant regimes.

4.3 Magical caves at The Knave

The limestone coast between Port-Eynon Point and Worms Head is fabulous. It is a privilege to present in particular Ram Grove and The Knave (Fig. 72), as they illustrate so well both the considerable age of initial cave development and the eventual impact of the coastal setting on cave drainage and habitation. Of the several caves in the vicinity of The Knave, two are especially interesting: Ogof Wyntog (aka Cunnington's Cave[2]) and Ogof Ffynnon Wyntog (Fig. 73). The caves at The Knave are likely to have developed initially *at least* some 10 million years ago, in Miocene times (23.03-5.33 Ma). The mapped passages in both Ogof Wyntog and Ogof Ffynnon Wyntog are predominantly phreatic (Figs 74 and 75), as are numerous less extensive remnants in the immediate vicinity up to around 40-45 m OD.

Phreatic passages form beneath a water table, fully flooded with water that dissolves the limestone around all contact surfaces; departures from an ideal cylindrical tube shape normally relate to some preferential dissolution along the bedding or fracture plane on which the tube initiated (e.g., Fig. 74). The key point here is that the phreatic passages in this fault-bound limestone 'block' can only have been created *before* the slades were cut deeply on either side and, of course, before formation of the sea cliffs in front where they are now revealed. Such an exposed upstanding limestone mass could not support internally any water table beneath which the known phreatic passages could form; cutting of the slades and cliffs must have drained *earlier-formed* phreatic passages. We applied the same argument for the Leather's Hole perforation of Great Tor at 35 m OD (Section 2.11).

We know that the marine planation that formed the 55-60 m platform, and thus truncated the heads of the slades, occurred in Late Pliocene times some 4-3 million years ago (Section 2.10). Since the slades must have been eroded to near their present depths before their truncation, they must have been formed at the latest by Early Pliocene times (5-4 million years ago). And, since the phreatic cave passages must have formed before any deep downcutting to form the slades, they must be older still, conceivably at least 10 million years old.

The phreatic passages must have been progressively exposed as the slades were cut and when the Late Pliocene planation removed uppermost levels (see below). The influence underground of the Late Pliocene marine transgression is unknown; most Gower caves must have been filled with water then and virtually stagnant for the duration. Final exposure of the cave system(s) occurred mainly in Early to Middle Pleistocene times (Section 2.12), less than 2.6 million years ago, when the coastal shelves and modern cliffs were cut back to their present form by marine erosion. Anglian ice covered everything about 450,000 years ago and Devensian ice and debris formed a partial cover around 23,000 years ago. Late Pleistocene periglacial freeze-thaw weathering degraded the cliffs so that talus (scree) covered or partially covered some caves (Section 3.16).

ALL OUR OWN WATER

*Figure 72. (**Top**) The Knave viewed towards west with Worms Head in the distance. The Knave (SS432862) is the feature left of centre with the north-facing steep grassy slope; the **red arrow** labelled '**a**' marks the position on the 'saddle' north of The Knave, at about 10 m OD, where there are preserved 'Patella Beach' superstorm-surge deposits (Section 2.14). The **red arrow** '**b**' marks the location, also at about 10 m OD, of the closely adjacent entrances of the two main caves here: Ogof Wyntog and Ogof Ffynnon Wyntog. (continued...)*

*The latter is the issue of a significant spring (ffynnon). **White arrows** locate some of the many phreatic cave remnants that exist near here at around 40-45 m OD. (Original photo courtesy of Geoff Williams). (**Bottom**) Ram Grove (SS429865) fault-controlled slade, showing the fault trace 'F' in shadow in the foreground with the cobble storm beach behind. The distant skyline is the south Gower platform edge at 55 m OD; the dry valley is forked at that far end, on either side of the prominent buttress, because the controlling fault splits there.*

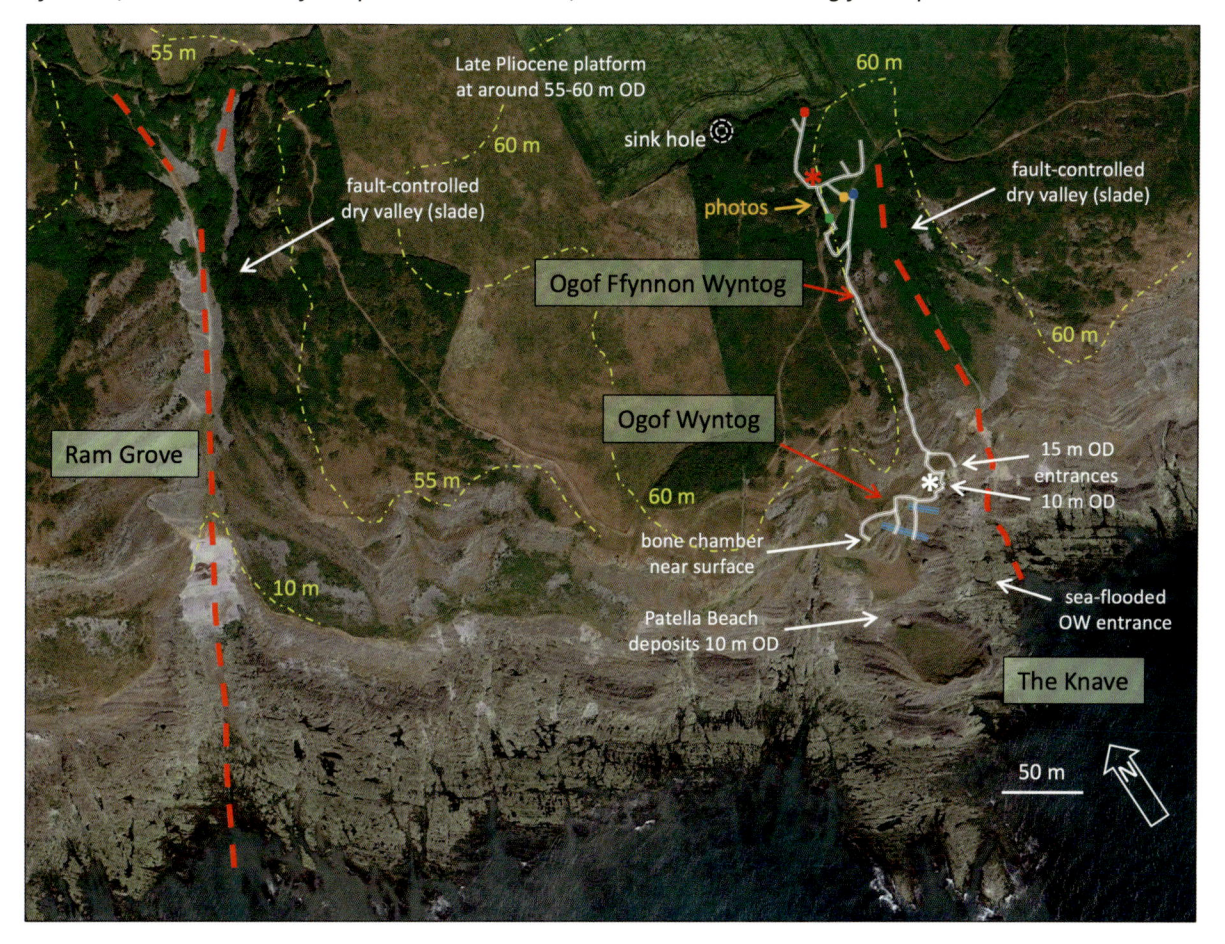

*Figure 73. Google Earth image of Ram Grove to The Knave with cave survey plans superimposed. The normally sumped rising (spring) of Ogof Ffynnon Wyntog (**white asterisk**) is at about 10 m OD. The cave drainage is down to the south and there exists no really substantial climb northwards before a rising from the water table at the **red asterisk**. However, there are phreatic loop passages below the main mapped feature, and passages climb >30 m higher in the northern parts. The linear distance from the entrances to the northern extremity (**red dot**) is ca. 240 m. A pronounced deep and sediment-blocked karstic depression (**double-dash circle**), more than just one of the many quarries[3], is a tantalising prospective surface connection to the northwest. The cave has considerable vertical extent with abandoned relics at high levels and persistently flooded lower levels. The locations of the three photos in Figure 74 are shown. The numerous gullies that cut the shore and cliffs reflect the abundance of fractures, both faults and joints, that intersect the limestone and influence its permeability. The cave passages tend either to trace parallel to limestone bedding, subparallel to the coast, or at right-angles, inland and guided by the fracture system.*

*The Ogof Wyntog lines are from the 1983 survey by Cwmbran Caving Club drawn by T Schurmans and K Jones; the **blue lines** are passages now accessed from below by the sea at high tide. The Ogof Ffynnon Wyntog lines are based on surveys by R Davies, W Naylor, P Dawson and R Philips drawn by R Davies in 1990, and by John Stevens, David Stevens and Paul Tarrant, drawn by John Stevens who kindly provided the data (pers. comm. 2021 all rights reserved; see also Stevens (2020)).*

Figure 74. July 12th Series in Ogof Ffynnon Wyntog; photos located by coloured dots in Figure 73.

*(**Top**; orange dot) Phreatic tube originally formed along limestone bedding. The bedding plane exposed in the walls, top left to bottom right, is the original surface along which the tube initiated.*

*(**Middle**; green dot) Original partly circular phreatic tube now vadose with a pebble-strewn active stream bed.*

*(**Bottom**; blue dot) Decorated alcove at the northern end of the long straight NNE-trending fault- or joint-directed passage.*

(Photos courtesy of Andy and Antonia Freem).

'Patella Beach' superstorm-surge deposits (Section 2.14) are preserved at about 10 m OD on the saddle between The Knave and the main limestone cliffs containing the Wyntog caves (Fig. 73). Importantly, practically identical Patella-bearing gravels occur within Ogof Wyntog (Fig. 75), resting on thick speleothem that must therefore be older, more than 123,000 years old. The gravel in turn grades up into overlying sand (Fig. 75) that now partly fills cave passages above and formed floors on which speleothem accumulated. A logical inference is that the raised sea level of Patella times, lasting several thousands of years around 123,000 years ago, caused passages below 10 m OD to become stagnant so that sediment accumulated within them. Here it is inferred that such sediment blockage is the reason why the spring of Ogof Ffynnon Wyntog is at about 10 m OD, with accessible passages upstream active near or above this water-table level and with part-filled phreatic loops at deeper levels.

The speleothem layer beneath the Patella deposits in Ogof Wyntog (Sp1 in Fig. 75 Bottom) can only have formed when the passage below 10 m OD was empty and dry, more than 123,000 years ago, and since the phreatic tube now accessed by the sea (Fig. 75 Top) is only a few metres lower, it too must have been drained by that time. These features indicate a pre-Patella cave drain that must now be below sea level, conceivably with phreatic passages lower still.

Sediments originally below and just above the Patella 10 m OD level (Figs 75 and 76) are largely 'missing' from the coastal cave passages, probably reflecting marine erosional undercutting since sea level reached near here, no more than 7000 years ago. The upper speleothem layers, on top of the sediments (e.g., Sp2 in Fig. 76), must predate the undercutting and their oldest age must post-date the sediment accumulation that initiated about 123,000 years ago. There is, as yet, no clear evidence of the waxing and waning of the Late Devensian Last Glacial Maximum, around 23,000 years ago (Phillips et al. 1994). Some glacial outwash from the Paviland Moraine reached the present shore in Ram Grove, but glacier ice did not cover the limestones containing the Wyntog – Ffynnon Wyntog cave system (Section 3.10).

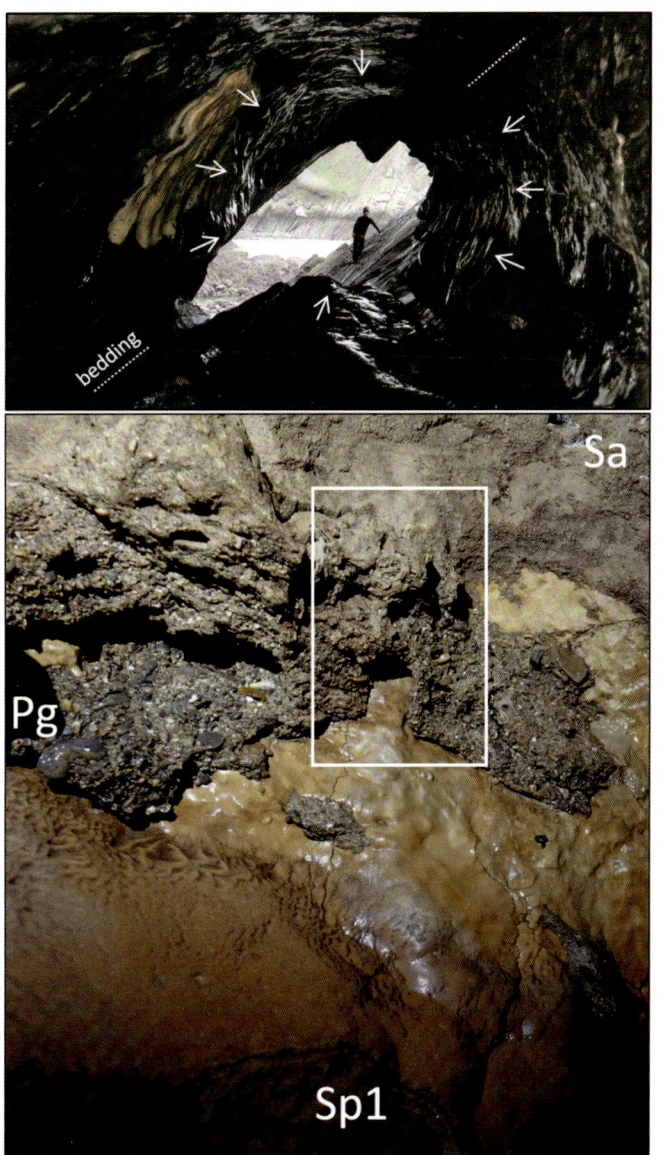

Figure 75. Ogof Wyntog.

*(**Top**) View outwards from the sea-breached passage, with the main top-right to bottom-left surfaces reflecting the dip of the limestone bedding. The rounded passage roof and walls (**white arrows**) are a remnant part of the original phreatic tube that penetrated horizontally along the bedding. The polished surfaces in the lower foreground are close to the maximum reach of energetic sea waves and are primarily due to abrasion by pebbles and cobbles.*

*(**Bottom**) Some 1.5 m higher than the top of the tube remnant (photo above), thick layered speleothem (Sp1) has remnants of Patella gravel (**Pg**) adhering to its surface, placing **Sp1** as certainly older than 123,000 years. The Patella gravel grades up into sand (**Sa**), which here is the bottom of deposits that partly fill the cave passages above. The **white box** is enlarged in Figure 76, which highlights the content of Patella (limpet) shell fragments.*

(Photos courtesy of Andy and Antonia Freem).

Figure 76. Ogof Wyntog.

*(**Top left**) Detail from Figure 75 with arrows pointing at two concave-up Patella shell fragments.*

*(**Top right**) Tall rift passage the floor of which is sand (**Sa**) underlain by the Patella gravel (Fig. 75). In the roof above the caver are passage enlargements and speleothem (**Sp2**) that formed above the sand, which is now partially washed and excavated out.*

*(**Bottom**) Successive speleothem layers that formed on the sediment that originally blocked the pothole linking down to similarly filled (but now sea-breached) cave beneath. **Blue and red arrows** mark completely collapsed layers, with **yellow and green arrows** marking respectively a skylight through a speleothem layer and its intact remnant with the original sand underneath.*

(Photos courtesy of Andy and Antonia Freem).

The northernmost reaches of Ogof Ffynnon Wyntog are being explored and extensions farther north than the underground rising (red asterisk in Fig. 73) are relatively new finds that may ultimately connect to the surface. The southern parts are likely to have originally been hydraulically connected with Ogof Wyntog; the present spring and sump at 10 m OD are products of the Patella-times blockage of passages lower down. The present sea-breached entrance to Ogof Wyntog clearly was initially a phreatic tunnel (Fig. 75), and, given the known considerable age of the phreatic system, it is logical to interpret that the comparatively recent return of sea level has only served to enlarge original passages, and not formed them.

The present cliffed coastline and inshore shelf were cut in several stages during Quaternary interglacial episodes (Section 2.12), although the submerged small cliff at some -15 to -25 m OD, could have formed in Late Pliocene times. Thus, the presently exposed caves almost certainly once continued southwards. In light of this it is interesting to consider the origin of an unusual 100 x 120 m diameter rock basin in the platform

CAVE DEVELOPMENTS

at about -10 m OD (Fig. 77), just offshore from the caves. Quite speculatively, this might register a cave collapse like those forming the 'daw pits'[4], e.g., in Bishopston Valley (SS576892 and vicinity), and/or a substantial spring embayment. The present spring, Ffynnon Wyntog (Windy Spring), releases water according to rainfall, but water presently 'leaks' to lower levels and an original system discharge lower down could have been considerable.

Before rising sea level reached the rock basin that is now offshore, which was probably around about 8000 to 7000 years ago, Stone Age men, women and children probably enjoyed a fresh-water pool there, close to and overflowing onto the shore – Pwll Ffynnon Wyntog. Artefacts and animal bones reflecting their presence were abundant in the nearby caves (now excavated), including Ogof Wyntog, originally just a short stroll away from the spring-fed pool, and Deborah's Hole (SS4338886301). Such occupation both predated and post-dated the Late Devensian glaciation, given that the Paviland hominid remains found nearby date at between about 34,000 and 33,000 years ago (Jacobi and Higham 2008). The Last Glacial Maximum ice did not extend quite this far, although the habitability here would have reflected the waxing and waning of the freezing periglacial tundra.

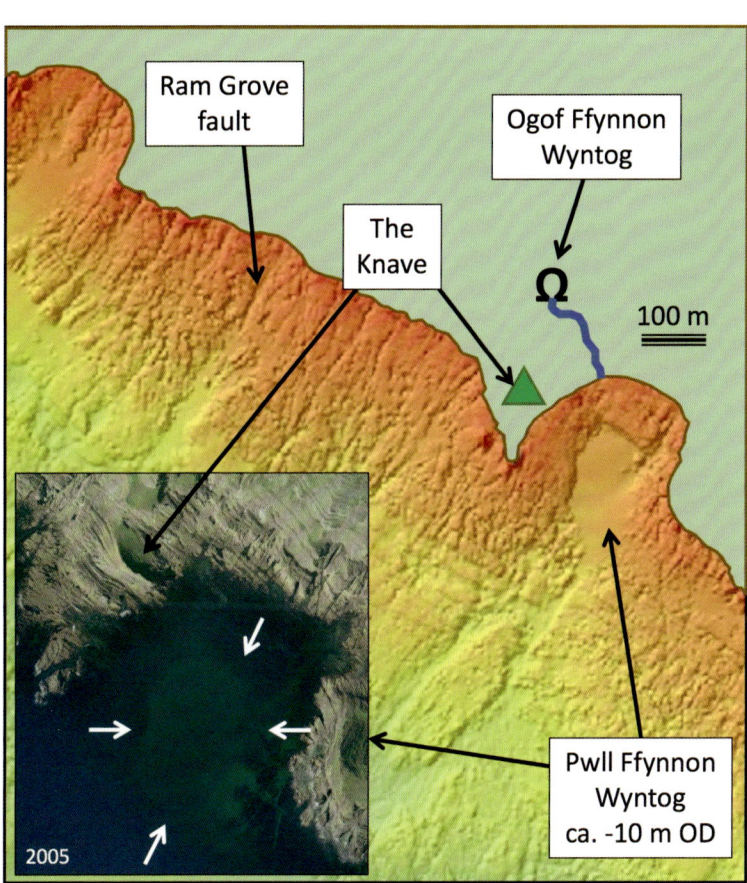

Figure 77. Sea-floor image showing speculative inference of a fresh-water pond, Pwll Ffynnon Wyntog, fed by the cave spring and ultimately lost to sea level rise perhaps some 8,000 to 7,000 years ago. The present sediment infill is about 10 m below Mean Sea Level and the depth to the rock base beneath it is unknown. It is a most unusual feature of the seabed and it is tempting, given its location, to link it genetically to the nearby cave system as possibly originating as a collapse pit and/or a substantial resurgence. Gullies eroded down the offshore slope reflect the abundant fractures that cut the limestones. (Base image multibeam bathymetry, at 2 m x 2 m resolution, from The Admiralty Marine Data Portal, kindly processed and provided by Gareth Carter. The inset image is from Google Earth © 2020 The GeoInformation Group NASA.)

The limestone sea floor in the vicinity of The Knave extends some 850-1,000 m offshore, where it becomes covered in sand (Fig. 78). The position of the contact with the poorly permeable Triassic-rock cover is unknown, although likely to be nearby (Section 2.4). Here it is deduced that the phreatic cave passages at The Knave formed initially (inception) beneath a karst surface that must have existed above the present surface. The original limestone massif must have extended out to the Triassic-rock cover rather like now, but at a higher level and with the contact farther north, because of its southwards dip. We understand that owing to *regional uplift*, which persisted until some 7 million years ago (Fig. 39), the broad valley of the ancestral River Severn and Gower were continuously eroded subaerially, until the Late Pliocene sea advanced over them. Thus, some thickness of limestone and its Triassic cover is likely to have been removed. Moreover, the upper phreatic passages require a water table well above their level at 40-45 m OD, certainly higher than about 50-55 m OD.

Figure 78 shows the bathymetry off The Knave and the limit of limestone exposure on the sea floor. We know that the unconformity at the base of the Triassic strata is topographically irregular and that it was originally close above the limestones (Section 2.4). Although the contact with the Triassic rocks would have been farther north than now, the limestone must have been sufficiently thick to support the phreatic system evident today *and* the water table originally above it. An attempt to explain this is in Figures 79 and 80.

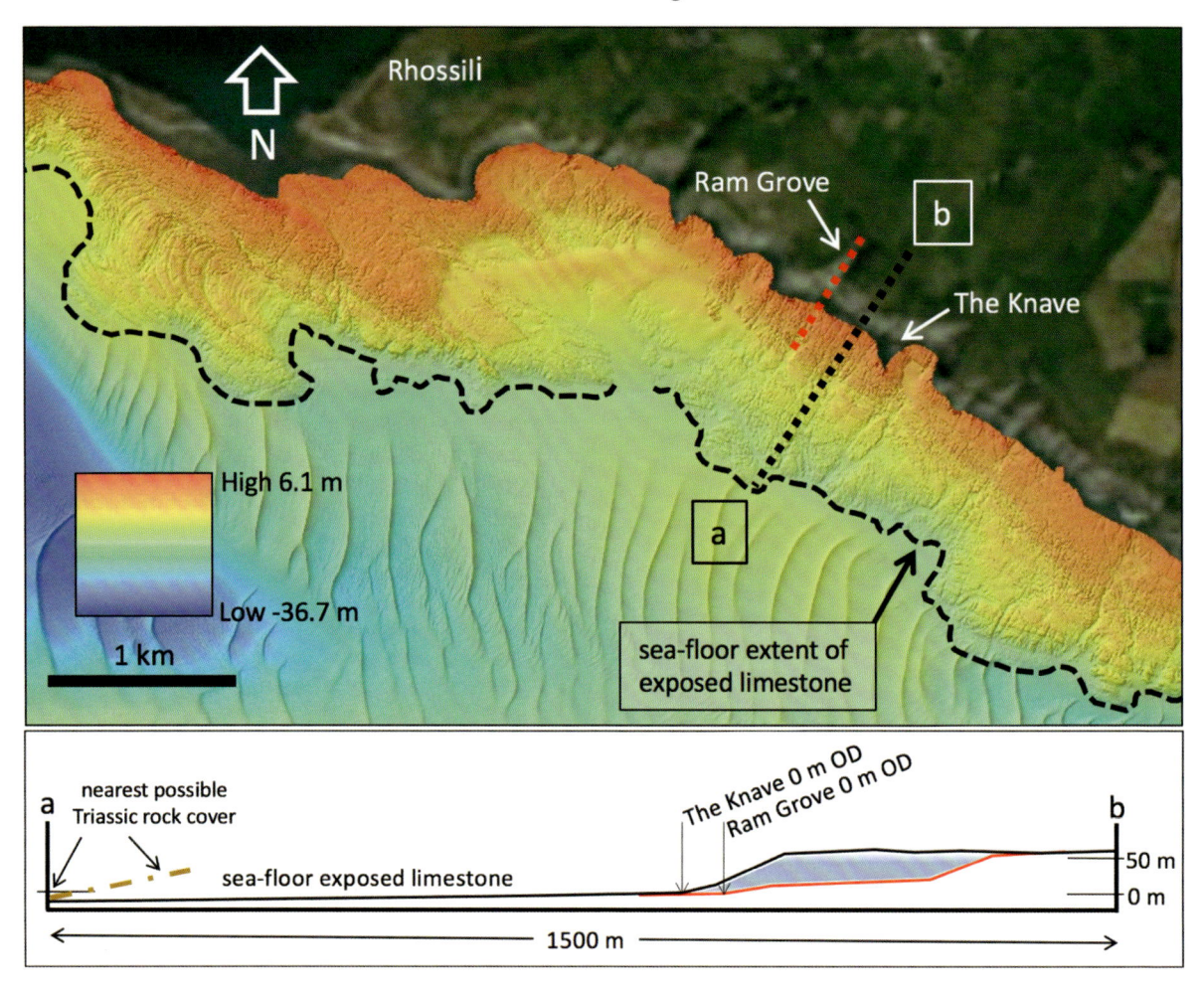

*Figure 78. Coastal profile at and near The Knave. Limestones continue as a gently sloping platform offshore 850-1,000 m distant from the cliffs. The caves extend at least 240 m inland. The on-land platform behind the cliffs is at 55-65 m OD, but farther inland it is covered by the Paviland Moraine. The true-scale cross section (**a-b** in bottom panel) shows the profile along Ram Grove slade in **red**, emphasising that the **blue-shaded** body of rock, which contains phreatic cave passages, could not support any elevated water table when the slade existed nearby. The same, of course, applies to the space in front of the cliffs; the phreatic passages must predate both the deep cutting of the slades and the cutting back of the cliffs here. (continued...)*

Triassic strata are known to cover the limestones offshore; the section shows their nearest possible occurrence and the fact that any original cover would slope gently up landwards (Section 2.4). Depths are relative to the level of the lowest astronomical tide: LAT. (Base image multibeam bathymetry, at 2 m x 2 m resolution, from The Admiralty Marine Data Portal, kindly processed and provided by Gareth Carter).

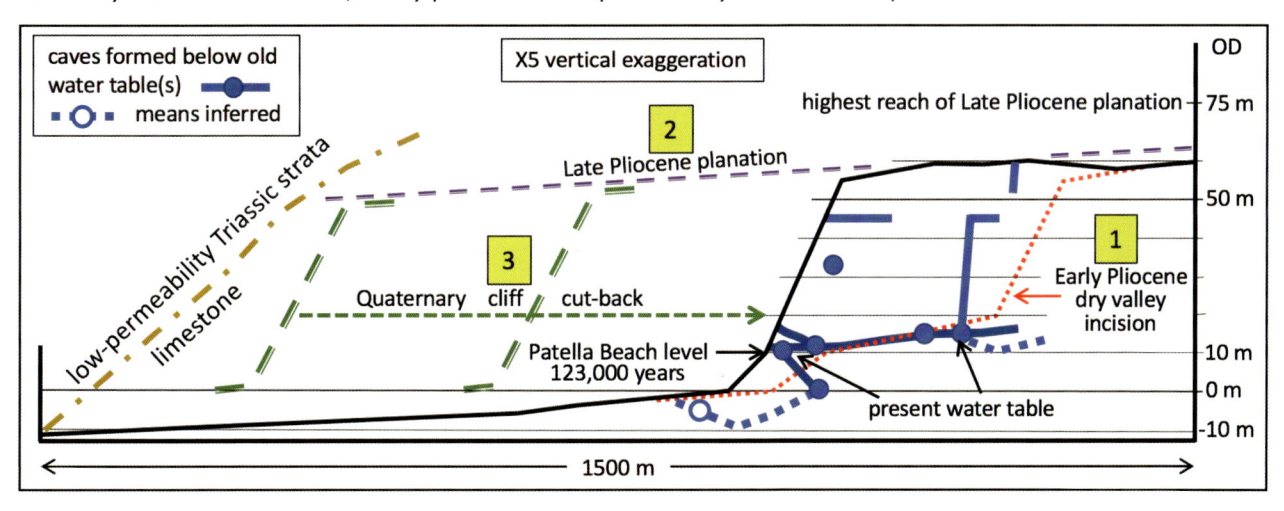

*Figure 79. Schematic diagram of three successive developments (**boxes 1-3** with processes beneath) that must have occurred <u>after</u> formation of the phreatic cave passages near The Knave. This is the same cross section as in Figure 78, vertically exaggerated x5 for clarity; this steepens angles but does not alter the significant observations. The present water table (sump and risings) level in Ogof Ffynnon Wyntog is close to the bottom profile of Ram Grove. The cave passage levels are shown approximately as they exist, although they are schematically drawn with the circles representing phreatic tubes trending mostly along bedding subparallel to the coast, into the plane of the cross section (see Fig. 73). The inferred contact between the limestones and the unconformably overlying Triassic strata is the closest it can be in the sea floor; its position is uncertain because sand covers the sea floor beyond the limestone (Fig. 78).*

The relatively late incursion of the sea to about 10 m OD, 123,000 years ago, caused blockage of deeper passages although these were partly washed out by Holocene sea level rise (since about 7,000 years ago). It is probable that caves existed beneath what is now Mean Sea Level (0 m OD). This is suggested by the offshore rock basin named Pwll Ffynnon Wyntog and it is required by the speleothem Sp1 beneath the Patella gravels in Ogof Wyntog (Fig. 75), which had to be in drained passage when it formed.

*The main take-away from this diagram is that the existing phreatic passages, **shown blue**, require restoration of the limestone thickness to a level that is now 'in space' along the Trias–limestone contact permeability barrier (**brown dash-dot line**). The implicit downcutting of the surface (see below) relates to uplift that ended about 7 million years ago, placing the age of the caves as most likely to be 10 million years old or older, of Miocene times (Fig. 39).*

Figure 79 emphasises that the phreatic passages near The Knave cannot have formed after the deep erosion of their host limestones to form the slades and the present cliffs. The water table within Ogof Ffynnon Wyntog lies close to the level of the valley floor of Ram Grove, which controls the water table less than 400 m away. The limestone is now wholly permeable in three dimensions, especially at right angles to the coast along fractures (Fig. 73) and subparallel to the coast along bedding planes. Thus, the level of any spring issuing on a cliff above sea level must be from a perched water table governed by a permeability barrier. Like many coastal springs on Gower, Ffynnon Wyntog lies at about 10 m OD because lower reaches of the original cave became infilled with sediment when sea level rose to 10 m OD in Ipswichian 'Patella times'. Some spring water, however, has found its way down to lower original passages that have been invaded by Holocene seas.

Figure 80 shows two stages of the likely continuous development of erosion in which both the (karst) ground surface (S1 and S2) and the water table beneath it (W1 and W2) were 'lowered' respectively on and within the limestone, in Miocene times. This erosion was in response to regional absolute uplift (Sections 2.7 and 2.8), so in theory the karst surface and water-table levels may have remained at the same absolute levels while migrating downwards on and within the actively uplifted limestone. The Early Pliocene cutting of the slades (Fig. 79) would drain all of the caves allowing speleothem formation, except after 123,000 years ago when the lowermost levels were blocked (see Sp1 in Fig. 75).

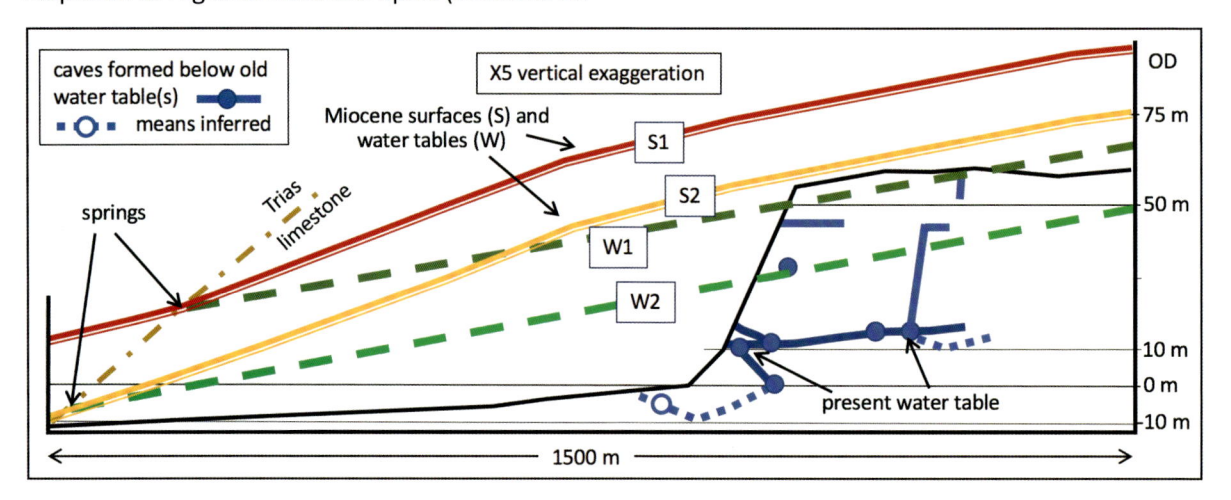

*Figure 80. Miocene erosion relative to the limestone in which phreatic passages formed beneath water tables, here related to two contemporary surfaces that were above the presently existing ground. The two stages depicted are 'snap shots' of the ground surface (labelled **S**) and associated water table (**W**) that would have occurred within a more or less continuous progression, in which phreatic passages become drained and left 'high and dry'. While the diagram suggests that the ground surface was physically lowered, from S1 to S2, we know that there was regional absolute uplift, so that the caves may have formed at greater depths within the limestone massif in response to the whole system being uplifted. The surfaces **S1** and **S2** are schematic average levels, only notionally smooth, whereas in reality they would have been typical subaerial karst.*

The marine planation that formed the 55-60 m platform clearly resulted from sea level rise *relative* to features so far discussed. We know that regional absolute subsidence was the main cause of this, with the marine transgression peaking in Late Pliocene times, 3 or 4 million years ago. We do not know what erosion might have occurred as the sea advanced over the Miocene land surface represented by S2 (Fig. 80), although the offshore submerged cliff at around -15 to -20 m OD elsewhere could have been cut then (Section 2.12). Subsequent Quaternary cutting back of the cliffs exposed the caves as we see them now.

4.4 Green Cwm and the Last Glacial Maximum underground

Llethrid Swallet Cave[5] and Tooth Cave (Fig. 81) are well known for caving adventures, although Llethrid was repeatedly blocked and eventually lost to us after about 2005, because sediment that had banked up in the stream by the entrance had been released and obstructed the rather tight passage inside. More on that later (Section 4.12). The two caves are parts of a single system in the mainly dry valley of Green Cwm. The age of the caves has been poorly understood but suffice it for now to say that they are in parts probably of the same initiation age as the caves at The Knave (above), that is, most probably at least 10 million years old. We shall consider the evidence for this, but perhaps the most interesting aspect of their history is the clear impact of the Late Devensian glaciation of Gower on both caves. Welsh Ice advanced over the caves and its subsequent retreat (recession) is recorded underground.

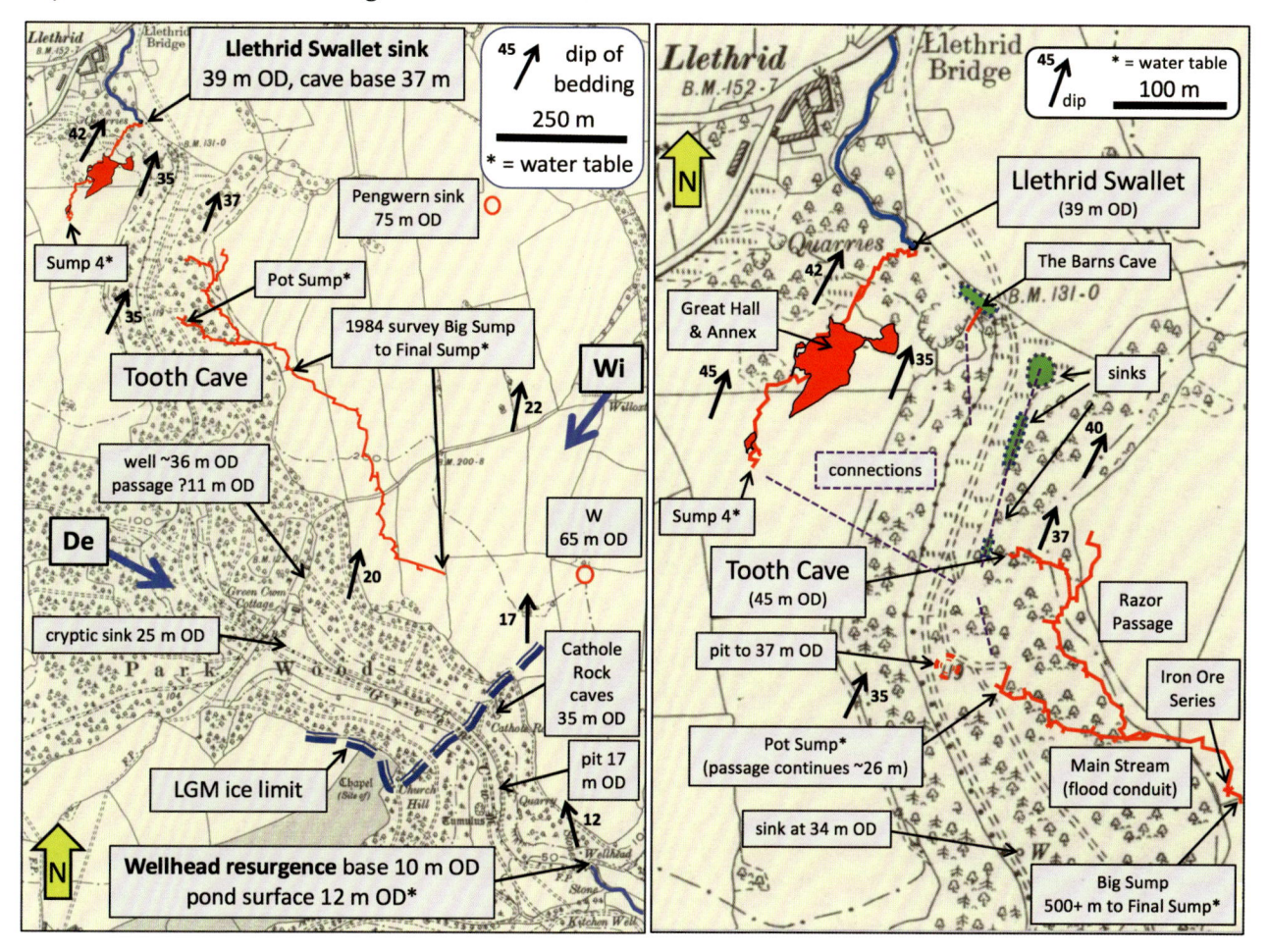

Figure 81. Green Cwm Cave System: Llethrid Swallet – Tooth Cave – Wellhead resurgence. (Left panel shows the entire cave system and the right-hand panel shows details of the upper reaches). In **red** are the Baynton and Jorgensen survey of Llethrid (January 1965) and the University College Swansea Caving Club and O'Reilly line surveys of Tooth Cave (1965-1966 and 1984). The sink to spring straight-line distance is 1.68 km and the vertical distance between cave floor levels, end to end, is 27 m. This drop amounts to an average slope of <1°, but much of it is within Llethrid cave, near the entrance, such that over most of the distance to the spring the water-table height difference amounts to roughly 10 m or a slope of only ~0.3°. The connections are all proved by dye testing, including the confluent flows (**blue arrows**) from Decoy sinks (**De**) and Willoxton sinks (**Wi**). The hydrogeology is discussed in Chapter 5. The Last Glacial Maximum (LGM) ice southern limit (**dashed double blue line**), near Cathole Rock, is a line mapped by the author and discussed in Chapter 3. '**W**' is Willoxton Dig[6] (SS53939028), which is a gently inclined metre-scale phreatic tube truncated by the plateau surface. The dips of the limestone bedding derive from original British Geological Survey mapping at 1:10,560 (see Strahan 1907a). (Base map is Ordnance Survey 1:10,560 Glamorgan XXII.SE published in 1900, reproduced with the permission of the National Library of Scotland).

The Green Cwm valley bedrock profile, its sediment floor and the caves were all modified during the Last Glacial Maximum (LGM), which peaked about 23,000 years ago (Chapter 3). The indicated connections between Sump 4 in Llethrid and Pot Sump in Tooth Cave (Fig. 81) probably in part follow the bedding strike, in general south-eastwards across the valley trend. The pattern of main passages involves trends broadly southwest against the dip of the limestone bedding, and broadly southeast along the limestone bedding strike[7]. The line of active floodwater sinks near the north end of the valley, and the apparent offset of the top contact of the limestone, seem likely to follow a fault or fault zone (Fig. 82).

The Main Stream Passage in Tooth Cave classically reveals its geological control by appearing in plan view rather like a staircase (Fig. 81). Some at least of the passage segments that trend broadly southwest exploit the numerous fractures (faults and joints) trending that way. Clearly most of the known cave is not under the valley floor; its confluences with water from Decoy sinks and Willoxton sinks are proven, but their locations are not[8]. Similarly, the last few hundred metres to Wellhead resurgence is unknown territory.

Until recently the location of the Tooth Cave Final Sump also was unknown. A survey from Big Sump in that direction by Andy Freem with Val Bednar, during the 2018 drought, had to be abandoned half way down owing to toxic concentrations of CO_2 that affected other team members near the Final Sump. However, plotting by Paddy O'Reilly of his 1984 survey with Blake Farren now provides the most up-to-date fix for the known downstream extent (Fig. 81). The inferred water-table surface is generally no more than ~16 m below the level of the valley floor in the northern part of the cwm and only a few metres in the southern end, which is one reason why the normally dry valley is prone to flooding after sustained heavy rainfall (Section 5.10).

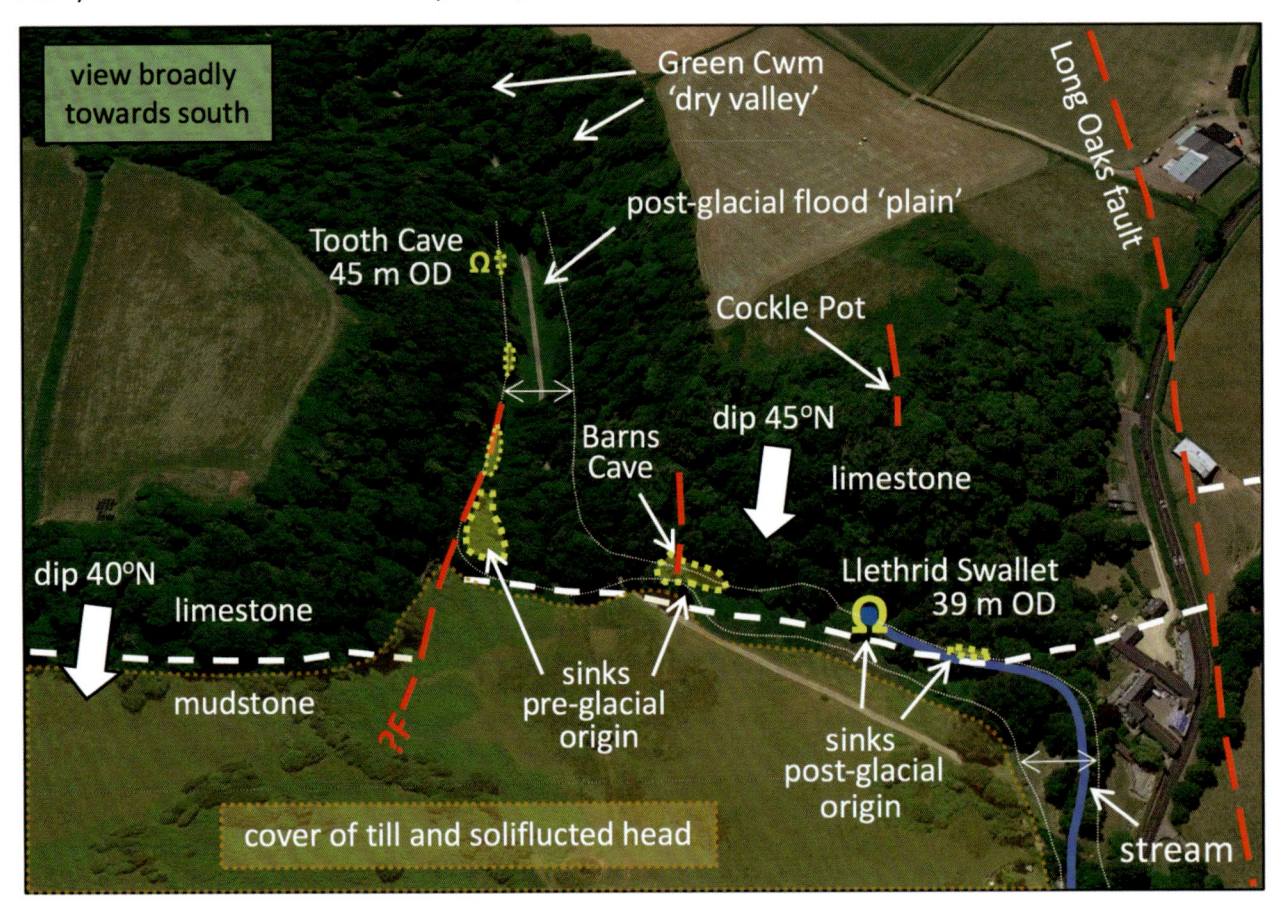

Figure 82. (Page opposite). Google Earth image of the northern end of Green Cwm viewed towards south-southwest. Llethrid Swallet, or sink, takes the normal stream flow (SS531911) and cavers have previously diverted the stream temporarily into a sink 40 m farther upstream[9]. During floods, water overflows the swallet and sinks in several places near the head of the cwm and as far as the sink in front of Tooth Cave entrance. Barns Cave is where digging found phreatic passage blocked with glacial outwash debris. Cockle Pot fault-controlled rift yielded marine shells some 0.5 million years old contained within Anglian glacial debris (Section 4.12). In extreme floods, water continues at the surface in the 'dry' valley all the way to the vicinity of Wellhead resurgence (Fig. 81). Green Cwm is floored by glacial outwash and erratics with a thin cover of recently accumulated sand, silt and organic debris. At its maximum extent, about 23,000 years ago, the Late Devensian Welsh Ice completely covered this location, reaching 1.5 km farther south as far as Cathole Rock (SS537900).
(***Above left***) *Incised floor of Green Cwm 650 m south of Llethrid Swallet. Occasional floods in the otherwise dry valley have cut this channel exposing the erratic boulders and glacial outwash debris that form the valley floor.*
(***Above right***) *Erratic quartzite boulders from the north crop of the South Wales Coalfield in the channel bank shown in the above left image (SS531904).*

Glacial debris of the LGM has dramatically influenced the geomorphology and cave development in the northern end of Green Cwm (Fig. 82). A cover of till, poorly draining and widely solifluected on slopes, blankets most of the mudstones and sandstones (Marros Formation) that overlie the limestones. The strata dip towards north-northeast at 35-55° and appear offset by a fault or fault zone along the east side of the cwm. Numerous sinks occur in the limestone, taking normal flows (Llethrid Swallet) and floods of water. The floor of Green Cwm, all the way at least as far as the vicinity of Parc le Breos Neolithic chambered tomb (SS537898) and Wellhead (Fig. 81), mainly comprises a flood plain of fluvioglacial outwash that broadly forms a flat profile cut into by flood-water channels (and modified by forestry tracks) (Fig. 82). The few exposures of the outwash show it to be at least 2 m thick, locally overlying a clay-rich till. Large north-crop erratic boulders of quartzite are embedded in the outwash and were delivered by the Welsh Ice, which extended as far as Cathole Rock (Fig. 81).

Figure 83 shows the 3.5 km² Pengwern rainfall catchment that drains into Llethrid Swallet and Green Cwm today (Section 5.6). The same basin would have gathered glacial outwash as Welsh Ice of the LGM advanced from the northwest across it, and also as it receded when first Pengwern and then Welsh Moor recessional moraines were formed (Section 3.14). We know that between those times the ice advanced widely across Green Cwm, shedding erratics both into the valley and on the high flanks above, as far as Cathole Rock (SS537900). We will consider in detail the influences within the caves of these changing conditions, but here it is reported that the overall effect *at the surface* was to choke pre-glacial sinks and deflect the post-glacial drainage to the western margin of the debris, where the modern sinks have formed. This explains why the former cavers' entrance passages into Llethrid cave, and the stream-diversion sink, are of rather tight immature passages ("unpleasant, narrow and wet") that contrast with the main passages and large ornamented chambers deeper within. On this interpretation, Llethrid cavers' entrance passage is relatively young, probably mainly developed in the last 20,000 years or so.

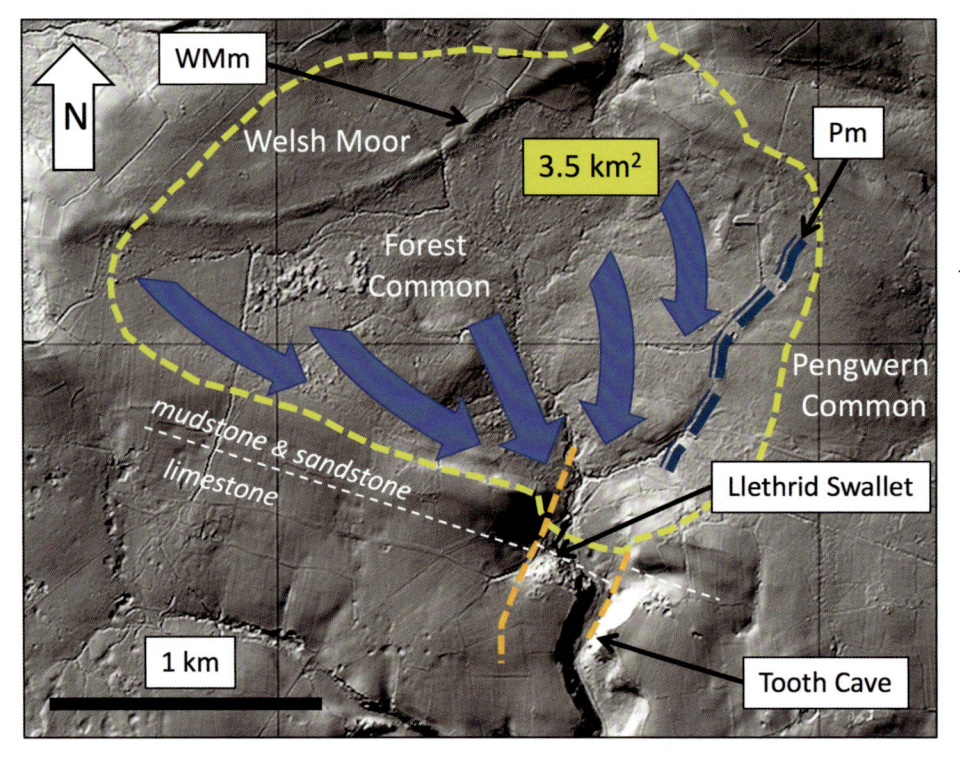

*Figure 83. LiDAR shaded relief image of Pengwern catchment of approximately 3.5 km² (Section 5.6); **blue arrows** show drainage directions. The same basin would have focussed glacial outwash towards Llethrid and into Green Cwm during both advance of the ice and its regression. Pengwern recessional moraine (**Pm**) and Welsh Moor recessional moraine (**WMm**) are shown, the latter far more prominent than the former. (LiDAR image uses base data © Natural Resources Wales).*

Insight into the blockage of a pre-glacial passage leading into the Green Cwm System derives from digging at the first substantial flood-bypass sink downstream from Llethrid Swallet (Barns Cave[10], Figs 82 and 84). The dig entrance is at the contact of valley-floor flood-plain sediment with limestone outcrop. It penetrates several metres down a narrow fault zone into a rift that drops farther down to passage that is in the roof of a tunnel of at least 1-2 m scale and mostly filled by debris (bottom unseen).

The debris is a poorly sorted mixture of variably rounded cobbles and pebbles with a predominantly sand-silt matrix; the coarse grains are a typical mix of local Pennant sandstones with quartzites, conglomerates and brown sandstones originating in the north crop of the South Wales Coalfield and delivered here by the Welsh Ice (Chapter 3). It is explained below, using better exposures in Tooth Cave, that this debris is fluvioglacial outwash that became remobilised and entered the caves catastrophically as a debris flow. The partly blocked passage is of a scale larger than the Llethrid Swallet entrance passage and, obviously, is pre-glacial. Digging some 10 m along the roof of the part-filled passage led to a choke of large boulders with matrix of distinctive red-purple sandy mud, like that found in the large chambers in Llethrid, 70-100 m farther west (Fig. 81). Open but tight draughting passages extend sharply down some 7 m from the 'red choke' and take flood water into drains that are too tight to follow but do not here reach down to the water table, which must be deeper down than about 22 m OD.

Figure 84. Excavation at Barns Cave.

*(**Top**) Near the surface, limestone blocks were broken out to widen a fault zone, beneath which a near-vertical rift became accessible.*

*(**Middle**) The rift bottoms out in the roof of a moderate sized phreatic tunnel, within which is a thick fill of glacial debris that had to be excavated to gain access.*

*(**Bottom left**) In situ glacial outwash debris including rounded cobbles and pebbles embedded within and grading up into sand and silt, which is laminated at the top beneath a roof pocket (scale is 10 cm).*

*(**Bottom right**) The excavated 10 m passage leads to a choke where red sandy mud fills between metre-scale boulders. The red mud is visible in places between the scaffold.*

(Middle and bottom photos courtesy of Antonia and Andy Freem).

We understand that before the Late Devensian advance of the Welsh Ice, water that drained off the Pengwern catchment (Fig. 83) entered the Green Cwm caves at the top of the north-dipping limestones. It is not clear yet how the beautifully decorated caverns of the Llethrid Great Hall and Annex (Fig. 85) were connected to the surface in pre-glacial times; the original Llethrid cavers' streamway entrance passages are too immature for that role.[11] Although close to the modern valley floor now, the pre-glacial phreatic Barns Cave passage must have formed long before the valley bedrock became cut down to the present level. Llethrid chambers are old.

The decorated chambers show substantial enlargement by roof collapse, both before and after considerable speleothem formation (Fig. 85), reflecting substantial drainage and dissolution. The sinks farther east in Green Cwm (Fig. 82) are presumed to have been similar to the Barns Cave entrance passage, according to evidence provided below concerning Tooth Cave. The level of the metre-scale Barns Cave passage is at about 30 m OD, substantially below the nearby ancient phreatic tube entrance to Tooth Cave, at 45 m OD. Evidently, much of the Green Cwm cave system here is very old.

Figure 85. Llethrid Great Hall.[12] (a) Maurice Clague Taylor on a thickly decorated boulder pile. Photographs were taken with long exposures illuminated by burning magnesium metal strips. (b) Detail of the slope shown in (a), with Marjorie Taylor for scale, obviously, for the photo, keeping as still as the surrounding stalagmites. (c) Huge fallen boulder near the northeast corner of the chamber; person for scale circled. Amongst known caves of Gower, this chamber is vast, almost 100 m from end to end (307 feet in the Baynton and Jorgensen survey of January 1965, and 320 feet measured by The Taylors in May 1956). These boulder falls are comparatively young, whereas elsewhere in the chamber the numerous snapped, fallen and subsided formations, many themselves thickly overlain with speleothem, register a considerable history of progressive subsidence. The original and currently active streamways are somewhere underneath. (continued...)

*Figure 85. Llethrid Great Hall continued. (**d**) The Church and Steeple shows subsidence with tilting of the broken-off thick column, shamefully desecrated by the author, then aged 14. Regrettably, from the first discovery onwards the splendid formations were routinely trampled, leant upon or fondled by cavers almost invariably muddied from the flood-prone chamber floor and passages. Early archive photos preserve some of the splendour; later ones show the need for conservation and a substantial clean-up, which is one objective of the current efforts to recover access (Section 4.12). (**e**) The Mushroom, here in clean condition with a youthful John Harvey, was rendered a dark muddied disgrace. (**f**) The Great Curtain or The Drapes is an unbroken, delicately thin, banded sheet several metres long. It hangs from a limestone bedding plane that dips northeast and forms much of the chamber roof (Fig. 81). (**g**) The roof of The Annex also is beautifully adorned. Photos (a) and (b) are from the SWCC archive of The Taylors; (c) by Paul Tarrant; (e-g) SWCC archive, (e) by David Hunt, (f) by Seaton Phillips, and (g) by Jem Rowland.*

As the Welsh Ice advanced on the Pengwern catchment (Fig. 83), fluvioglacial outwash would have been flushed into Green Cwm and probably into some open sinks. The rock floor beneath the middle of the valley is not exposed, so this initial phase is conjectural. However, substantial blocking of the cave passages only occurred later, during the glacial recession (see below). As well as fluvioglacial debris, the ice delivered metre-scale erratics that were dumped into the cwm. As the ice reached some 1.5 km beyond the sinks, there will have been a phase during which meltwater with sediment moved beneath it, both sub-glacially along the valley floor and in some cave passages.

Ultimately, during the recession, catastrophic inundation by debris caused cave passages to be substantially filled and the original cave entrances to be buried to several metres depth. This deflected the remaining stream, then a misfit, to the western valley margin, where it found new limestone weaknesses along which the rock was dissolved to form the relatively immature Llethrid Swallet entrance passages.

4.5 Tooth Cave: a stygian realm of curiosities

Tooth Cave[13] has 1.5 km of accessible passage[14] but is prone to flooding and should be avoided in all but settled dry weather. A branched and notoriously uncomfortable entrance series leads to a main flood-bypass passage between constricted diver-resistant sumps in the water table (Pot Sump and Final Sump; Fig. 81). Dye testing shows that slow-flow passages below the water table continuously connect Llethrid stream to the Wellhead resurgence, but after heavy rainfall flood-water bypass via Tooth Cave is relatively fast (Chapter 5). The slow-flow passage(s) are interpreted to be phreatic and possibly also vadose conduits that developed when the system base level was as much as 25 m below modern sea level (-25 m OD). These subsequently became flooded and infilled during sea level rise (Section 4.2).

Tooth Cave phreatic-tube entrance (Fig. 86a) is 220 m south of Llethrid Swallet, at 45 m OD and 5 m above the valley floor outside. The only permanently 'dry' sections are close to the entrance, including Bone Chamber (Fig. 86b) and the nearby choke and rift, and also the high-level phreatic passages of the Iron Ore Series (also known as Aven Series[15]), which, remarkably, is in and above a Triassic-sediment-filled fault zone (Section 4.8). Razor Passage (Figs 81 and 86c) involves challenges to poorly protected knees and elbows, in addition to occasional sump development and blockage by mobile gravel.

The main flood-bypass passage (Main Stream Passage; Figs 81 and 86d-h) is mostly gently sloped, but near horizontal at about 24 m OD, and is obviously within a few metres of the water table. So called 'Big Sump' only dries in droughts but can be dived[16], while passages farther downstream near the 'Final Sump' can have toxic levels of CO_2, probably due to decomposing leaves. Added to this, the main passage clearly floods to its roof where leaves become lodged, so, all-in-all, penetrating the cave requires some planning and resolve. However, set against this, it should become apparent that this rather gloomy and threatening cave holds some splendid and curious records of the Late Devensian glacial episode, as well as the remarkable occurrence of sediments more than 200 million years old. These features are the focus here.[17]

Main Stream Passage is characterised by two features that record contrasting phenomena. Firstly, its phreatic-tube walls extensively are deeply and beautifully scalloped, reflecting relatively fast flow of 'aggressive' water, i.e., liable to dissolve limestone. Secondly, there are impressive large banks of cobbles (256-64 mm) and pebbles (64-4 mm) variably set in a granule/sand/silt/clay-grade matrix, widely covered by a veneer of dark brown to black iron-manganese hydroxides and with speleothem on top (Figs 86, 87 and 88). As we shall see, the banks reflect a single episode of catastrophic intrusion into the cave of the mixed debris, followed by comparatively normal processes of erosion and sedimentation.

In the cobble-pebble banks (Fig. 87), the large fragments are variously rounded, some irregular and others slabby, and commonly they form a touching framework with granules, sand, silt and clay-grade sediment filling the spaces in between. Elsewhere the large fragments are fully supported within the finer matrix. Taken together the pebbles and cobbles, which dominate the deposits, are well sorted in terms of grainsize (excluding the 'fines') and the assemblage is typical of coarse-grained, water-sorted and transported debris found in rivers. Small-pebble and granule gravel (8-2 mm) is sparse in patches and large boulders are conspicuously rare[18].

The large fragments and their sorting are closely similar to those of the stream sediments in the catchment of the caves, but here there is finer grained sediment mixed with the coarse and also in a grainsize-graded top that transitions up into sand, silt and clay (mud)[19]. Locally there is a considerable thickness of precipitated dark-mineral coating overlying the fine-grained top layer, showing that this remained unreworked. The cobble and pebble-rich deposits generally lack stratification (there is no distinct layering) but locally show flow-alignment imbrication fabrics, mostly at a shallow angle to the floor but locally steep (Fig. 87, bottom panels). The deposits half-fill some passages, either broadly mounded or flat-topped as in some side passages where the fine matrix continuous with the covering layer is especially well developed (Fig. 87 top panels). The banks are remarkably stable and predate substantial flowstone and stalagmite speleothems. Patches of pale fine sand or silt are everywhere the most recent deposit (Figs 86e and f, and 87 top panels) and represent 'normal' flood transport and deposition.

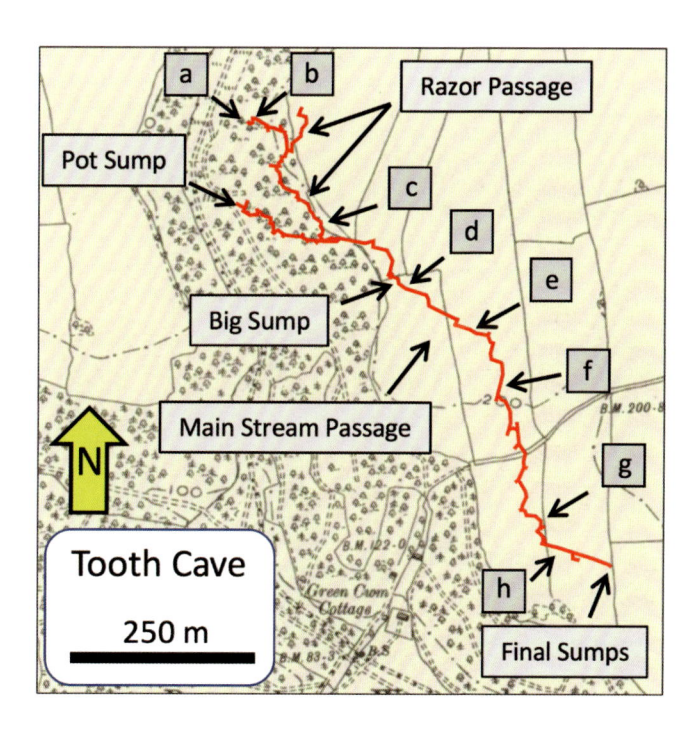

Key to photograph locations in Figure 86 (next page).

ALL OUR OWN WATER

Figure 86. Main features of Tooth Cave passages. (Photos courtesy of Andy and Antonia Freem).

(a) Gated phreatic-tube entrance at 45 m OD, with pendant 'tooth' near the middle.

(b) Bone Chamber is a few metres in from the entrance and several metres down, close to the valley floor. During substantial floods into the valley, a sink is active outside and at one time a small stream could be seen at the bottom of a narrow rift; a tight passage extended towards the sink, but the access was filled with excavation debris (Baynton 1968-1969 p. 27, and John Harvey pers. comm. 2021). The **white arrow** points at a layer of fluvioglacial material that entered the cavern as a debris flow (it was undercut by manual excavation). This debris both rests on and is overlain by fallen angular blocks, while the stalagmites and flowstone are on top of all. Evidently block fall occurred before and after arrival of the glacial in-wash and the calcite formations formed more recently. The chamber floor, several metres in diameter, was excavated in 1962 when Bronze Age human remains were discovered in a thin 'earth' layer beneath the speleothem (Harvey et al. 1967), which was partly disrupted for the dig. Humans originally accessed the chamber via the vicinity of the fallen blocks, which are parts of the same pile that forms a debris apron beneath the cliff outside. The stalagmites record growth in no more than about 4,100 years (start of Bronze Age; see below).

(c) A relatively open part of Razor Passage where intermittent (flood) stream flow and caver traffic have excavated a trench through a partial fill of in-washed fluvioglacial debris.

(d) Downstream of the intermittent Big Sump, flood water has eroded a channel into in-washed fluvioglacial debris. The stalactites show a partial coating of an amorphous iron-manganese hydroxide, which, where missing, reflects partial dissolution and/or wear by sediment in suspension (corrasion).

(e) Typical section of Main Stream Passage with substantial banks of cobbles and pebbles with finer sediment matrix and with pale partial cover of flood-water-deposited sand and silt.

(f) Downstream the passage is generally larger than elsewhere. Here the section known as The Nave shows extensive upwards phreatic dissolution along the fracture that guides the passage here (NNE-SSW). The eroded banks of cobbles (left) show that the passage was once substantially filled with the debris.

(g) Towards the Final Sump, confluent side passages coincide with a zone of intricate dissolution forms, probably reflecting occurrence of irregular zones of less-soluble dolomite and possibly also mixing of waters with different dissolved mineral contents, which causes enhanced dissolution ('mixing corrosion' of Bögli 1980, pp. 35-37).

(h) Near the Final Sump the passage splits into narrow rifts with beautifully scalloped walls. There are in fact two accessible but tight sumps and in drought they can be dry; to date these constrictions have not been passed. This is the lowest accessible part of the known cave and, after prolonged drought, it is prone to accumulation of toxic concentrations of CO_2 in the stagnant air.

Figure 87. Poorly sorted cobble and pebble-rich debris with graded fine-grained top and local cover of speleothem. (Photos courtesy of Andy and Antonia Freem).

(a) In Main Stream Passage the banks of debris are significantly mounded but typically cohesive, reflecting a (sticky) clay component within the fine-grained matrix. There is abundant evidence of erosion by floodwaters to produce channels with steep sides (e.g., at the distant location of a caver), whereas the mound surfaces are stable and widely coated in a veneer of iron-manganese hydroxide.

(b) Here the coarse debris has partly filled a side-passage and has a smooth surface layer of silt and mud. Patchy pale sand-silt flood-water deposit rests on top. The fill surface to roof space is up to about 1.5 m.

*(c) Eroded cobble-pebble bank, about 40 cm high near the downstream end of Razor Passage, showing the upper surface overlain by stalagmites and flowstone. The **red box** highlights a steep 'log-jam' fabric of slabby fragments, which is typical of en masse transport and deposition, here with mass-flow from the right having pushed against and thus steepened an original imbrication.*

*(d) Eroded bank of poorly sorted cobble-pebble deposit with fine matrix. There is a weak imbrication fabric of slabby pieces (**red lines**), aligned top right to bottom left, and locally a steepened orientation (fragments indicated by **white arrows**). The fabric records mass flow from left to right, near the junction of Razor Passage with Main Stream Passage.*

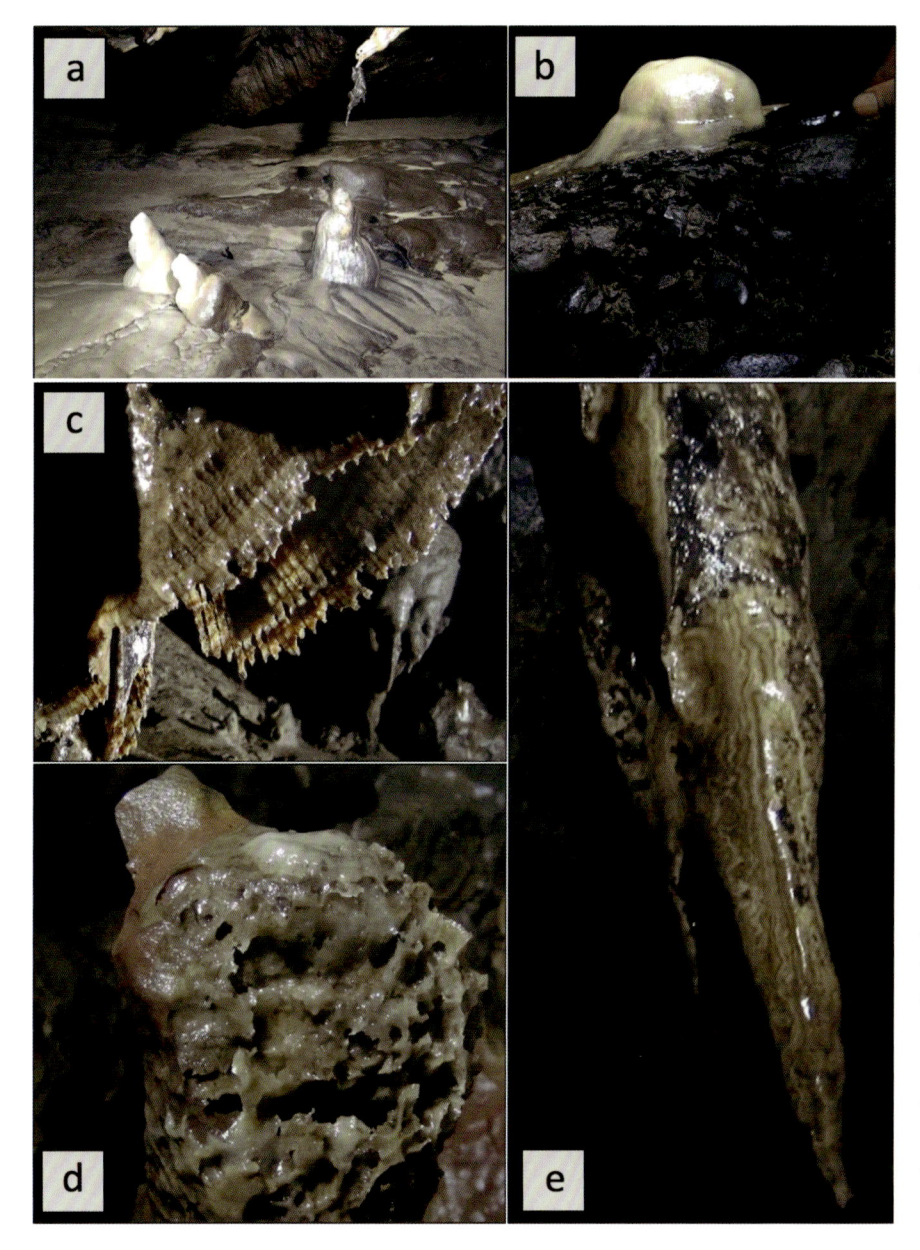

Figure 88. Speleothem in Main Stream Passage. (Photos courtesy of Andy and Antonia Freem).

(a) Clean flowstone and stalagmites thickly cover a cobble-pebble bank. The stalactite just showing has caught a leaf, showing that flood water reaches the roof. Front of view is approximately 1.5 m wide.

(b) Clean stalagmite formed on soft iron-manganese hydroxide veneer that coats the in-washed fluvioglacial material and is several millimetres thick.

(c) Fretted partly dissolved curtain; field of view approximately 20 cm.

(d) Fretted boxwork of partly dissolved stalagmite, 12 cm wide.

(e) Stalactite about 25 cm long showing some black mineral coating with maximum corrosion and/or corrasion of the upstream-facing side. Flood water in this passage has a suspended-sediment load and this asymmetrical feature could represent wear (corrasion) due to this.

The cobbles and pebbles in Tooth Cave are predominantly of tough sandstones and conglomerates, with some representative of local Pennant sandstones and many being erratics typical of the northern coalfield outcrops of the Old Red Sandstone and Carboniferous strata (Fig. 89). The coarse debris seen in stable banks evidently was deposited in a discrete, one-off episode followed by more normal processes. There is no presently active process that has the energy to transport such large stones so far into the cave, and also no obvious presently open point of entry for this size of material into the known passages. The general lack of stratification in the cobble and pebble-rich banks, the occasional imbrication and upright jamming fabrics, and the variably mounded forms all indicate rather catastrophic transport and dumping. The finer grained, poorly sorted sediment matrix and cover of the cobble-pebble banks show that it arrived with the coarser material, with partial expulsion upwards of fines during the rapid dumping and then transition to contrasting tranquil settling conditions. Matrix of this nature (e.g., Fig. 87d) cannot form by later infiltration and so is an original feature. The steep banks eroded into these deposits record subsequent stream erosion, although the bank tops widely were not reworked or covered by eroded material.

Most of the cobbles and pebbles are coated with brown to black, amorphous iron and manganese oxide and hydroxide veneers, some of which are thin and tough while others are thicker and softer (Fig. 89). In places on top of the banks the mineral veneer is 2-3 mm thick and forms a continuous drape. Similar deposits occur in places on the cavern walls and on some stalactites. Passage walls near Pot Sump, which is the top inlet to Main Stream Passage, are variously coated. Some surfaces are beautifully scalloped clean limestone, some have a patchy ferruginous brown coating and/or a muddy veneer, and others have soft black layers and drapes. The coated surfaces here are widely scratched; some are enjoyed by browsing snails (detailed discussions below). The harder veneers are identical to those on cobbles in the bed of the stream that enters the caves, above Llethrid Swallet, where they indicate oxidizing and basic (hydroxide forming) conditions typical of fast flowing, oxygenated waters.[20]

Figure 89. Coatings of iron and manganese oxides and hydroxides in Main Stream Passage.

(a) Cobbles collected from the passage show a range of coating types from mm-thick and earthy to very thin and shiny, as if polished. The broken counterpart fragments (b) show north-crop Marros quartzite (pale, white to yellow) and coarse ORS Brownstone. The shiny surface is of a thin veneer of haematite (Fe$_2$O$_3$) that cannot be scratched with a steel blade and is difficult to scratch with quartz. The veneer on the 'shiny' stone on the right completely coats it, while the stone in the lower middle has a similar thin complete haematite coating beneath a thicker encrusting layer on the one side. (continued...)

*The thicker coatings are softer, friable iron hydroxides, mainly goethite and/or amorphous limonite, respectively FeO(OH) and FeO(OH).nH$_2$O, with black amorphous manganese hydroxide MnO(OH). (**c**) A cobble bank top surface is thickly coated, essentially veneered or draped over, with iron and manganese hydroxides with its steep eroded face less thickly coated and locally with trapped pale flood-water sand or silt. (**d**) This coating of fine to amorphous iron and manganese hydroxides is soft, slightly plastic but readily breaks. (Lower photos courtesy of Antonia Freem).*

In one or two places in the downstream reach of the cave, the rapidly dumped cobble deposits overlie thinly bedded to laminated fine sands and silts that record an earlier episode of sediment settling in tranquil water (Fig. 90). These deposits display concentric colour bands due to post-depositional migration of iron and/or manganese hydroxide(s), as commonly occurs by diffusion within or at an interface of pore-waters during drying; technically they are akin to Liesegang rings.

*Figure 90. Thinly bedded to laminated sediments, roughly 50 cm thick in the far downstream reach of Main Stream Passage[21], are older than the cobble deposits, which lie over and against them (**white arrows**). The photo is skewed; the layering is actually horizontal and that which is marked by the **red brackets** was originally continuous proving the cobble deposit to be younger with an erosional base.*

*Curved colour bands crossing the sediment laminations, highlighted by the **white dotted line**, are concentric fronts of Fe-Mn enrichment probably formed during drying of the sediments. (Photo courtesy of Andy and Antonia Freem).*

Here, *based on the observations made so far*, is a summary of apparently successive conditions and processes involved in the development of Tooth Cave. These are inferences and there are uncertainties and substantial gaps to be filled. Some problematic issues are discussed further below, and the overall evolution of the entire Green Cwm System is considered later.

(1) There was a long period of phreatic passage creation, conceivably initiating at least 10 million years ago. Successively lower passages were formed beneath what is now the phreatic cave entrance and Iron Ore Series, which respectively are at 45 m and 38 m OD, while the base level migrated down to below present sea level, conceivably to some -15 to -25 m OD (Sections 2.5 and 4.2), probably then with an unrestricted stream exit.

(2) During the Ipswichian, around 123,000 years ago, the 10 m OD high stand of sea level (Section 2.14) led to tranquil, low-energy sedimentation that formed laminated, fine sand

and silt deposits within what is now the main flood-bypass passage several metres above 10 m OD. This probably happened over some few thousands of years when lower reaches of the cave became stagnant as water backed up due to the higher base level. Passages lower than 10 m OD were substantially infilled. When sea level subsequently fell, the sediments in the flood bypass passage dried and were extensively eroded by renewed flood-bypass activity. Secondary redistribution of iron and manganese oxides and hydroxides formed concentric bands in the sediments. Passage walls in places not actively corroded to form scallops would be coated in thin veneers of haematite and some Fe/Mn hydroxide.

(3) As Late Devensian Welsh Ice advanced onto Gower, about 23,000 years ago, fluvioglacial sediment including rock flour ('glacier milk') must have been washed into Green Cwm, but evidence for this is not exposed. As ice flowed over the cave and penetrated 1.5 km into Green Cwm, it is probable that sub-glacial melt water carried sediment into the cave.

(4) During the glacial recession, accumulated fluvioglacial outwash debris was catastrophically remobilised from the Pengwern catchment (Fig. 83) and deeply penetrated underground via open sinks as a debris flow. This caused the wall scratches at Pot Sump, as discussed below, and it formed a remarkably extensive, poorly sorted cobble- and pebble-rich deposit, which also affected the Llethrid main stream entrance – now Barns Cave. The debris buried cave entrances and left the valley floor much as it is today. The ultimate cause of the catastrophic remobilisation is speculatively considered below.

(5) As more 'normal' post-glacial conditions re-established, iron and manganese oxide and hydroxide precipitate veneers formed thickly on top of the coarse sediment banks as they became cut by channels taking flood waters. Channel sides locally have mineral veneers suggesting that the active channelling diminished in time.

(6) Speleothem including extensive flowstone and stalagmites formed on top of the banks and on their precipitated chemical veneers. As sea level rose close to its current position by about 7,000 years ago, any deep passages remaining open downstream from Wellhead would again have become stagnant as the Pennard valley filled with alluvium.

(7) In Bone Chamber speleothem covered bones of Bronze Age humans. Such development is ongoing, although some early formed stalagmites and stalactites now show partial dissolution in the main flood-bypass passage.

(8) Modern floods involve limited erosion of the (known) cobble and pebble banks; fine sand, silt and some mud are washed through to the undersized resurgence at Wellhead (Chapter 5), and minor amounts are deposited underground from the waning flood flows. There is no active plastering of the passage walls with mud, so that many mineral crusts and speleothem surfaces remain perfectly clean.

4.6 Debris flow, scratches and snails

This section explains the inference of a debris flow and its deposits in Tooth Cave, and hence in Green Cwm too, and it proposes a causative post-glacial catastrophic event. Bizarre and highly unusual scratches localised in the vicinity of Pot sump are explained in terms of debris-flow dynamic behaviour, attributing them to momentary forceful contacts made by fragments passing through in the debris flow. This 'explanation' was hard to come by for the author, so heaven help the reader. Diverse observations have been assembled that, hopefully, restrict the hypothesis presented, but Occam's razor and parsimony are in there too, as well as snails.

The cobble- and pebble-dominated debris that occurs widely within Tooth Cave and partly blocks the Barns Cave passage, near Llethrid (Section 4.4), has poor sorting (i.e., a wide-ranging mixture of grain sizes), lacks bedding, contains internal imbricate fabrics, and its matrix grades up into a fine-grained top. These features together are typical of aqueous debris flow deposits. And, from what is accessible to see, there appears to be only the one such sedimentary unit. Fluvioglacial outwash entering underground driven by a melt-water river could not form such a singular poorly sorted deposit graded up to a fine-grained top.

Debris flows can be immensely fluid, fast and penetrating. They commonly advance in series of energetic waves with considerable momentum, and can persist for tens of minutes, depending on the source, pathway and slope.[22] The phenomenon inferred here is described as catastrophic because it constitutes a distinctly energetic event of finite duration, involving rapid transport and dumping of coarse material, followed by transition to tranquil settling of fines. Mounds commonly form when flowing debris piles up against material in front that is moving more slowly, as commonly happens with spontaneously formed waves, or has stopped. Debris flows moving down solid phreatic tubes will experience minimal frictional resistance along contacts if there is water there, and the normal slowing of subaerial flows when they spread out on open ground will not occur underground, so considerable penetration is to be expected.

From the considerable penetration of the debris into the caves and into Green Cwm (it entered Tooth Cave Bone Chamber from the cwm), it seems inevitable that there must have been a one-off causative event that mobilised a huge amount of debris. Lakes impounded by glacial debris in front of receding ice sheets are normal and one may readily imagine that a lake outburst, or a sudden release of subglacially ponded water (a jökulhlaup), could have flushed accumulated fluvioglacial debris from the Pengwern catchment into the sharply constricted entrance to Green Cwm (Figs 82 and 83).

Detailed speculation regarding the specific site(s) and nature of entry of the debris flow into Green Cwm is unwarranted, but within 300 m of the location of Pot Sump there are several active flood-water sinks (Fig. 82), all of which have taken considerable water recently, e.g., in the flood that reached Wellhead through the cwm on 13th December 2020 (Section 5.10). Now they are choked with glacial debris, sand, silt, mud and rotting vegetation, and they drain slowly, but before the Late Devensian glaciation they would have been open and are the most likely sites where the coarse-grained glacial outwash debris entered underground.

The 30 m-long arcuate cliff that contains Tooth Cave entrance, with the semicircular depression around the sink there[23], resembles a partly buried steep-walled collapse pit like those that remain unfilled in a similar setting in Bishopston Valley (e.g., Lower Daw Pit at SS575890). Suffice it to say that the vertical distance of more than 10 m from the valley floor down into the caves will have conferred significant hydraulic head (pressure) on the underground flow, with any thickness of the debris flow moving into the constricting cwm significantly adding to that motive force.

What of the scratched walls near Pot Sump? Could they be a record of the debris flow? The known upstream limit of Main Stream Passage is ~25 m beyond Pot Sump (also known as Top Sump). Cave divers have found that the passage, about 3 m wide, continues down with the dip of the limestone bedding but becomes constricted by a sloping floor of sand and gravel (see below). The sump gets its name from the normal 4-5 m steep drop to the water level in dry weather, although this fluctuates as much as 10 m and was once observed overflowing to form a small lake in Main Stream Passage[24]. In flood the water coming through the sump can fill the passage completely to the roof and from dye testing it is known to connect with Sump 4 in Llethrid (Fig. 81), although that is not likely to be the only upstream connection.

In places on the steep walls of Pot Sump, and on the roof and walls nearby in Main Stream Passage, there are numerous, variously orientated long scratches (up to tens of cm), some in parallel sets and crossing surface irregularities (Fig. 91). Here the limestone widely

ALL OUR OWN WATER

has brown coatings of mud and mineral, mainly goethite[25], although coatings are absent nearby. C*rucially, the parallelism of scratches in sets, their continuous tracing over uneven surfaces, and their multiple directions, have to be explained.* They are most easily envisaged as having been produced by rock fragments momentarily forced against the walls while moving. We examine the scratches more closely below, but, up front, they cannot have been produced by rock fragments simply colliding with the walls as occurs on coastal rock platforms where rocks are tossed up and dragged back bouncing and rolling by storm waves (Section 2.6). That abrasion damage is made by random-hit percussive spot impacts, even where fragments are trapped by gravity at the bottom of potholes; long scratches do not form there.

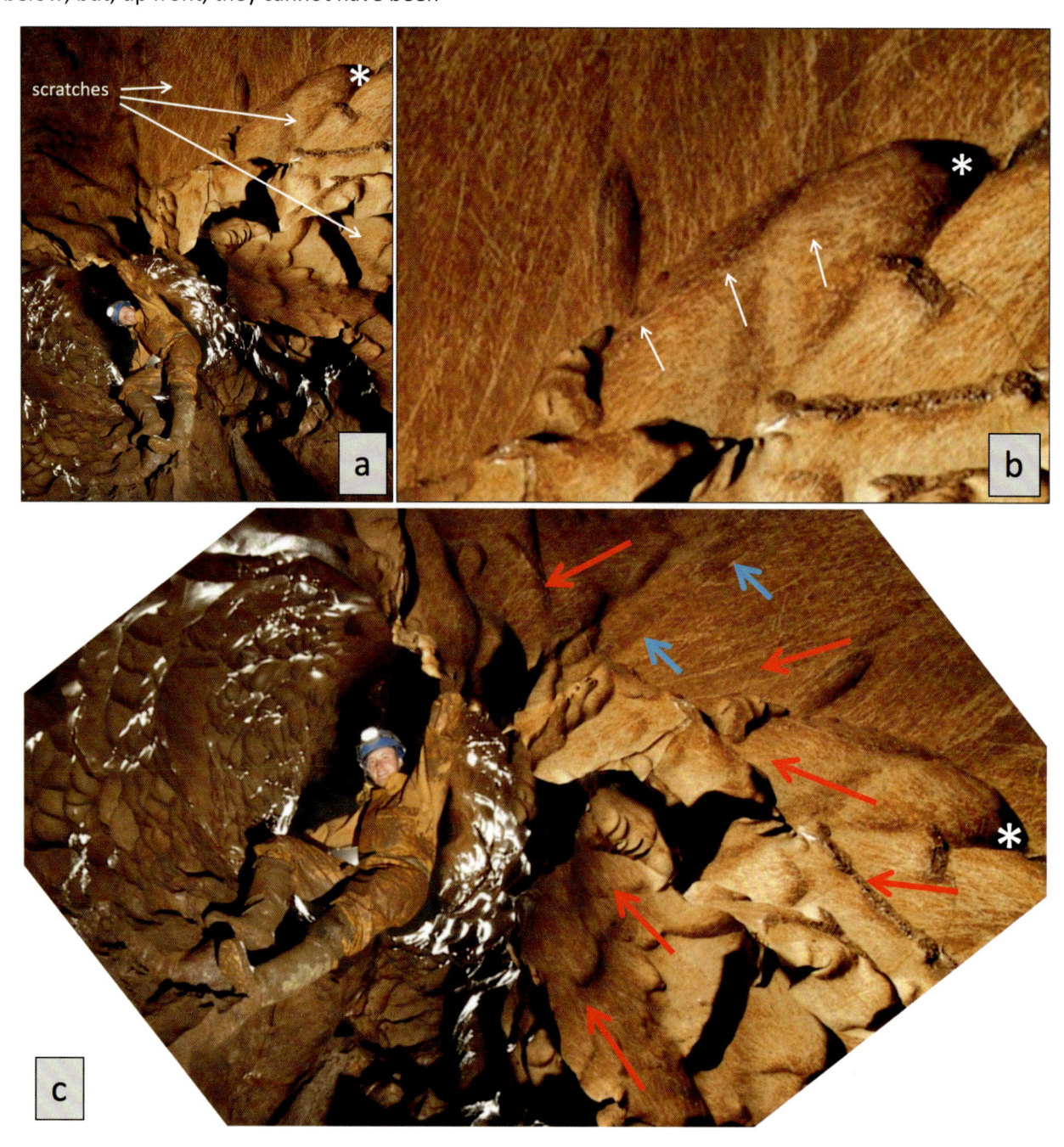

Figure 91. (Page opposite). View from Pot Sump looking steeply up the 4 m-deep irregular tube with scratches over variously orientated surfaces; near-horizontal Main Stream Passage is behind the caver.

*(a) Part of the original photo and (b) an enlargement of the scratched wall; the **white asterisk** marks the same place. Several long scratches, such as the one indicated by **white arrows**, trace continuously over curved surfaces.*

*(c) Rotated crop of the original photo, attempting to restore the steeply upwards perspective. The **red arrows** show the locally predominant trend of scratches (there are considerable variations), which tend to converge from the right-hand side upwards in the pot. The wall opposite (left of caver) is not so obviously scratched and is beautifully scalloped. The **blue arrows** mark a calcite vein. (Photo © ogof.org.uk, with permission kindly provided by Brendan Marris).*

Figure 92 illustrates most of the key features of the scratches. Scratches close together in perfectly parallel sets, absolutely straight or becoming slightly curved, are common. Most telling, perhaps, are the sets of concentric scratches showing that the damage was done in unison during a small amount of rotation. These can only have been made by the damaging pointed 'tools' being forced against the surface while moved round together. And the scratches are many and variously orientated, suggesting a protracted history of damage events involving many 'tools' with considerable directional freedom. The deep scratches can only have been achieved by hard and sharp 'tools' that are rock fragments; although leaves and small twigs occur in the passage, there are no larger fragments of wood.

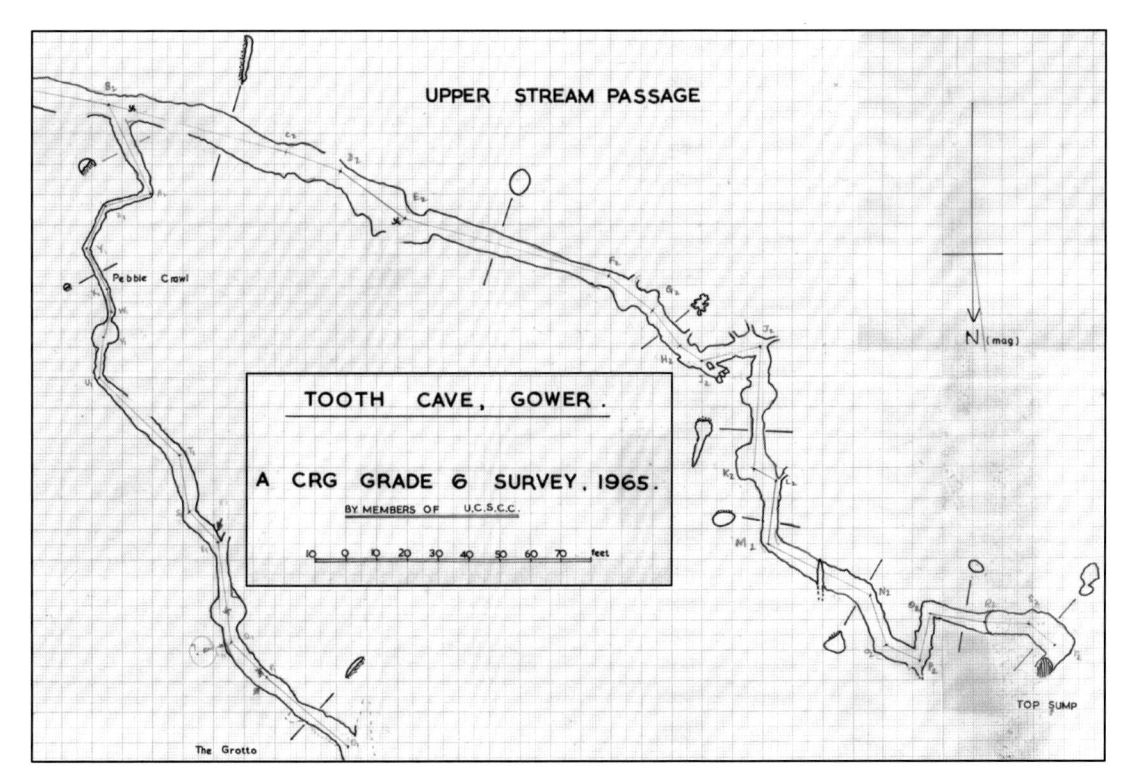

Part of the original drawing by Paddy O'Reilly of the 1965 survey, showing the lower reach of Razor Passage (left) and the Main Stream Passage to Pot Sump ("Top Sump").

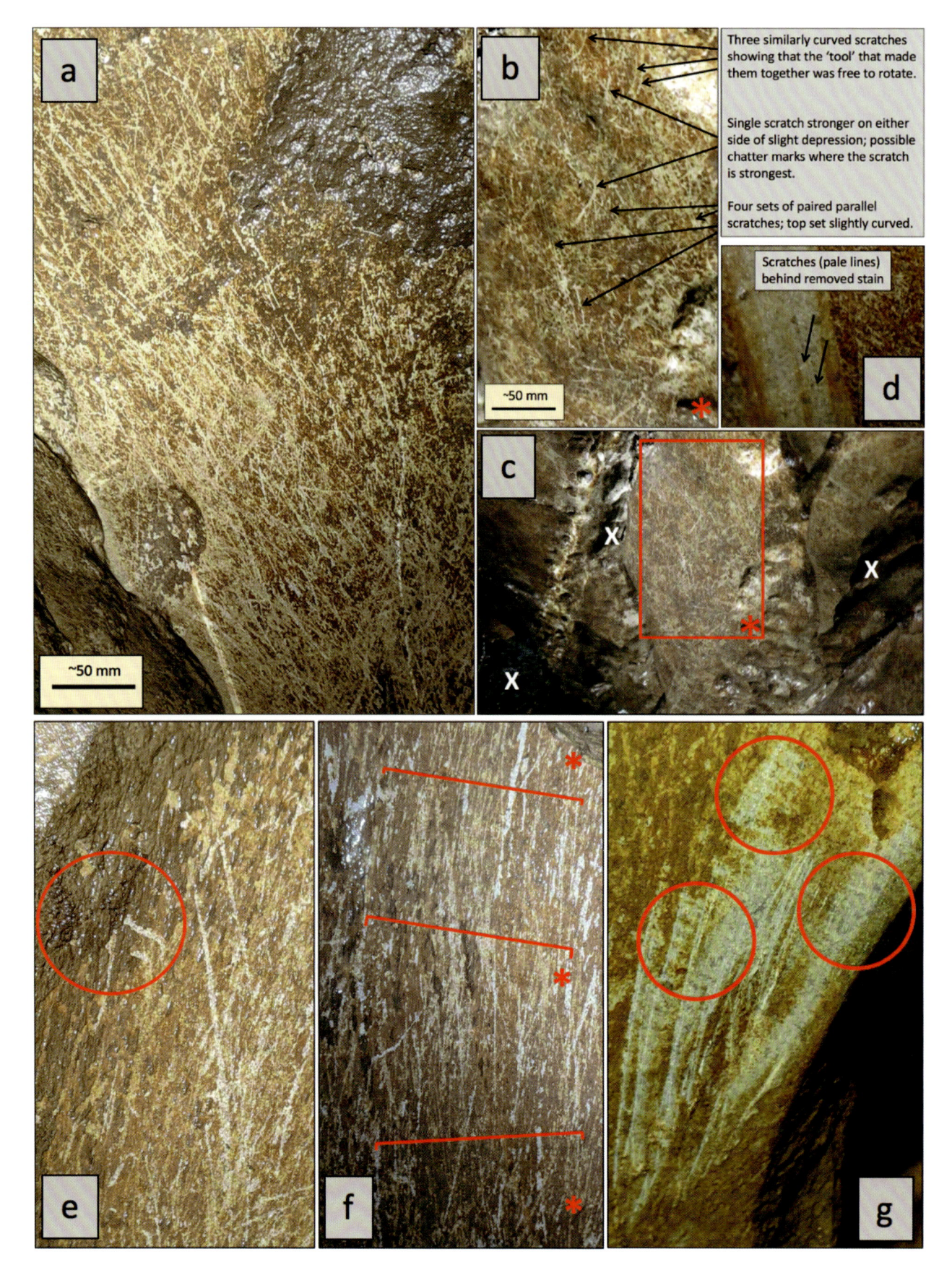

Three similarly curved scratches showing that the 'tool' that made them together was free to rotate.

Single scratch stronger on either side of slight depression; possible chatter marks where the scratch is strongest.

Four sets of paired parallel scratches; top set slightly curved.

Scratches (pale lines) behind removed stain

~50 mm

~50 mm

144

Figure 92. (Page opposite). Detail of the nature and location of scratches in the passage roof and walls near Pot Sump.

(a) General view of a scratched wall surface.

*(b) Detail from the area of the **red box** in panel (c); the **red asterisks** are in the same place. The key features to explain are itemised in the top-right panel. The scratches must have been made by 'tools' capable of damaging the limestone, moving singly, or in pairs, or in threes, or more. They trace up to about 15 cm long and in places are curved, requiring that the 'tool' or 'tools' were firmly held while they were forced round singly or together across the surface to be scratched. Some of the deepest scratches show chatter marks, which are chain-link-like strings of damage pits typical of forceful shearing contact with stick-slip motion. A prime example is the prominent pale scratch in the lower right-hand side of panel (a). Some scratches vary in severity according to occurrence over prominences and recesses, or simple curvature. The 'tools' apparently could adjust to undulations, thus remaining in contact with the rock, albeit with varying force. There are a few pale 'dots' that could be percussion marks[26] recording sharp impacts, for example those that damage the mud veneer in the small pocket at lower left in panel (a). There are thousands of scratches.*

(c) Scratches are most prominent on rock faces fully exposed in the passage wall and are lacking in relatively protected positions, e.g., marked X. In this vicinity, the coating on the walls ranges from thick mineral deposit, showing black with label X lower left, to patchily absent revealing grey limestone.

(d) The coating nearby could be rubbed off to reveal the most pronounced (deepest) scratches behind.

*(e) Scratches are of variable severity. Many appear simply to score a mud veneer; some reveal the limestone surface stain and others score into the limestone forming the pale marks. Such marks are in places covered by the pale brown muddy veneer (e.g., **red circle**) while elsewhere they cut that veneer.*

*(f) Pale brown veneer is scored by fine parallel scratches in bundles (**red brackets**) that slightly change direction; **red asterisks** trace two of the more prominent parallel scratches. Later scratches are superimposed.*

*(g) Caver-made chatter marks (**red circles**) form series of parallel stick-slip marks across the downwards scraping direction made with a steel blade in scraping off the scratched surface material for laboratory analysis.[27]*

Figure 93, on the next page, is a diagrammatic model proposed to account for the scratches at Pot Sump. It is in part based on diver-derived details. It may be difficult to imagine how the upper entrance to Green Cwm may have looked before the debris-flow inundation that choked the sinks there and left the valley floor level. For those who know the Bishopston Valley downstream from Bishopston Church (SS577893), where much of the valley floor is naked limestone, the multiple drains and large sink sites there might be considered as somewhat similar to how Green Cwm was before the event, quite capable of taking part of any highly mobile debris flow underground.

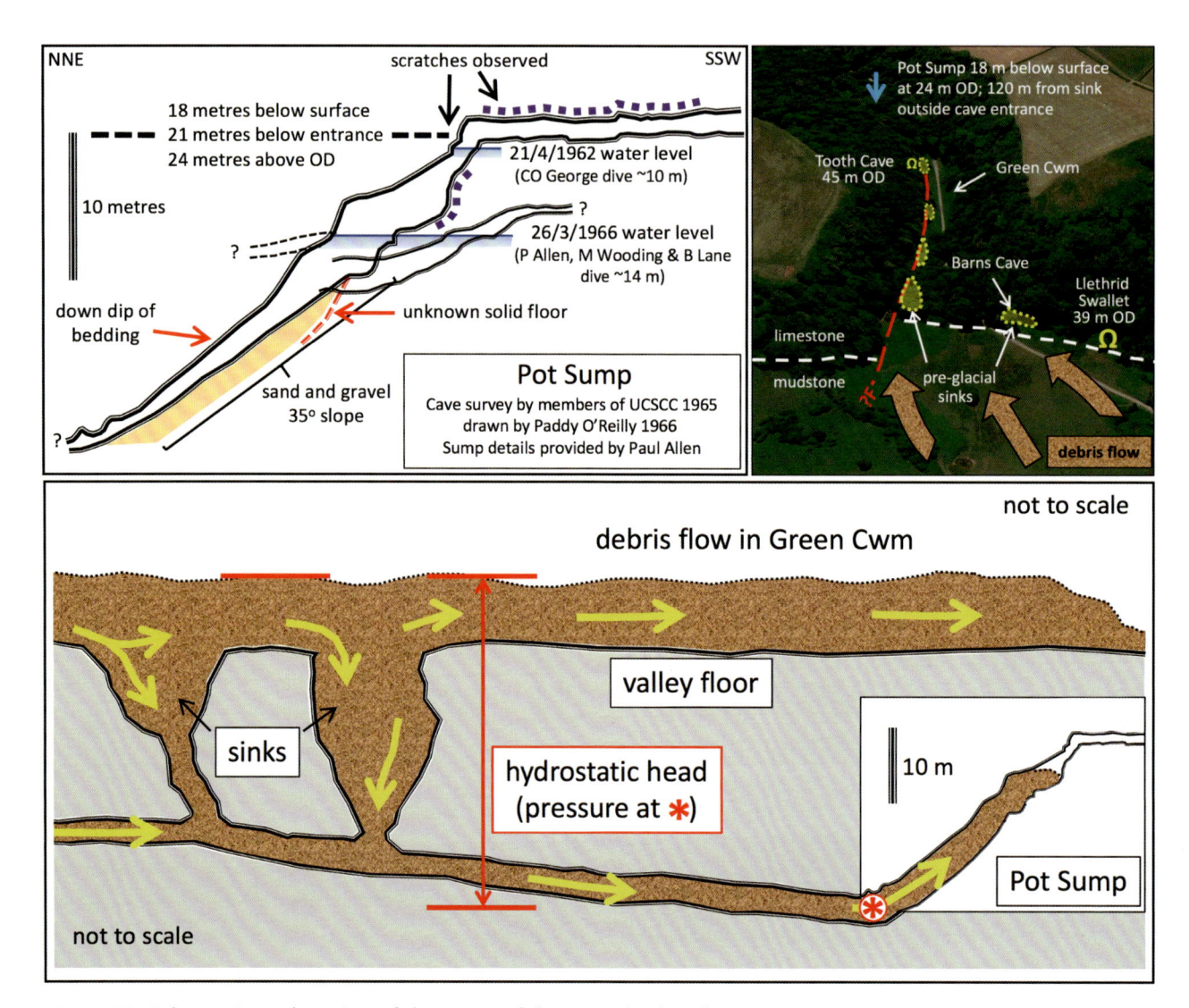

Figure 93. Schematic explanation of the cause of the scratched walls at Pot Sump.

*(**Top left**) Detail of Pot Sump derived by divers and cave survey (UCSCC is University College Swansea Caving Club). This diagram infers that the observed sand and gravel floor of the sump is a repose slope of trapped sediment and that the phreatic passage there is unlikely to have been permanently so constricted. Such banks of trapped sediment are common features of the low rising part of phreatic loops. Paul Allen recorded that the connected emergent passage (to the right) had "leaves and soil debris on the floor".*

*(**Top right**) Shows the line of sinks that would have been open to incursion of the debris flow, conceivably like the 'daw pits' in Bishopston Valley (e.g., Lower Daw Pit at SS575890). Now the fluvioglacial debris forms the wide flat valley floor.*

*(**Bottom**) Hypothetical not-to-scale scheme indicating the catastrophic incursion of the debris flow into Green Cwm and the cave passages underneath. This is an ideas-in-principle diagram. The underground flow is shown at the instant it is about to be forced out of the top of Pot Sump, where the scratches will be formed by it. This is at 24 m OD while the original valley floor would have been at about 37-38 m OD. Together with the (unknown) thickness of the debris flow at the surface, immense hydraulic pressures would have developed and driven the underground flow. Debris flows can be impressively fast and fluid (see end note [22]).*

Scratched-wall key features and inferences are as follows. (1) Scratches are of variable severity but normally form lines centimetres to tens of centimetres long showing that the responsible 'tools' briefly remained in contact with the wall surface. (2) Scratches involve a brown mineral stain on the limestone and also a veneer of brown mud that covers the stain. The mud is variably scratched but also is plastered on top of scratches. This shows that mud plastering accompanied the scratching. (3) Dark brown to black mineral layers in places cover scratched surfaces, as they do covering the top of the cobble and pebble banks. This places the scratching in the same time frame as the catastrophic debris flow.

The simplest explanation is that the one-off catastrophic debris flow plastered the walls with mud while transported rock fragments did the scratching. This would have happened during at least several minutes and probably during tens of minutes while the flow was hydraulically forced up out of Pot Sump. Mud-bearing debris flows normally coat the surfaces they flow against; the plastering can be missing behind obstacles where the flow locally separates from the surface.[28] Explaining the scratches is less simple.

For many decades it has been known that *dense granular* flows, like the debris flow considered here, do not move (slide) like a rigid body against its contacts but flow rather like fluids, rapidly changing shape, forming waves and splashing (see end note [22]). 'Dense granular' means that the grains, in this case our cobbles and pebbles, widely are touching some of their neighbours while the fluid behaviour shows that the contact arrangements must be changing and the grains to an extent must be free to rotate.

High-quality photoelastic material and sophisticated experimental arrangements now allow us to see a simplified representation of what is happening. Rather than there being uniformly transmitted pressure, as in a continuous liquid like water, forces are transmitted along chains of touching grains, and these arrangements change rapidly as the mass deforms. Thus, there are transient 'force chains' within the granular flow (see explanation box).

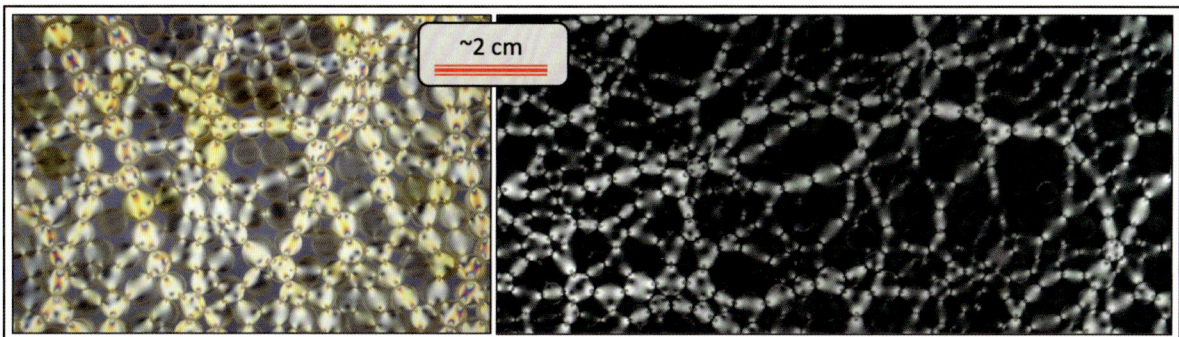

Force chains in photoelastic granular material. These images show 2-D arrays of stress-sensitive discs that only transmit (polarized) light when stressed; non-stressed discs remain uncoloured (left) or dark (right). They show an instant when force is communicated along chains of discs bearing most stress. As the mass is deformed different chains are activated. [Images by Karen E Daniels (Daniels 2017) Open Access Article distributed under the terms of the Creative Commons Attribution Licence 4.0 (http://creativecommons.org/licences/by/4.0/)]. Estep and Dufek (2012) show that in granular flows relatively strong stresses may be transmitted locally by force chains that cause erosion of substrate; adjacent domains move more freely and the whole behaves as a fluid.

For our purpose regarding the scratches, it is tentatively inferred that transient force chains are what momentarily held moving rock fragments against the passage walls to leave their marks. The scratch patterns graphically reveal the motions of the transported particles, broadly in the general flow direction but with considerable local changes of direction and with rotations (Fig. 92), and with rapidly changing force-chain configurations and forces during flow. We cannot know the absolute durations of the rock-fragment contacts with the surface being damaged, because we do not know the flow velocity at all, let alone near the surface. Nevertheless, we may infer variable *'brief' durations* from the variably short length of the scratches (nowhere more than a few tens of cm and mostly much shorter), using the knowledge that countless contacts occurred in each small area here while the flow reached at least ~850 m down Main Stream Passage. The chaotic nature of the damage implies some non-trivial velocity. Debris-flow velocities of 15 metres per second (m s^{-1}) occur, but we should consider more time than the 1 minute such a velocity would entail for that distance. Perhaps about 1-1.5 m s^{-1} is reasonable to envisage an energetic flow of roughly 10-15 minutes duration at Pot Sump. This, of course, is one or more of vague, subjective and tenuous, but it gets worse.

It seems possible that the constriction at the top of Pot Sump passage is implicated in the scratch localisation near there. Some surfaces seem to have been 'out of the way' of the debris flow and farther downstream the passage tends to widen, which may have lessened the effects by reducing the flow velocity. Also, the record is far from complete. It is evident in places that the more normal storm-water transport of sand and silt has worn some original surfaces (corrasion), while the widespread occurrence of scalloping suggests that aqueous corrosion has modified surfaces as well. Perhaps most difficult to comprehend is the preservation of the mud plastered on surfaces and scratched during the flow. Nevertheless, mud plastering is not a feature of the normal flood-bypass behaviour, which typically leaves similarly ancient mineral precipitate and speleothems clean, with only waning-flow residues of sand and silt on top.

As far as this author is aware, neither extensive debris flows in caves, nor the graphic record of their granular motions, have been described before.[29]

And the snails, from the sublime to the ridiculous. We now know that **the brown coatings on the cave walls and on cobbles and pebbles in the passages are extremely variable (Figs 89 and 92), from absent to thick, from extremely hard to plastic and even possible to wipe off. Now we find that in places they are browsed by snails.** Some coatings have been removed by small fresh-water snails (molluscs of the Physidae Family), suggesting nourishment, although perhaps not directly. Figure 94 shows scratched oolitic limestone, in Main Stream Passage near to Pot Sump, where the exposed ooliths – near-spheres of calcite about 1 mm in diameter – are generally coated by a thin brown stain, samples of which include goethite[25]. This stain is missing along somewhat sinuous feeding trails and in places the culprit browsing snails can be found. The scratches are just like those nearby (Figs 91 and 92), showing physical damage to the limestone where the stain has been removed. On the inference (above) that the scratches were formed by stones in the debris flow, during the glacial regression sometime after 23,000 years ago, how is it that the same scratched stain appears to interest snails now? The variability of the wall coatings probably involves locally differing physical and chemical conditions, although the apparent food association here suggests some biological involvement.

Figure 94. (Page opposite). Grazing molluscs have removed this superficial brown stain on oolitic limestone near Pot Sump. Scratches show limestone mechanical-damage grooves where the stain is also removed, indicating that the stain is ancient. The freshwater snails doing the grazing, one here within an aqueous film (inset; a few mm long), are of the Family Physidae, known to devour organic material including micro-organisms. Here is the apparent paradox of the scratches being perhaps around 20,000 years old, yet the stain is being browsed today. Brown organic slime films and drapes occur in places nearby suggesting a food source, but that is a stygian curiosity too much to contemplate and involve here.

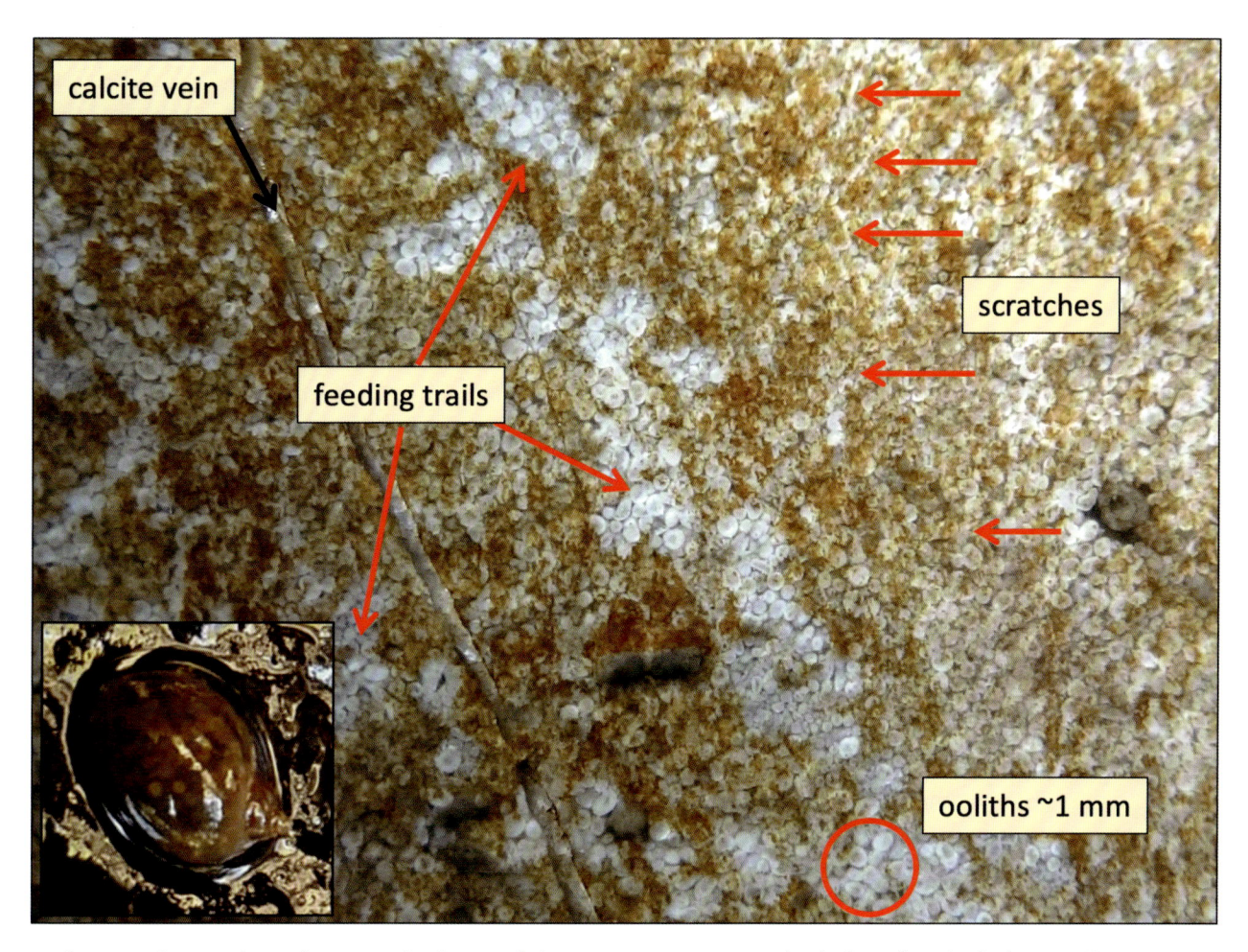

[Figure labels: calcite vein; feeding trails; scratches; ooliths ~1 mm]

Factors that perhaps bear on the issue of the age, nature, and possibly the changing nature of the stains, are as follows. Thin coatings of haematite appear to be the first formed; they stain the body of the fresh rock (Fig. 89 Top right) and occur underneath softer hydroxide coatings on stones; this is a common order. They are likely to have formed on walls and stones at the same time, long before the glacial climax. Thin coatings of goethite also developed. The later and softer, thick chemical veneers, which formed on top of the cobble-pebble piles and locally on walls, are or were probably mainly brown goethite, although some veneers and stains are probably an amorphous mix of both iron and manganese hydroxides. This substantial Fe/Mn mineral precipitation occurred after the debris flow that invaded the caves. Speleothem deposition was continuous after the substantial Fe/Mn chemical activity, so it seems that the mineralisation was most marked after the glaciation on Gower. Conceivably this reflects the new exposure of bare rock and glacial till before substantial vegetation recovered.

Evidently, the Green Cwm caves were significantly impacted by the Late Devensian glaciation, but what is the evidence that they started to form many millions of years before the Pleistocene? What happened after their inception while the marine planation occurred, long before the Last Glacial Maximum? Next, we consider the long-term evolution of the limestone massif around Green Cwm, but that will not be the end of it. Before leaving the Green Cwm System, we must explain an ancient record of global-scale continental fragmentation within Tooth Cave, perhaps with a surface manifestation, and involving a dead mouse.

4.7 Inception of caves in the limestones around Green Cwm

There is a big bit of Green Cwm cave history missing in the story so far. Relative rise of sea level by about 3 million years ago (Late Pliocene) must have completely flooded the entire existing system as the marine platform was cut above it (Section 2.9). Sea level fell back in Pleistocene times, from about 2.6 million years.

We established that caves in the limestone massif that ultimately formed cliffs at The Knave, on the south Gower coast (Section 4.3), probably initiated at least 10 million years ago. The case there hinged on the knowledge that phreatic passages, which only develop beneath a water table, must predate the valley sides where they are now exposed. And the valleys must have been formed before the platform that sharply truncates their original upper reaches, there at around 55-60 m OD. If valleys are to form, they must have head-water catchments, for which now along the south coast there is no substantial remainder; it is the 'headless valley problem' we considered in Section 2.10.

Because the main platform of Gower, rising gently from around 40 m to 75 m OD, was cut some 4-3 million years ago, it follows that the valleys must be older, at the very least 5-4 million years old. The regional uplift that caused substantial erosion persisted until some 7 million years ago (Fig. 39), and, as the phreatic cave passages exposed in the valley sides must be much older than the deep cutting of those valleys, we reasonably infer cave initiation "at least 10 million years ago".

Whereas the slades on south Gower probably had head catchments mainly in the proto-Gower limestone massif, the valleys of Green Cwm, Ilston, Bishopston and Caswell are through-going with large head catchments on the less permeable strata to the north, above the limestones, namely on the mudstones and sandstones of the Marros, Coal Measures and Pennant formations. Regarding Green Cwm, it seems most probable that some south-draining forerunner of the Pengwern catchment (Fig. 83) existed at least 10 million years ago. This drainage direction is probable because the northern perimeter is in high ground of the relatively tough Lower Coal Measures, above the softer Marros mudstones to the south. However, the general ground surface 10 million years ago must have been well above the present general surface and the ancestral Green Cwm cannot have been cut as deeply as it is now.

On the platform about 350 m northeast of Green Cwm and Cathole Rock (Fig. 81), the entrance of Willoxton Dig at 65 m OD (SS53939028) is the truncation of a metre-scale phreatic tube that has no remaining surface drainage leading to it; its water table must have existed at least several metres above it with the original ground surface higher still. Unlike some collapse sinks in the platform, this gently inclined phreatic cave cannot have formed by upwards propagation of a cave void towards the surface. Similarly, Tooth Cave entrance passage and caves at Cathole Rock, at around 45 m and 35 m OD respectively, and the Iron Ore Series at about 38 m OD in Tooth Cave (see below), must also register water tables that existed at or near or higher than the platform surface above them now at 60-75 m OD. By the same token, Llethrid and Tooth Cave passages, now close to the modern valley floor, must have been formed farther beneath a contemporary surface than they are now.

Overall, the rather ill-defined picture must be that *at least* a few tens of metres of rock have been removed from the flank areas of Green Cwm and even more from any original valley floor. Thus, in what we deduce to be *at least* 10 million years, the Green Cwm cross-valley profile has both lowered overall and deepened. At the same time, early formed phreatic passages must have been successively abandoned as phreatic drainage worked deeper. Early formed chambers such as Great Hall and the Annex in Llethrid then enlarged as they were undermined and expanded by roof collapse.

There is no way of telling how thick any Jurassic or even a Cretaceous rock cover of Gower might have been; quite reasonably we expect a Jurassic cover at least (Section 2.4). We know from Lundy Island that at least 1 km thickness of rocks was removed from the region during uplift through Paleogene and Neogene times, from about 62 million years ago (Section 2.7 and Fig. 39). Before that, the Triassic erosion and unconformity had cut across the limestones so that the former limestone massif could not have reached very high above the surface now remaining; we find vestiges of Triassic sediments widely on Gower (Section 2.4 and Fig. 10). Thus, we are constrained to infer that limestones in which caves could again form (since Late Triassic times) were only uncovered 'late' during the regional uplift but probably well before 10 million years ago. Put another way, cave formation in Gower limestones could not occur during Jurassic and Cretaceous times because they were buried, and it could only re-commence once that cover had been removed, at least partially, by Paleogene to Neogene uplift. For the moment, cave formation from "at least 10 million years ago" and not *very* much earlier is about as good as it gets.

Presently it remains unclear what might have happened inside the existing caves in Late Pliocene times, 4-3 million years ago, when marine planation cut that surface above Green Cwm. Since then, the area has endured two glaciations that must have modified any earlier surface deposits, and 'fresh' water must for a long time have flushed through the caves drained of any sea water.

4.8 Remarkable record of a desert above

Before leaving Green Cwm, there is one more wonder to contemplate. In Tooth Cave, Main Stream Passage is interrupted by a boulder pile just upstream from the intermittent Big Sump. A short scramble up through the boulders leads to a comparatively large and tall but narrow chamber from which a laddered ascent originally lead into some small high-level crawls, passages and chambers. As youngsters we knew this as the Iron Ore Series (Fig. 81) on account of the red colour there of the cave walls, fallen boulders and sticky mud.

The tall chamber has two parallel walls of limestone, roughly 2 m apart, practically vertical and trending 25-30° east of north. These, plus fallen blocks below, show calcite vein mineralisation indicating location on a fault. Between the fault walls there is a succession several metres thick of sub-horizontally bedded red micaceous sandstones and siltstones (Fig. 95).

These sediments are similar to other Triassic rocks on Gower, but they show clear evidence of accumulation in water, with some ripple forms on bed surfaces and grain-size-graded beds with lobe-and-cleft bases. If the contemporary surface above was primarily arid desert occasionally affected by flash floods, it is sensible to assume that the sediments in this Tooth Cave fault-fissure were washed in during floods. The graded beds are typical of rapid deposition from mixed-grainsize suspensions in which the coarser grains settle first (e.g., see Section 5.14). The symmetrical ripples then probably relate to sloshing of the floodwater within the fissure between the confining walls. Draughts sufficient to generate waves capable of forming symmetrical ripples seem unlikely.

This remarkable occurrence, which was coincidentally intersected by Tooth Cave, is closely similar to other fault zones that contain red sediments and are commonly exposed in Carboniferous limestones along the coasts of south Pembrokeshire and Gower. They are interpreted as Late Triassic in age when desert conditions existed widely 220-210 million years ago (Section 2.4). A closely similar sedimentary succession is in Red Chamber (Fig. 10d).

*Figure 95. (**Top left**) Bedded purple-red fine-grained micaceous sandstones, with thin siltstone partings, fill between the fault walls where they are slightly deflected upwards at the original sedimentary contacts. Alteration of the original iron oxide content of the sediments has rendered parts of them coloured cream.*

*(**Top right**) Symmetrical ripple forms in fine sandstone with cream siltstone partial drapes, on a block fallen from the fault-fissure infill.*

*(**Bottom left**) Cleaned samples of the fine sandstone showing calcite-filled fractures, which show that there was some ongoing deformation across the fault infill after the sediments were lithified to form solid rock.*

*(**Bottom right**) Basal surface of a micaceous sandstone bed showing lobes, bulbous downwards, with intervening pinched clefts. Such beds show grainsize grading, coarsening upwards from medium to fine sand and silt. The pattern on the bed base is characteristic of sedimentary loading of a sand layer into an underlying, yielding finer deposit, although it could possibly reflect original existence of a thin microbial slime mat that had formed and wrinkled before the sand was deposited (e.g., Davies et al. 2016). The small 'pimples' on the surface were rhombic dolomite crystals, ~ 1.5 mm in diameter, that grew within the sand during its cementation to form rock and are now replaced by ferroan calcite[30]. (Top photos courtesy of Andy and Antonia Freem).*

Nigel Woodcock and colleagues at Cambridge University show that when the vast Pangaea supercontinent broke apart, in Triassic times, the Carboniferous rocks became extended – effectively pulled apart – so that earlier-formed faults trending around north-south across South

Wales opened up to form deep fissures that commonly breached the surface (Woodcock et al. 2014). Triassic-age deserts with seasonal rainfall that formed flash floods produced red sediments on the eroding uplifted land and in places this poured via wadis into the opening voids. On the evidence that the Iron Ore Series fault infill extends down to about 27 m OD, a few metres above Main Stream Passage, and that the *platform* surface above now is at about 65 m, the red sediments are likely to have penetrated down at least 40 m from the original desert surface.

4.9 One weird depression

There is a conspicuous semi-circular embayment in the east flank of Green Cwm, roughly 120 m in diameter and with steep inwards facing slopes (Fig. 96). The LiDAR survey picks out a distinct straight lineament almost 200 m long trending north-northeast across the site. The online Ordnance Survey map (© Crown Copyright and database right 2021) here shows 'Workings (dis)', but no previous maps show any such thing, and the site has no such recent disused development. It would appear that the features have been misidentified. High on the southeast side of the lineament is a bank comprising glacial erratic boulders and cobbles, best interpreted as representing a temporary ice margin of the retreating LGM ice (Fig. 81).

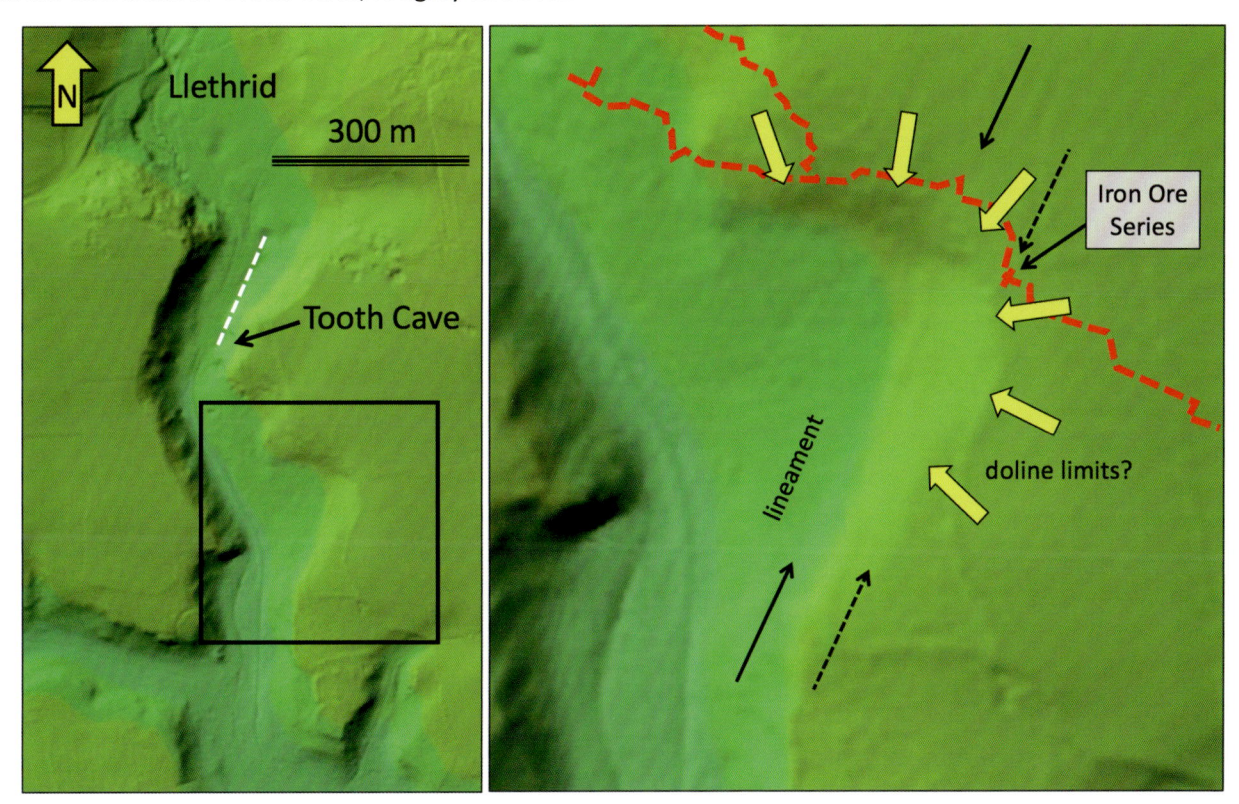

*Figure 96. Suspected ancient doline possibly related to faults that are suggested by topographic lineaments. The semi-circular feature on the east flank of Green Cwm is conspicuous as an anomaly, lying across distinct straight topographic lineaments that trend 25-30° east of north, parallel to faults in this vicinity, including the Triassic sediment infilled fault intersected by the Iron Ore Series in Tooth Cave, which also has a measured trend of 25-30° east of north. The **white dashed** line by Tooth Cave entrance is along the line of active sinks there and is interpreted as a fault or fault zone (Figs 81 and 82). The **red dashed line** marks the underground position of Tooth Cave. The sharply defined edge of the depression is taken to mark its truncation by the Late Pliocene platform-forming erosion here. (LiDAR image uses base data © Natural Resources Wales).*

The University College Swansea Caving Club (UCSCC) 1965 survey of Tooth Cave, replotted by Paddy O'Reilly in 2021, places the Iron Ore Series (aka Aven[31] Series) and Triassic sediment filled fault (Section 4.8 above) along the east flank of the surface depression (Fig. 96). This author's recollection of the high Iron Ore Series was of a frightening series of muddy crawls in places beneath unstable roof. Given its location, it seems too great a coincidence not to relate the surface and underground features – as a doline, a collapse-related depression, centred on a fault zone. From underneath, the Triassic fissure walls are practically vertical and so would project directly upwards to the east margin of the feature at the surface. Given that the platform at about 65 m OD here was cut in Late Pliocene time, 4-3 million years ago (Section 2.10), and that it sharply truncates the depression, the doline probably is a remnant of a truly ancient landscape feature.

Because Tooth Cave is so prone to flooding, the Iron Ore Series was, in the 1960s, stocked with emergency rations[32] and, because of the height of the aven, it was also considered as a possible site of rescue from the surface. To this end there was a radio-location survey here (and at Llethrid), but the findings of so long ago have not come to light (John Harvey pers. comm. 2021). It has nevertheless been remembered that closeness of the Iron Ore Series to the surface was evident during the 1960s exploration, from occurrences of tree roots and a dead mouse.

4.10 Caves lost in time

We have established that in the vicinity of the extensive caves of Gower, at The Knave and in Green Cwm, some considerable thickness of originally overlying rock has been removed since the initial phreatic passage development (inception). Restoring that land surface in 'the mind's eye' may be a little tricky, but perhaps it is more difficult to appreciate the huge volumes of cave-containing rock that have been removed in forming the coast and the main limestone valleys, notably Pennard Valley to Three Cliffs Bay and the coast beyond. The evidence for this is isolated and fragmentary, but nevertheless it is clear that Gower caving would once have been far more of an adventure, in far bigger systems, many millions of years ago.

Perhaps consistent with the long established and relatively extensive catchments of Green Cwm and Ilston Cwm, which feed to Parkmill and then Pennard Valley, fragmentary evidence of originally extensive caves is found in and around Three Cliffs Bay (Fig. 97). The isolated phreatic passages of Leather's Hole in Great Tor show that the limestone there must have been within an extensive massif long-since mostly removed (Section 2.11). Those passages (Figs 26 and 27) are at 35 m OD and beg the question of how much rock has been removed from above and to the sides of Great Tor since Leather's Hole was below a water table.

The arch that penetrates the Three Cliffs promontory (SS538887), 800 m to the east of Great Tor, has part of a 2-3 m diameter phreatic tube in its roof, at about 10 m OD (Fig. 97), which must have been part of a passage within a limestone massif where now there is the scenic void of the main Pennard Valley and Three Cliffs Bay. And 1.1 km farther east, the cave known as Devil's Kitchen (SS548872) is another similarly scaled phreatic tube at about 10 m OD with a vertical section and a horizontal passage truncated at the cliffed coast (Fig. 97). Heaven only knows how this tube once connected beneath a water table.

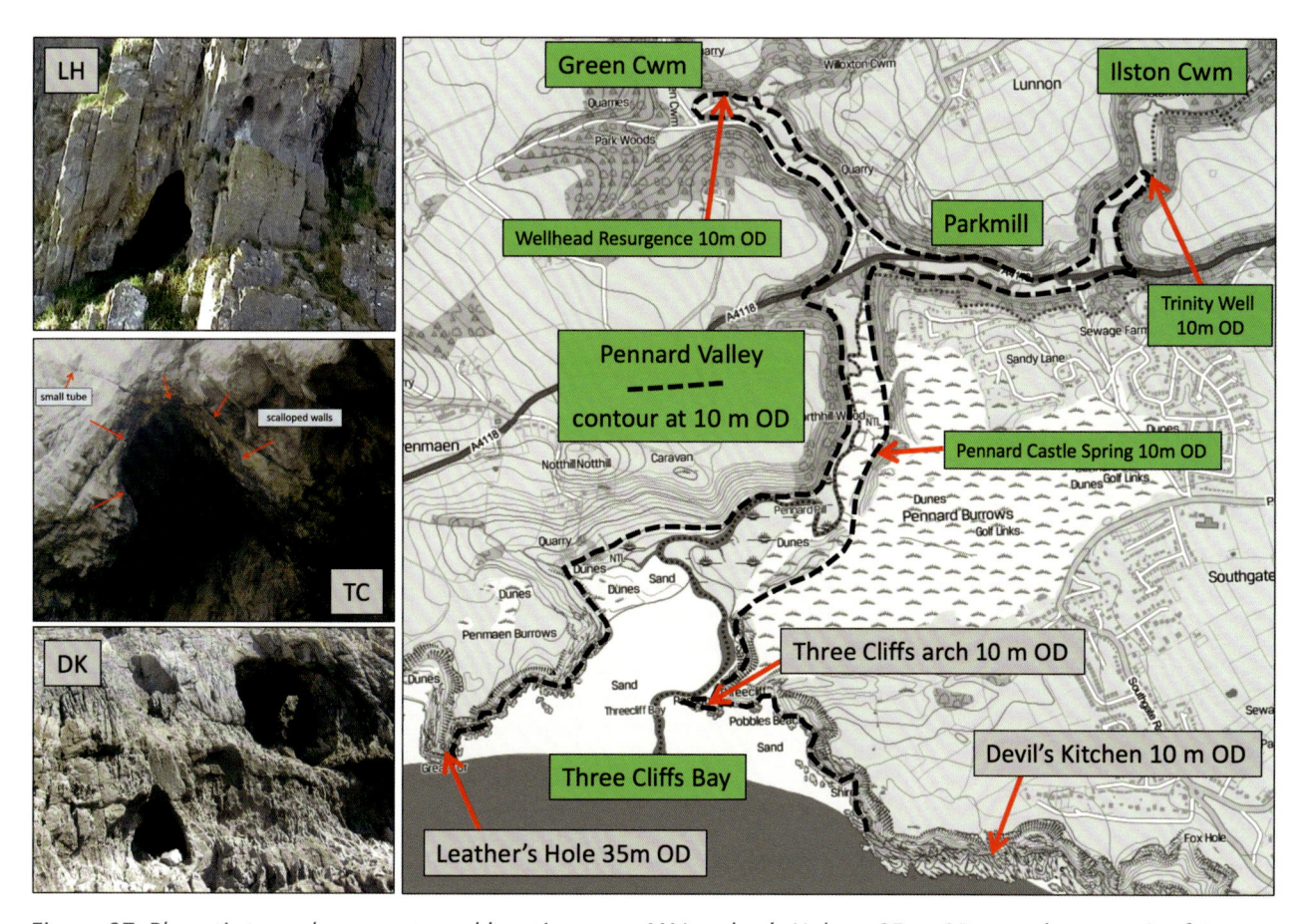

*Figure 97. Phreatic tunnel remnants and location map. **LH** Leather's Hole at 35 m OD near the summit of Great Tor (Section 2.11). **TC** Three Cliffs arch with partial tunnel remnant in the roof roughly 2-3 m in diameter, at 10 m OD. **DK** Sea-cliff opening of Devil's Kitchen Cave phreatic tunnel is top right, about 4 m in diameter at 10 m OD. This cave trends vertically upwards at ca. 3 m in diameter forming a skylight just visible at the back. Lower left is the Devil's Spring (aka Westcliff Spring). In contrast to the nearby practically circular phreatic tube, this smaller cave narrows to small fissures and has the classic A-shaped entrance reflecting marine erosion.*

The map shows the 10 m OD level at which caves at deeper levels are blocked and where misfit resurgences now occur (labelled). It is impossible to know how the three isolated phreatic tubes were once plumbed as parts of long-since removed cave systems within the original limestone massif of more than 10 million years ago. (Map © Natural Resources Wales).

Although there are many other caves in the vicinity of Three Cliffs Bay[33], the three mentioned here (Fig. 97) specifically begin and end in space so one must conjure a cave almost completely removed by erosion. This removal would have been a massively long time after the initiation beneath a water table. One has to envision that there once was no Pennard Valley and that, probably, there was some drainage southwards across this area at a higher level in a limestone massif now substantially reduced. Given that the slades and cwms were cut long before 5 million years ago, we must consider a missing landscape significantly above the presently isolated tubes. This perhaps had a drainage inherited from the Paleogene and Neogene peneplain evolution (Fig. 39) that would eventually cut down to the present system – a landscape where surface water on the limestones first percolated down to initiate phreatic enlargement of pathways beneath the water table of the time. One may sensibly guess that this was *at least* 10 million years ago, during regional uplift (see above and Section 2.16), and that gradually the surface drainage cut deeper as the water table base level migrated downwards. Similarly, the coast must have cut back as rock was eroded above. Viewed from the valley of the ancestral River Severn, now the Bristol Channel, one could expect there to have been several subaerial resurgences of streams from tunnels in limestones far in front of and lower down than what is now Three Cliffs Bay and the cliffed coast on either side.

4.11 Speleothem formation ages

Stalagmites, stalactites and flowstone are forming in Gower caves now. Fine formations in Tooth Cave Bone Chamber have been forming since the Bronze Age[34], which was from about 4,100 years ago, and those on the debris-flow deposits farther underground there (Sections 4.5 and 4.6) can only have developed since just after the Last Glacial Maximum some 23,000 years ago. In the ancient isolated passages of Leather's Hole, which perforates Great Tor (Section 2.11), speleothem ages are from just after the Anglian glaciation and late after the Last Glacial Maximum, probably Holocene. Nothing, however, has been sampled and found to reflect the many millions of years since the early development of the phreatic passages deduced from the geomorphology. This may be an issue of recognition, but the most widely used speleothem dating method is not applicable for such ages and 'older than' is the best to hope for.

In ideal circumstances U-series dating (aka U-Th dating, e.g., Ford and Williams 2007) can yield precise and accurate ages of speleothem materials less than about 600,000 years old. For this reconsideration of Gower caves an application was made to the Natural Environment Research Council for a pilot study in which it was hoped to confirm a few inferred relative ages and fix some absolute ages. Always an 'unknown' before any preliminary analyses, it unfortunately turned out that the five speleothem samples submitted contained rather little uranium and mostly too much thorium for all of them to yield really useful dates. Non-negligible amounts of initial Th, when corrected using an average continental crust U-Th isotope composition, result in large calculated age uncertainties. Table 1 presents analyses of the three comparatively viable samples, followed by a brief discussion of the age determinations.

Sample	AR=activity ratio				Measured, Tracer-Corrected					
	Sample wt (mg)	U ppm	232Th ppm	230/232 AR	232/238 AR	±2s (%)	230/238 AR	±2s (%)	234/238 AR	±2s (%)
TBC2	408.28	0.04	0.002	3.5	0.0126514	0.06	0.04448	0.90	1.1070	0.17
LH1	146.17	0.10	0.017	2.7	0.0570049	0.06	0.15211	0.53	1.1682	0.17
LH2	222.44	0.22	0.000	9382.7	0.0001116	0.14	1.04687	0.23	1.0497	0.13

Detrital-Corrected										
230Th/238U AR	±2s (%)	234U/238U AR	±2s (%)	Corr. Coef. 08-48		Date corr (ka)	±2s (abs)	Initial 234U/238U AR	±2s (abs)	Rho T-g0
0.0343	21.46	1.1082	0.51	0.278	TBC2	3.426	0.742	1.1092	±0.006	0.293
0.1098	29.67	1.1766	2.10	0.288	LH1	10.660	3.255	1.1820	±0.026	0.281
1.0469	0.23	1.0497	0.13	0.000	LH2	426.250	14.380	1.1652	±0.005	0.723

Table 1. U-series data on speleothems from Tooth Cave Bone Chamber and Leather's Hole. (Determined at the British Geological Survey Isotope Geosciences Facility at Keyworth, Nottingham[35]. The data are used with permission copyright BGS, UKRI.)

TBC2 is from a stalagmite that had originally been on top of Bronze Age human remains in Tooth Cave Bone Chamber (Fig. 98, next page); it was recovered broken from within excavated debris. Its determined age is 3,426 years old, which is appropriate (Bronze Age is ~4,100-2,750 years ago), but the analytical 2-sigma error of ± 742 years is large. Sample LH1, drilled from a hidden position and resin backfilled from wall flowstone in Leather's Hole, yields 10,660 years old, again with a large error of ± 3,255 years but consistent with a post-Devensian age (post-LGM), some time after complete deglaciation at about 18,000 years and conceivably Holocene (from 11,500 years). Sample LH2, broken from flowstone in a hidden high tube, gave an analytically more satisfactory result, consistent with its slightly higher U and lower Th concentrations. Its calculated age is 426,250 ± 14,380, which although analytically sound would be more robust if a closely similar age could be determined on a sample immediately adjacent to LH2 in the same specimen. This age places that speleothem growth in Leather's Hole at the end of the Anglian glacial interval (478,000 to 424,000 years ago; Lisiecki and Raymo, 2005) and with the 2-sigma error of ± 14,380 years it is conceivable that it developed as conditions warmed towards the Hoxnian interglacial. While it was hoped that this sample would prove to be older than the 600,000 years limit of the dating technique, it nevertheless records growth long before the Ipswichian interglacial of around about 123,000 years ago, which is registered nearby in the Patella notch and deposits (Fig. 26).

4.12 Recovering Llethrid

This final section is a little bit of recent caving history, for the record and with a couple of small surprises in the end. At some time during or just after the year 2000, entangled timber and wire holding back assorted debris was pulled out of the main stream sink at Llethrid Swallet. It had been supposed that the obstruction was causing bypass flooding into Green Cwm, but, regrettably, its removal released several 100s of cubic metres of stream gravel that had accumulated behind, which thus went into the sink. With washed-in tree branches and assorted rotting vegetation, the released debris blocked passages beyond the gated entrance and hence the cavers' way into the cave. There would be no hope of clearance of the original route without a massive effort to take the debris back out. And, of course, the periodic bypass flooding into Green Cwm continued as before (Section 5.10).

We know that the Llethrid entrance passages were formed mainly after the Late Devensian glaciation (peak 23,000 years ago) when the then misfit stream became displaced to the valley side (Section 4.4). Unsurprisingly, these immature passages were tight in many places and for the first explorations had to be chiselled to get through. The low crawls between loose rocks and tight squeezes with names like the 'Needle's Eye' and 'The Filter' speak volumes about the passage that became blocked. Perhaps more telling of the entrance passage was the dramatic rescue on Sunday 9th - Monday 10th February in 1964.

At 4.00 pm on the Sunday, leaving the Great Hall, Howard Butler fell down an awkward drop, broke a leg and damaged his ribs. He was only 120 m from the entrance, but it took 25 hours and 20 minutes following the accident to get him out, involving a monumental effort by more than 30 personnel of police, ambulance and mines-rescue teams, including explosives engineers, with three medics, and the Women's Royal Voluntary Service. Cavers were first on the scene, some who were at the time also in Llethrid and others in Tooth Cave. Members of the rescue team immediately came from Swansea, Penwyllt (South Wales Caving Club headquarters) and elsewhere, to assist. There were also several volunteer local school-boy cavers. Thyssen engineers stood by with large-bore drilling equipment, while food, stoves, hot water bottles, blankets, a tent and assorted medical and rescue kit were taken to the casualty, who was well-attended, splinted, treated for pain and sedated.

ALL OUR OWN WATER

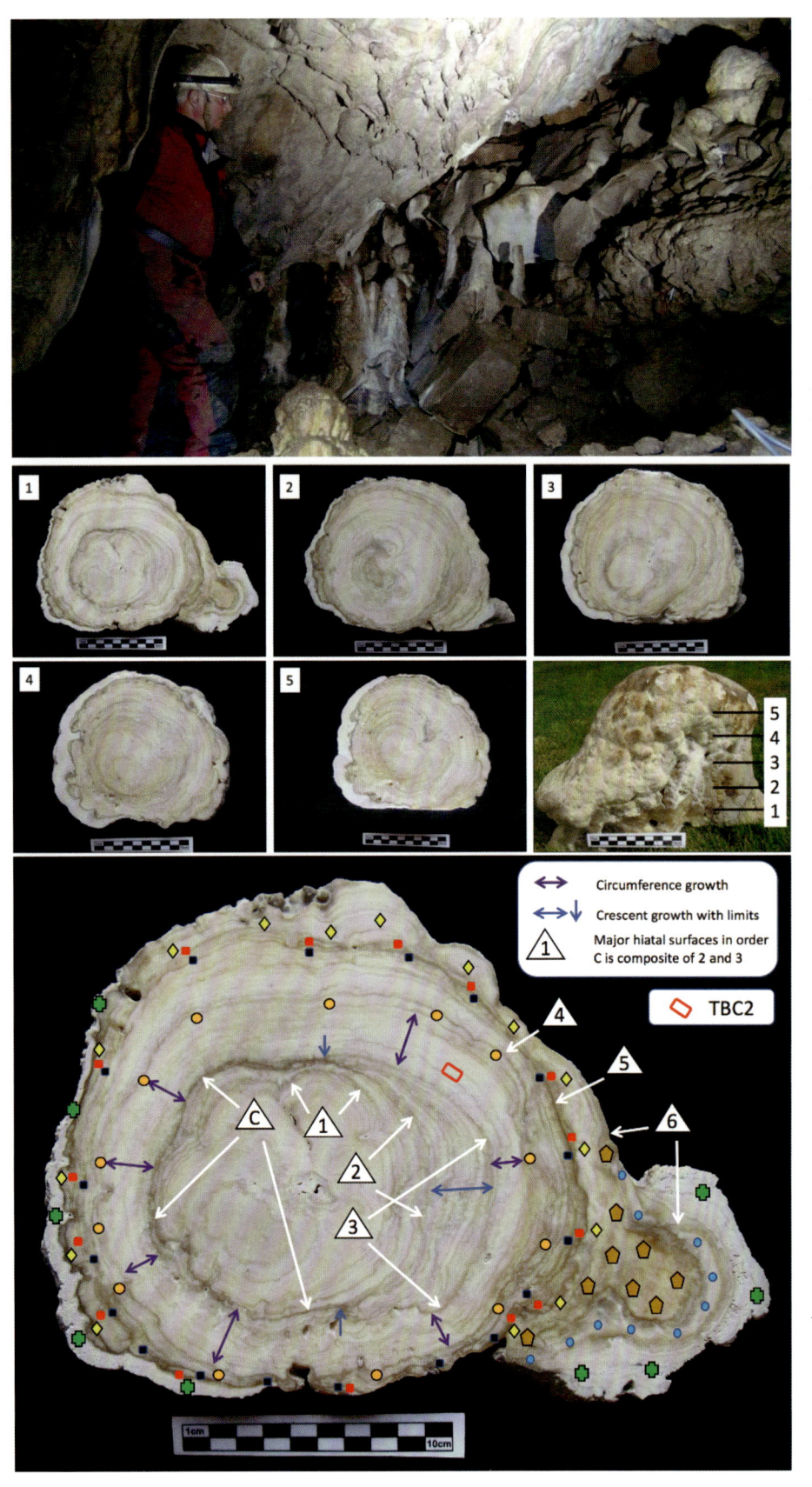

*Figure 98. (**Top**) Tooth Cave Bone Chamber showing in situ speleothem formations with part of the archaeological excavation site in the foreground.*

*(**Middle**) Stalagmite recovered from excavated debris and serially sectioned as indicated.*

*(**Bottom**) Lowermost section showing position of the dated sample TBC2 (sampled from the remaining basal counterpart). A sample TBC1 was taken from the oldest growth but failed to give any useful data.*

The various coloured symbols mark discrete outer growth zones. Recorded early growth of the stalagmite was irregular to asymmetrical (e.g., crescentic growth layers) with several darker and more irregular layers that show discontinuities ('hiatal surfaces'), following which growth was for a period more symmetrical (circumferential growth) before again becoming irregular. The variations from white to brown are mainly variations in crystallinity. Growth variations are commonly taken as proxies for climate and at this site in particular we know that it has long been close to the surface, sometimes open and probably occasionally flooded. Reading this record, however, is not going to be attempted here. Any takers?

158

Most of the long time that was taken before transporting the casualty out of Llethrid was to render the tortuous and tight passage near the entrance, in places only 30 cm wide, able to admit passage of the stretcher. In the event it was often pulled over people lying in water to keep it dry. The passage rendering involved pneumatic drilling and explosives, and afterwards the cave was temporarily closed. The entrance passages remained awkward, but the explosives had made some parts precariously loose and in need of some immediate and perilous sorting out. Nevertheless, in the following 40 years or so, hundreds of cavers, including many children, mine included, passed through to wonder at the unexpected spectacular large spaces and splendid formations inside.

As the research for this book got under way, several 'local' cavers considered recovering access to this jewel in the crown of Gower caves; modern speleothem dating and vast-space-lit photography beckoned, along with some local pride. On several criteria, including the caves with prehistoric relics, Green Cwm and its flanks constitute a Site of Special Scientific Interest. Hence, via the land owners Julie and Adrian Parker, cavers Andy and Antonia Freem and this author secured Natural Resources Wales permission to excavate a new access to recover Llethrid Cave. As described above (Section 4.4), the first 'dig' eventually became Barns Cave (named after the property Llethryd Barns). With numerous helpers[10], we made significant progress along a passage that must have been a pre-glacial stream entrance (see Figure 84). It seemed not far to go to break into the Llethrid Annex that leads into the Great Hall.

We had taken the only published map of the cave and carefully superimposed it on the map of the surface; apparently the Llethrid chambers were just inside the hillslope above the dig. The map had been published by both Oldham (1978) and Ede and Bull (1989), the latter clearly attributing it "after a survey by T. Moon, courtesy of South Wales Caving Club."

In February 2021, from the archives of Paddy O'Reilly, we discovered another map of Llethrid, surveyed and drawn in 1965 by Dick Baynton and Brian Jorgensen. Ironically, this author (PK) was present for part of that survey and indeed still retains copies of the hand-written sheets with all of the bearings, back bearings and distances measured. To our horror, that survey (reported as CRG Grade 5)[36] showed the cave trend rotated clockwise about the known fixed point of the entrance, by some 70° relative to our previous map. Instead of just about to break back into Llethrid, the Barns dig was some 60 m from the known cave! The Baynton and Jorgensen map orientation and measures could be verified as correct, because they matched the dip direction of the limestone beds in the roof of the Great Hall as well as the accurately measured length of that cavern, at nearly 100 m.

This author often went caving and climbing with Dick Baynton (Fig. 99) and Brian Jorgensen, and knew them well, but in 1966 they went travelling abroad, where, shortly afterwards, Dick developed leukaemia. Very sadly, Dick died in 1967 (see O'Reilly et al. 2021). His files were passed to Paddy O'Reilly and most of his extensive studies on Gower were published in the South Wales Caving Club Newsletters (Baynton 1968-1969). The Baynton and Jorgensen 1965 map of Llethrid, however, did not see the light of day, until February 2021. In what was a remarkable re-connection of the old timers who had caved on Gower in the 60s, we discovered from Terry Moon that his map, which we had been using to guide our dig, was never made as a proper survey but was a "back-of-an envelope sketch" showing how to navigate through the cave. He made it on his "first trip to Llethrid with three school mates, in 1960 or 1961"!

Fig. 99. Dick Baynton[37] in 1965 outside Pork Hall, the pigsty that we converted for accommodation in Stonemill (east Parkmill), close to the Gower Inn (photo courtesy of Brian Jorgensen).

In March 2021 we overlaid the Baynton and Jorgensen map on the LiDAR ground image and immediately found a strong draught in abandoned quarries where we started a new dig, downwards vertically above the Annex. My suggestion that we call it 'Far Side of T Moon Cave' was rejected as too silly and obscure. It became known as Cockle Pot (Fig. 82), a narrow fault-controlled fissure partly filled with debris from the surface.

Cockle Pot, unsurprisingly, is named for cockles excavated there. Initially it seemed likely that we were digging into a Romano-British midden; there are relics of this occupation elsewhere above Green Cwm and a hoard of Roman coins had been discovered *very* close nearby. Soon, however, it became clear that the shells were dispersed in freeze-thaw textured till-cum-soil to at least 4 m below the surface. Further, it also became clear that there are other marine shells present and that the assemblage does not represent any deliberate collection for food (Fig. 100b). The ground surface around there is littered with Devensian north-crop glacial erratics, but these are absent in the cockle-containing fissure fill.

Another surprising discovery. Setting aside Noah's flood, the shells can only have been delivered to Llethrid by the Anglian ice, which came from the northwest across Gower some 450,000 years ago. Anglian glacial debris is widely present on Gower, both beneath the much younger Devensian glacial deposits and remaining where the later ice did not reach (Fig. 42), but such 'fossiliferous' Anglian deposits are rare. Clearly the shell assemblage registers shallow marine to estuarine conditions, like those of the mouth of the Loughor Estuary and Carmarthen Bay today (Fig. 100). Evidently the Anglian ice must have picked up the material in crossing ancient equivalents of today's deposits, but at a time when sea level would have been nowhere near there. The shells must record marine conditions that predate the Anglian glaciation and so could be some 0.5 million years old. They tell us that the Anglian ice crossed relict shallow marine to estuarine deposits that had been left high and dry as tundra and freezing conditions advanced ahead of the glaciation.

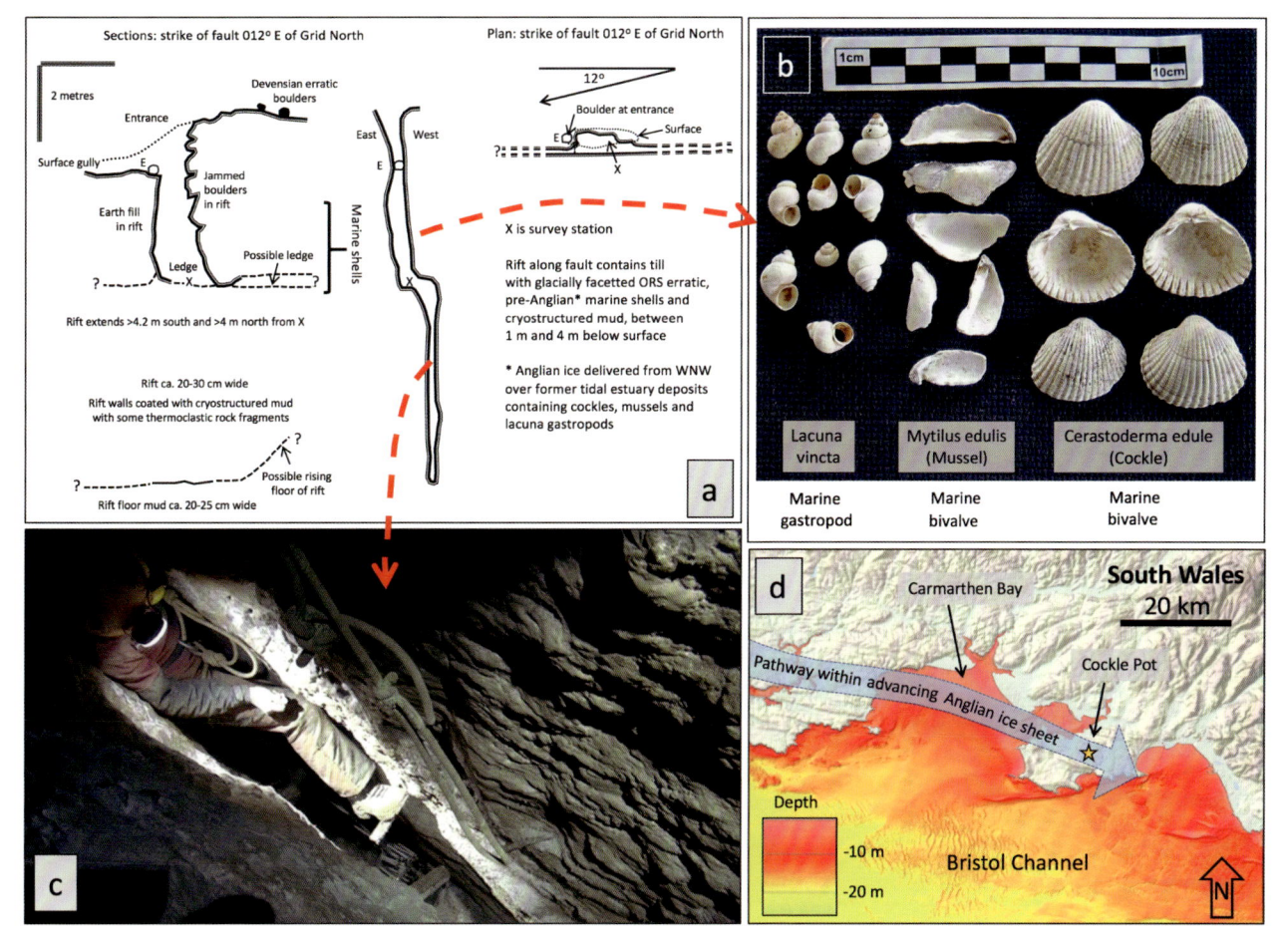

Figure 100. (a) Cockle Pot fault-fissure (SS5306591147; at 70 m OD) section and plan applicable June 2021.

(b) Representative marine faunal remains collected 1-4 m below the surface in Cockle Pot. The shells conceivably are some 0.5 million years old.

(c) The lower part of the fault-fissure is dangerously vertical and desperately tight, necessitating placement of footholds and use of 'cows tail' ties to an overhead safety line. A wire ladder hangs vertically beyond the caver, but at this stage it is beyond the reach of lovers of beer.

(d) Probable course of the Anglian ice that delivered the marine shells to Llethrid. The pathway shown is notional and a narrow part of an extensive ice sheet advancing on a broad front. At that time, 450,000 years ago, the coast would have been far to the southwest of here, with relict marine sediments subaerially exposed to the advancing ice not only in the Carmarthen Bay area but also widely across the valley that would become the Bristol Channel. (Base image from EMODnet 2019).

We have not given up on recovering access into the great and well decorated chambers of Llethrid.[38]

Caves Notes

[1] The submerged limestones off south Gower normally extend down to some 25 m below sea level, beyond which lie Triassic strata (see Chapter 2). However, the contact with the Triassic rocks is imprecisely located and the base level formed by it may range between -15 m and -30 m OD.

[2] Members of the Cunnington family explored several caves in the vicinity during 1908-1928. An inscription in what is now known as Ogof Wyntog reflects the death at age 27, in March 1918, of EC (Ned) Cunnington while tending to the wounded close to the front in WW1.

[3] Whereas many small quarries were opened for stone to build mostly straight walls around the cliff-top fields, in this case the depression is skirted around by the wall, which suggests a sink hole.

[4] Daw pits are named for the jackdaw birds (*Corvus monedula*) that enjoy the security of the steep walls, in places overhanging, to nest and roost. They form by cavern collapse that propagates up to the surface.

[5] Llethrid is sometimes rendered Llethryd, especially since the re-Welshing of some place names. Llethrid is the name on the maps used in this work and the original name used for the cave.

[6] Discovered by Bill Clarke in the 60s and dug for some 15-20 m with Roy Morgan, Glyn Genin and John Harvey. Abandoned still with good prospect after a mishap with gelignite. Rediscovered by the author.

[7] The dip of bedding is the slope directly downwards on the bed surface; the strike is the horizontal direction along the bed.

[8] It is tempting to wonder whether it is water input from Willoxton that causes the distinct zone of enhanced honeycomb dissolution shown in Figure 86g, which could reflect the mixing of waters with different dissolved mineral contents ('mixing corrosion' of Bögli 1980, pp. 35-37). When two H_2O-CO_2-$CaCO_3$ solutions saturated with respect to calcite but with different compositions mix, the non-linear equilibrium saturation relation Ca^{2+} vs H_2CO_3 causes the mixture to become 'aggressive'.

[9] This was done by the Taylors when they were working to penetrate the main sink (Taylor 1991). The Taylors were Maurice Clague and his sisters Eileen (shown below left, on the way to Wooley's Hole by Landimore with some serious ladders) and Marjorie (right). The right-hand image shows the early problem with penetrating Llethrid Swallet, which caused them to divert the stream. The Taylors' finding, exploration and description of so many caves on Gower in the 50s and 60s constitutes a monumental contribution. Many of us who took to caving in the early-mid 1960s followed and were inspired by them. These photographs are from an archive presently held at the South Wales Caving Club (Rowland 2019), which includes the two photograph albums of their exploration of Llethrid. There exists no better account of the cave and the wonder and excitement of its discovery than: *'Llethrid Swallet a photographic record by the Taylors of the best decorated cave in Wales (1956)'*.

[10] Much of the digging at Barns Cave (and eventually Cockle Pot) was done by Andy and Antonia Freem with the author, but the following people all made significant contributions (not all together): Luke Ashton, Val Bednar, Chloe Francis, Paul Hartwright, Duncan Hornby, Allan Richardson, Margaret Richardson, Jem Rowland, Gareth Smith, Paul Tarrant, Dawn Thomas, Glanville Thomas, Claire Vivian, and Alan Walsh. All of this was most generously supported by Julie and Adrian Parker, the land owners.

[11] At the time of writing, it is unknown how the pre-glacial Barns Cave passage at the 'red mud boulder choke' may continue towards the main Llethrid chambers. There remains potential for further passage and even decorated spaces. In the early years we were vaguely aware that the setting of the chamber formation must relate to more major passage than just the immature entrance series. Extension of Barns Cave towards the Great Hall of Llethrid would shed light on how the chambers developed.

[12] Llethrid Swallet Cave was first entered in the late 1940s by Phil Duncan and penetrated more fully by a SWCC team led by Don Coase, in 1951. Great Hall was discovered in 1953 and first entered by John Long and Michael Steward, with Eileen, Marjorie and Maurice Clague Taylor.

[13] I thought that Tooth Cave must derive its name from some special archaeological find, but a tedious literature search found nothing. The entrance passage was already named Tooth Cave when, in 1961, the extensive system was discovered by breaking through thin speleothem to access what is now Bone Chamber, where human bones (including teeth) and various artefacts were discovered and soon excavated (Morgan and Harvey, 1961; Harvey et al. 1967). Somewhat annoyingly, while recently surveying the entrance area, my wife pointed out to me the obvious: "Surely Tooth Cave is named for the prominent tooth that dangles in the entrance!" (Fig. 86a and below). More prosaic but perhaps less annoying I was told of the finding of a horse's tooth in the entrance passage. It was first named in Welsh, Ogof y Dant, and I am now at last assured that the name derives from both the finding of a horse's tooth and the pendant feature in the entrance (John Harvey pers. comm. 2021).

[14] Surveyed length is 1494 m excluding very minor loops and side tubes and the unsurveyed Iron Ore Series (pers. comm. Paddy O'Reilly 1st April 2021). Tooth Cave was originally surveyed in 1965 as far as Big Sump by Paddy O'Reilly with Terry Moon, Colin Fairbairn, Susan O'Reilly and other members of the University College Swansea Caving Club, with additional contributions from Richard Baynton. This was first published in Speleotawe 65 by UCSCC. The line survey was also included as an illustration in the SWCC 21st Anniversary Publication in 1967. The section beyond Big Sump to the Final Sumps was eventually mapped in 1984 by Paddy O'Reilly with Blake Farren from University of British Columbia. In 2021 these two surveys were completely replotted by Paddy O'Reilly, in accord with magnetic declination of the time and with an accurate fix on the entrance, with contributions from Peter Kokelaar, Eric Leinberger and John Harvey. This was published as an illustration in the SWCC 75th Anniversary Publication in 2021 without full details. Passage profiles and notes have subsequently been added and will be submitted for publication along with the history of discovery and exploration, surveys and features of the cave, in O'Reilly and Kokelaar (202x).

[15] To us as local cavers in the 1960s, this was always known as the Iron Ore Series (Fig. 81) after the characteristic ochreous red-staining mud there, which we did not understand at the time. Like so much in this cave of mysteries, we were blissfully unaware then of the interesting features we passed by.

[16] Its level and hence length are variable, commonly too long to duck under safely.

[17] For more general and original accounts see Morgan and Harvey (1961), Harvey et al. (1967), Baynton (1968-1969), Ede and Bull (1989).

[18] Boulders that do occur are not part of this deposit but relate to local collapse, e.g., in Main Stream Passage beneath the Iron Ore Series.

[19] Clay is used here as a grainsize term; clay minerals are present, as determined by X-Ray Diffraction (Section 5.8), but some is of other minerals finely ground to 'rock flour'.

[20] These mineral precipitates are typically amorphous mixtures of red-brown Fe-rich oxides / hydroxides and black amorphous Mn-rich oxides / hydroxides. They tend to precipitate where surface waters with dissolved Fe and Mn enter shallow, fast-flowing, well oxygenated waters of relatively high Eh and pH. The Fe oxides tend to precipitate more readily and so typically form coatings under more Mn-rich layers (Hale et al. 1984).

[21] For those familiar, just upstream of Mud Hall and the Christmas Cake. See recently drawn map by Paddy O'Reilly (drawn at the time of writing but to be published with additional details in O'Reilly and Kokelaar 202x; map to be filed with SWCC).

[22] Examples of debris flows are shown at https://www.youtube.com/watch?v=Fsh5E9m3PrM and https://www.youtube.com/watch?v=5ZKoIZHKRUM In these videos the catchment slope is steeper and in the first case the large rocks are much bigger than applicable to the Pengwern catchment, although the slope (lacking vegetation) and valley constriction at Llethrid would focus a fast flow into Green Cwm. The considerable mobility and forceful advance of the high concentrations of wet debris are relevant, however, especially underground. The second video of the channelised flow may well represent the nature of the debris flow that plunged into the Green Cwm caves. Note how towards the end of the video the once highly mobile debris has stopped to leave a dense thick, mounded deposit of large rocks in fine grained matrix; very little water drains from it.

[23] The rock face containing Tooth Cave entrance continues down behind the talus slope at its foot. The western side of the Bone Chamber is a talus (thermoclastic) breccia pile continuous with that outside; Bronze Age people entered the cave there. Thus, the wall would have been at least some 8-9 m high with the chamber entrance at the base. The post-glacial debris flow entered there as the floor of the valley built up to its present level.

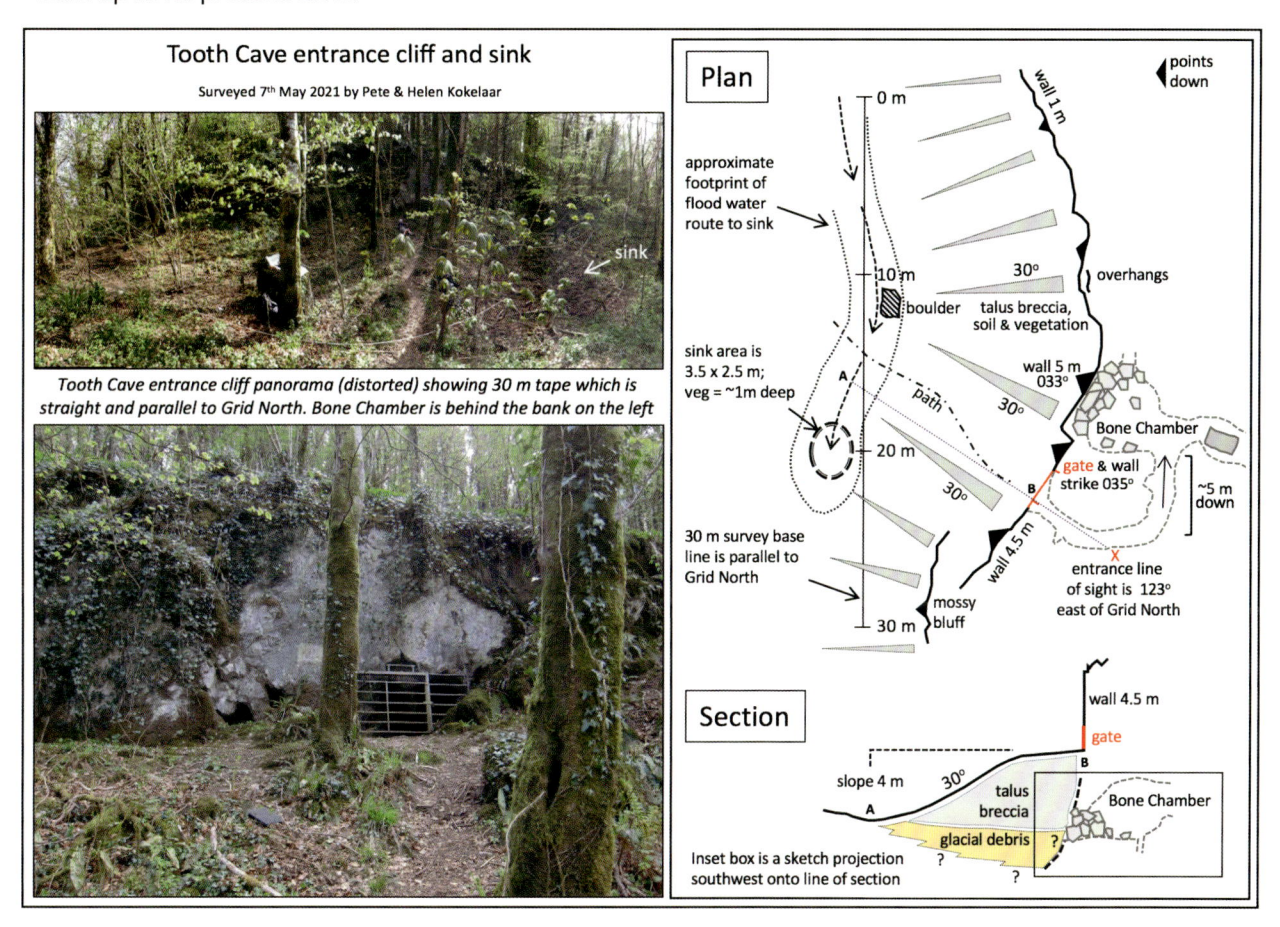

Tooth Cave entrance cliff and sink
Surveyed 7th May 2021 by Pete & Helen Kokelaar

Tooth Cave entrance cliff panorama (distorted) showing 30 m tape which is straight and parallel to Grid North. Bone Chamber is behind the bank on the left

[24] Baynton (1968-1969) reports this observation made by M Coburn.

[25] Scraped off stain contains goethite (FeO(OH)), proved by X-Ray Diffraction, while X-Ray Fluorescence shows a trace amount of manganese. (Analyses kindly undertaken by James Uttley at the University of Liverpool, July 2021).

[26] Percussion marks on rocks (e.g., Wilson and Lavé 2013) are single-strike points of damage, commonly showing a central pit with surrounding concentric microfractures. They typically occur on high-energy fluvial and storm-beach bedrock surfaces, as well as on the surfaces of the associated cobbles and boulders.

[27] Antonia Freem, Phill Thomas and Tomasz Zalewski obtained some of the photos and all of the samples at Pot Sump, while Dave Dunbar and Luke Ashton took measurements in the Iron Ore Series, on 24th June 2021. The author's knees are immensely grateful to them.

[28] Mud plastering due to the Montecito debris flow of 2018. The yellow arrows mark corner 'obstacles' where the flow continued downstream but separated behind them to leave locally undamaged surfaces (photo courtesy of Nico Gray).

[29] Flying a kite while treading thin ice seems a suitably muddled metaphor for this hypothesis. The author has studied granular flows for decades, including those that occur on the Moon (Kokelaar et al. 2017), but this occurrence of scratches is the most bizarre of records and possibly unique. As an illustration of the struggle to arrive at the current hypothesis, here is the previous painstaking drawing in which ice boulders from the glacial maximum were invoked to explain the scratches. The problem here was that two bizarre events would be involved. Occam's razor favours just the one.

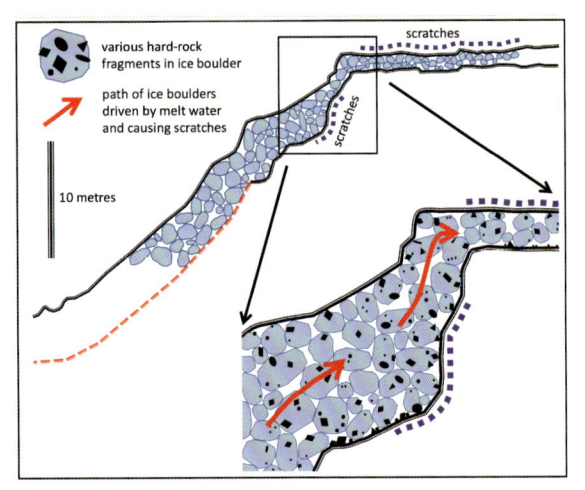

[30] Sections of the rock were kindly prepared and microscopically examined by Nigel Woodcock and Tony Dickson of Cambridge University. The sections show that the sediment became rock by permeation of Ca- and Mg-bearing fluids, from the adjacent dolomitic limestones, to form a dolomitic cement with scattered dolomite rhombs. The rocks were subsequently bleached in places (to cream colour) and the dolomite altered.

[31] In caving terms an aven is a cavern or passage high up in a cave roof. As is the case in Tooth Cave, avens commonly reflect early phreatic passage development with subsequent abandonment as base level is lowered and new passages form lower down.

[32] Items remained until they were removed recently when the fault-wall orientation was measured (see note 27 above); they were of little nutritional value, although a swig of the remaining methylated spirits might have been tempting in extreme circumstances.

[33] Many caves in the modern cliffed coastline have obviously been opened through coastal erosion; these commonly show roughly A-shaped entrances because the marine erosion is greater low down and diminishes upwards. Many are formed on weaknesses that are fractures in the limestone (faults or joints), and some lead back to solutional tubes. None of the examples listed here are of those types and because they continue into space (unlike blind tubes) they must be remains of truly ancient phreatic passages.

[34] The formations overlie Bronze Age human remains (Harvey et al. 1967). Both on this archaeology and on bats hibernating here the cave is an SSSI and has seasonal entrance restriction.

[35] The author is immensely grateful to Diana Sahy for providing the analyses.

[36] Cave surveys are graded according to achieved accuracy. Grade 5 was at that time basic for accurate bearings and distances. See Judson (1974) and https://en.wikipedia.org/wiki/Cave_survey

[37] Dick Baynton's tragically early demise in 1967 robbed us of a talented cave scientist and amusing friend. In his teens the author spent many muddy, abraded and freezing hours with Dick exploring Tooth Cave and elsewhere on Gower; his enquiring and systematic approach to hydrological field studies substantially influenced the style of the author's subsequent career as a field geologist. Hopefully this book reflects some of Dick's great legacy (Baynton 1968-69; O'Reilly et al. 2021).

[38] Setting the pages for printing of this book was ongoing in August 2021, with repeated attacks on the access to Llethrid affording interesting and energetic diversions. Triumphant inclusion of modern photographs of the vast Great Hall, illuminated by powerful LED lamps to show its previously unseen scale, seemed to have eluded me in the end...

Barns Cave dig

Breakthrough! Llethrid Great Hall, 18th August 2021, Andy and Antonia Freem with Gareth Smith.

5 SINKS AND SPRINGS

5.1 All our own water: an introduction

It may have been my first brush with geology. In school we discussed how Gower's main water supply from a spring at Parkmill (Wellhead), and other springs, are fed by rain that fell miles away on the northern skyline of the South Wales Coalfield, in Breconshire and Carmarthenshire. Many of the older inhabitants of Gower, and Dŵr Cymru (Welsh Water) engineers, not all of whom went to the same school as me, would tell you that the water 'goes down' in the limestone hills between Carreg Cennen Castle in the west and Black Mountain to the east and 'comes up' feeding springs on Gower. If you were offered any seemingly plausible reason for this belief, setting aside Celtic wizards and mystical forces, it would probably be that in all living memory of severe droughts on Gower the main springs never ever ran dry and so must have an unseen supplementary input.

Through the industrialisation of Britain and certainly by the time of World War 2, the demand for coal and steel meant that the geology of the South Wales Coalfield was comprehensively mapped for its mineral wealth and structure. Important coal-bearing strata of north Gower, thoroughly mined underground in the belt from Llanmorlais and Penclawdd through Killay and Black Pill towards Swansea, were known to reappear at the surface north of the Loughor Estuary, in the long belt from the valley of the Gwendraith Fawr near Kidwelly in the west across to the 'Heads of the Valleys' in the east (Fig. 101). These south and north outcrops[1] of the coalfield strata define the main down-fold, or syncline, which dominates the rock structure across southern Wales.

Just as the coal-bearing strata of north Gower overlie Carboniferous limestones that are exposed to the south (with intervening Marros strata; Fig. 101 section A-B), so too the north-crop coal-bearing strata lie above limestones that form the hillsides farther north (Fig. 101 section C-D). Those limestones are well known for their caves in the belt from Carreg Cennen to Black Mountain, and farther east around upper reaches of the Swansea Valley. Thus, it has seemed plausible that water entering the limestones along the north crop could be driven by the hydrostatic head (pressure due to higher elevation) down towards the lower springs in the same rock unit on Gower, via the syncline beneath the estuary. This potential source certainly was 'known' in the 1940s and seemingly became the accepted wisdom regarding Gower water supply from then onwards.

It is rather too easy, with the benefit of hindsight gained from more recent discoveries, to 'knock' early ideas by pointing out original errors. Rather than doing that, this Gower project was initiated to find out positively how our spring water is actually derived. Evidently the project developed some greater scope. As the title declares, the springs are found to be of 'our own water', from rain that fell on Gower. There turns out to be nothing exceptional about Gower water, but there has been no modern synthesis of the hydrogeology of the peninsula and the discovered details constitute fascinating 'case studies', at least so for the author, especially as they are intimately bound to the evolution of the peninsula (described in the previous chapters).

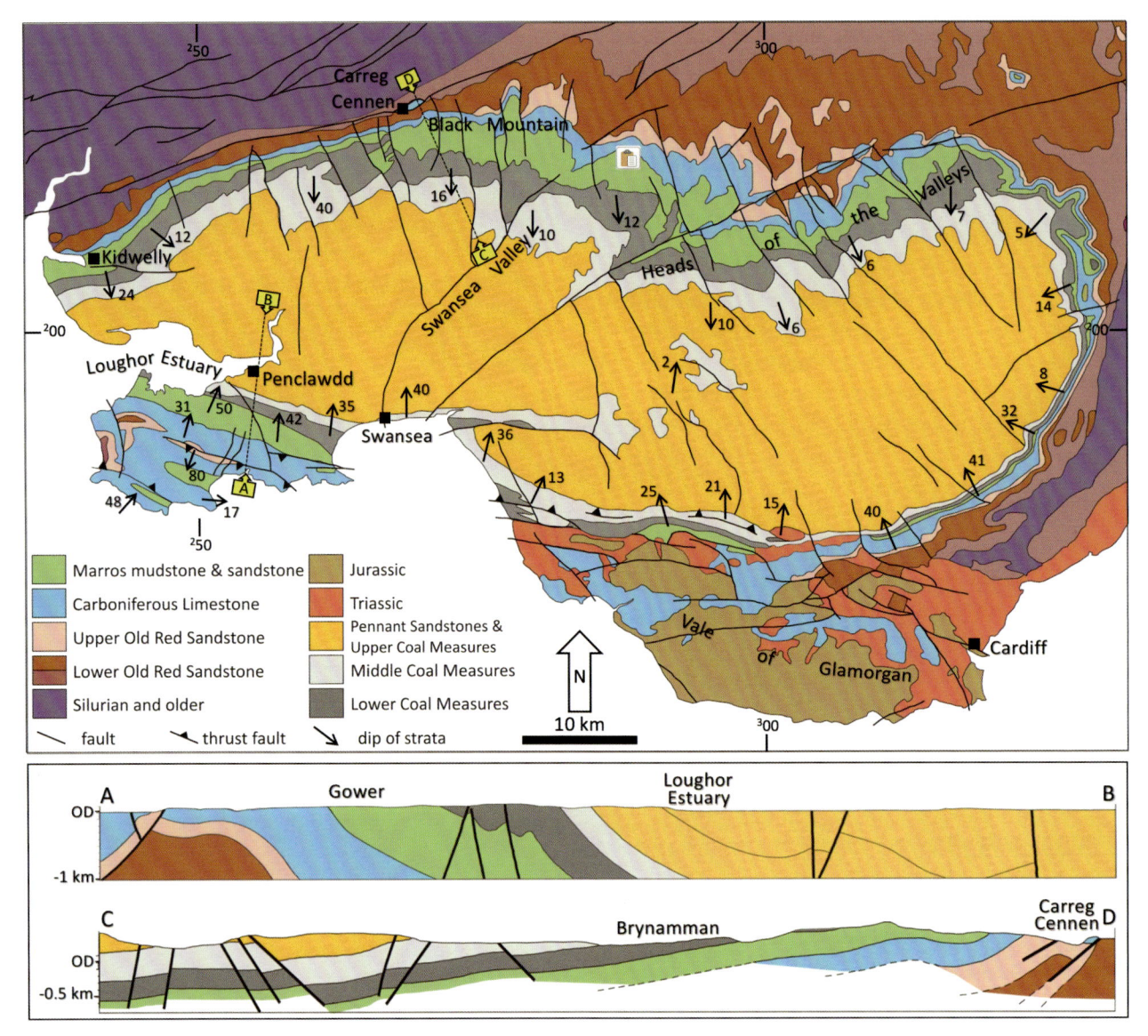

Figure 101. South Wales Coalfield map and cross sections[2]. The oval pattern on the map defines the syncline, with coal measures and underlying strata around the edges all dipping inwards (see dip symbols and cross sections). Coal Measures and coal-rich Pennant Sandstones are the main part of the southern limb of the syncline across northeastern peninsular Gower, along and south of a line between Penclawdd and Swansea. Gower limestones continue beneath the Marros beds and Coal Measures (section A-B) to reappear along the 'north crop' (section C-D); thus, the rock sequence is mirror-imaged across the coalfield syncline and it is easy to see why it was thought that the north and south crops of the limestone were hydraulically connected beneath the coalfield. Strata above the limestone must together reach more than 3 km in thickness under the middle of the coalfield, as evident from the thicknesses shown in section A-B.

The following sections will first present a key to the subject locations and data collected, followed by some grossly simplified views of spring types and behaviours, adding a view peculiar to the seaside setting of Gower subterranean drainage. Then aspects of the main historic water supply are described, before a review of the case for an exotic, northern supply. It should become evident that the original observations on Gower water were fair and that it is quite understandable why they were interpreted as reflecting the exotic northerly source. The alternative explanations, although simple, are perhaps interesting, but, before going there, it seems worth explaining briefly how this study came about.

As a boy living and caving on Gower, I became interested in the water flow underground, with a rather contrary view reflecting a mix of optimism and sheer ignorance (Fig. 102). When I retired and returned with some understanding of geology, the topic still vexed me sufficiently to dare mention my misgivings about the northerly source of the water. The scornful dismissal I received included a remark along the lines: "If you don't believe me, ask Peter Sambrook who was the engineer at Wellhead Water Treatment and Pumping Station for 34 years. He'll tell you." Peter did tell me, and we became friends regularly revisiting local highlights together (Fig. 103), even after I convinced him that Gower water comes only from Gower.

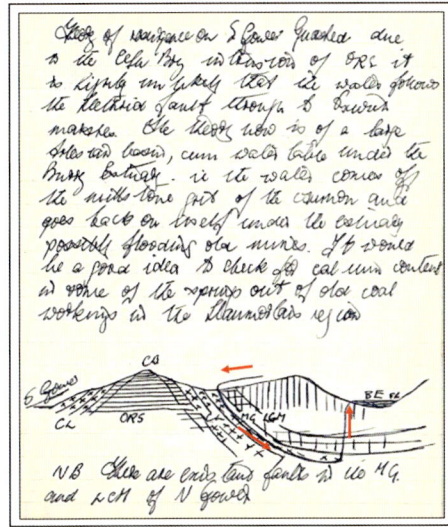

Text reads:

"Theory of resurgence on S Gower quashed due to the Cefn Bryn intrusion of ORS it is highly unlikely that the water follows the Llethrid fault through to Oxwich marshes. The theory now is of a large Artesian basin, cum water table under the Burry Estuary. ie the water comes off the millstone grit of the common and goes back on itself under the estuary possibly flooding old mines. It would be a good idea to check for calcium content in some of the springs out of old coal workings in the Llanmorlais region

Cross section

NB There are existant faults in the MG. and LCM of N Gower"

In the section: CL is Carboniferous Limestone, CB is Cefn Bryn, ORS is Old Red Sandstone, MG is Millstone Grit, LCM is Lower Coal Measures, BE is Burry Estuary and SL is sea level. The proposed course of the water, unlike anything conceived before or after, is the line to which the red arrows are added.

Figure 102. Page from the author's Log Book written in 1964-65 (at age 15) concerning the fate of water at Llethrid. Evidently even then I had my own contrary ideas. The emanation of waters from mines at Llanmorlais is, just perhaps, as fanciful as a north-crop origin, but springs from limestones along the south side of the estuary are a reality. The "Llethrid fault" of my notes is the Long Oaks Fault on geological maps; it extends from Llethrid though Cefn Bryn to Oxwich marshes, via Nicholaston Woods where I was at the time digging to explore a secret cave[3]. What blissful innocence!

Figure 103. (Top) Peter Sambrook, former engineer at Wellhead, most kindly contributed unwritten information, documents, photographs, stimulus, company and great humour for this project. Whereas his teaching of water divining worked for my wife, Helen, whose dowsing rods swung despite her not knowing exactly what water we were trying to detect, I simply could not get the same rods in the same place to understand that I too was seeking water.

(Bottom) Wellhead Water Treatment Works and Pumping Station at the time of opening in 1954 and in 1990.

5.2 Sinks and springs of central Gower

This chapter focuses on central Gower, essentially including the vicinity of the main water supplies and associated cave systems that have been studied.[4] The author is familiar with the delights of Bishopston Valley (SS5788), where sinks, caves and springs are numerous, but these are not discussed here. They are well described in general by Baynton (1968-69) but, compared to Green Cwm, they lack detailed maps and information. The sites studied are located in Figure 104, with further information and photographs in the appended gazetteer. Text references to features discussed are made with the site number in square brackets, for example the springs at Wellhead [14] and Holy Well [6], which have been studied in detail.

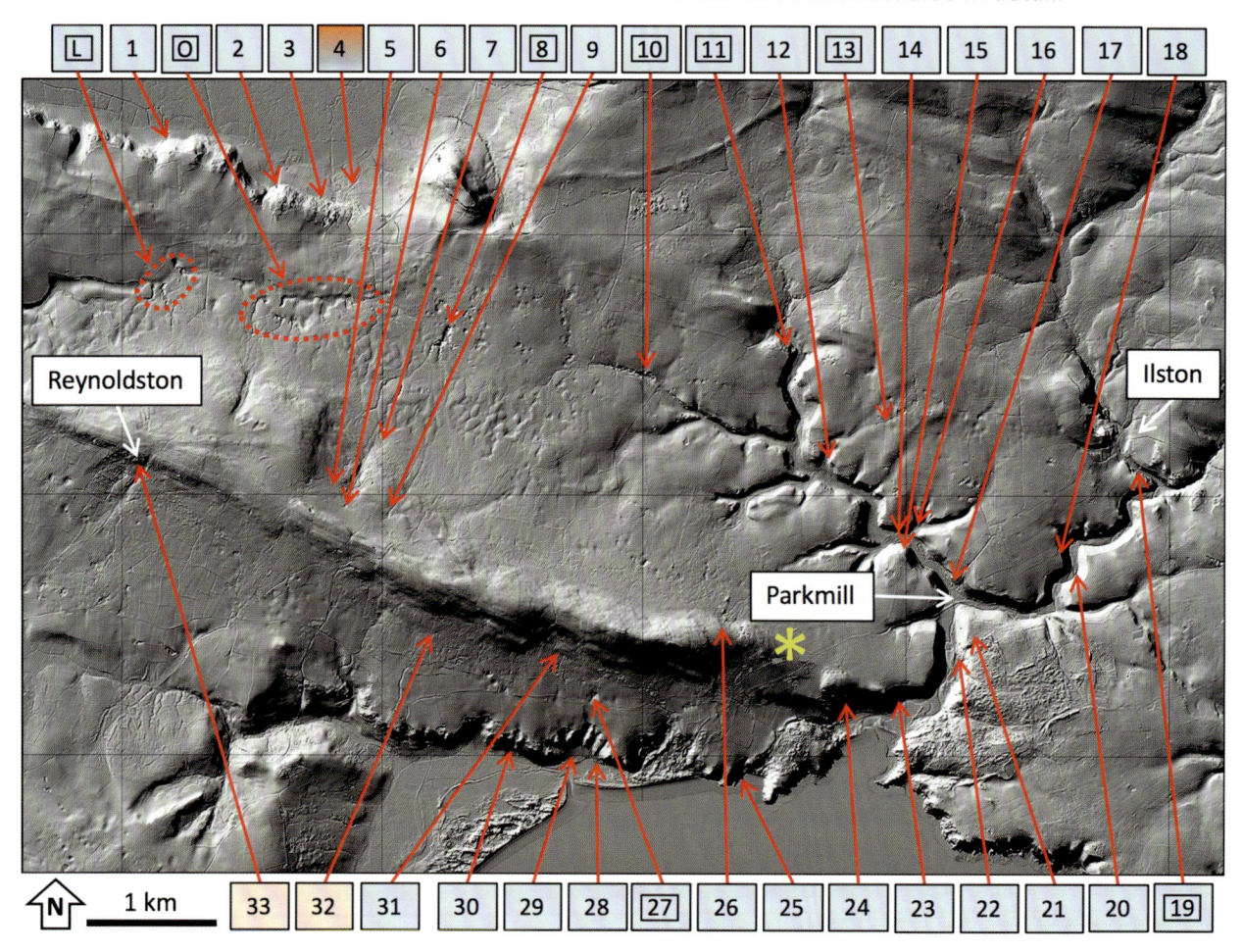

Figure 104. Shaded-relief LiDAR image locating the sinks and springs of central peninsular Gower that are studied in this work; some have known names, other are named here by location. Location references to the Ordnance Survey National Grid are given in the gazetteer in the appendices. **Blue boxes** *indicate a limestone setting;* **buff** *indicates conglomerate and sandstone;* **brown/blue** *indicates mudstone on limestone.* **Double boxes are sinks; single are springs**. *The* **yellow asterisk** *locates Penmaen weather station.* L *indicates Llwyn-y-bwch sink cluster and* O *indicates Oldwalls sinks cluster. 1. Leason. 2. Staffal Haegr. 3. Llanrhidian Church. 4. Butter Well / Saint Illtyd's Well. 5. West Well. 6. Holy Well. 7. Hot Well. 8. Moormills sinks. 9. East Well. 10. Decoy sinks. 11. Llethrid Swallet (sink). 12. Former well at Green Cwm Cottage. 13. Willoxton sink. 14. Wellhead resurgence. 15. Kitchen Well. 16. East Wellhead. 17. Parkmill Heritage Centre and Lunnon boreholes. 18. Lunnon East. 19. Killy Willy sink. 20. Trinity Well and Killy Willy Rising 200 m to the north. 21. Sunnyside borehole. 22. Pennard Castle (aka Saint Mary's Well). 23. Sambrook. 24. Notthill West and East. (continued...)*

25. Tor Bay. 26. Parc le Breos pond inlet. 27. Nicholaston sink. 28. Nicholaston East. 29. Nicholaston West. 30. Nicholaston Woods. 31. Parsonage. 32. Perris Wood. 33. Reynoldston Post Office. (Image base data contains Natural Resources Wales information © Natural Resources Wales and Database Right).

The prominent ridge from near Reynoldston towards Parkmill is Cefn Bryn, largely an outcrop of low-permeability quartz-pebble conglomerates and brown sandstones of Devonian age (Old Red Sandstone; see geological maps, Figs 3 and 101). The majority of water features on either side of Cefn Bryn are in Carboniferous limestone, except for Butter Well [4] in Llanrhidian, which is mapped as close to limestone but possibly lies on overlying low-permeability Marros mudstones. Broadly speaking, the limestones north of Cefn Bryn dip northwards and those immediately to the south dip southwards.

In terms of data collected, many of the springs in the central Gower area (Fig. 104) have been repeatedly measured in the field (up to 32 times) for total dissolved solids (TDS), temperature and pH, over a period of 12 months during 2018-2019. Wellhead resurgence [14] and Holy Well [6] have had water and local-air temperatures recorded at 30-minute intervals for one year and water samples from 13 representatives of the springs have been chemically analysed. The latter samples were collected late during the great drought of 2018, which maximized the intrinsic compositional signatures rather than having these variably diluted by rapid flushing with new rainwater. The 2018 drought fortuitously also allowed recognition of styles of discharge response, particularly when it rained afterwards. Importantly, it allowed repeat measurements of the discharge from Holy Well [6] and adjacent springs [5 and 9] renowned for never failing to flow. As shown below, had the 2018 drought persisted much longer, they would have dried up. We shall see that the historical persistence of the springs relates to the frequency and amount of rainfall on Gower.

Dŵr Cymru (Welsh Water), via Ian Murphy, generously supplied access to many years' worth of chemical, temperature and turbidity data recorded at the Wellhead water treatment installation [14], much of it by Peter Sambrook. Borehole data are also presented; one hole drilled close to Parkmill in 2019 penetrated to 50 m below sea level and recovered fresh water from the limestone. Rainfall records for 1962-2009 from the nearby Penmaen weather station[5] (Fig. 104) have been used, with more recent data from several sites on Gower linked to the 'Weather Underground' website.

In keeping with the objective of understanding Gower's water, main aims have been to characterise and explain the diverse spring types and behaviours that occur, and to find whether local rainfall alone can account for the well-known persistent discharges. To help the reader to know where this work leads, before we work through the evidence, here is the most significant finding: *It is certain that the early-envisioned supply of water from the north crop of the coalfield is geologically extremely improbable and, more importantly, that there is absolutely no need to invoke it.* So, here we go…

5.3 Sorts of springs

Springs are where water emerges from underground[6]. We will not dwell long on the topic, but it is perhaps worth knowing that spring waters in general can have any one or a combination of origins. Some 'primordial water', retained deep in planet Earth as it formed (e.g., Hallis et al. 2015), slowly migrates upwards and typically is released at volcanoes. Other water is expelled from deeply buried 'wet' rocks, a large amount originating as rain or sea water that circulated deeply underground driven by heat within the Earth's crust to emerge at hot springs, commonly laden with dissolved chemical elements. Here, fortunately, we can focus on rainfall onto land and its pathways of sinking into the ground and eventual return to the surface. Despite the peninsular situation of Gower, sea water has not been found to have penetrated inland as any saline 'wedge' beneath fresh water; forgive the pun, but the outwards flow of fresh water appears to keep the sea water at bay.

It is simplest initially to think in terms of two end-member types of springs. In nature various hybrid types are the norm, but it is an interesting and instructive feature of the hydrogeology of Gower that near-end-member types exist as well as different hybrids. Accounts of these different types are presented below, but here are the end members (Fig. 105).

A *conduit-flow spring* is the emergence of water that has flowed through underground passages generally larger than a few centimetres in diameter (>5-15 mm; Ford and Williams 2007), commonly much larger. Limestone cave passages initially form fully flooded with water and typically develop as tubes that enlarge by slow dissolving of rock around all contact surfaces. These so-called 'phreatic' passages may ultimately drain partly to form underground streams with open air space, known as 'vadose' passages. The water flow in these two types of conduit primarily depends on hydrostatic head (pressure) and slope, reflecting height differences between the points of recharge (water in) and discharge (water out), and on hydraulic conductivity, or 'ease of passage'.

On Gower we will need to understand phreatic-conduit flow with a relatively tranquil throughput. As a thought experiment, imagine a pipe (conduit) about 2 km long, with some variations of metre-scale-diameter and with a sink-to-spring slope of less than 1°. Now imagine adding water at the sink due to runoff from a passing rain shower. This added flow input would raise water levels and thus constitute an addition to the hydrostatic head, creating a 'pressure wave' that passes along the 2 km pipe and in time would emerge as additional output at the spring. Any tracer dye added at the sink during the rain-water peak influx would not travel directly to the spring but would mix and become to some extent dispersed in water pre-existing in the conduit (see below). It is common for systems involving phreatic-conduit flow to have relatively 'old' water pushed out of the spring before any added new water arrives there, as we shall see.

In reality (in the Green Cwm System), the just-described 2-km long phreatic conduit actually has several free-surface openings along the way and is overlain in parts by vadose passages. Adding to the thought experiment, imagine now that the rainfall was heavy and persisted for sufficient time for the phreatic throughput to be overwhelmed. The phreatic conduit then could not take the full flow, which consequently bypassed via higher vadose passages. Like open streams that include flood bypass routes, the vadose conduit throughput would be relatively rapid, and any added tracer dye would emerge relatively quickly. Again, any 'old' ponded vadose water, e.g., in an underground lake, is likely to be flushed out in the front of the main flood of new water. Conduit-flow springs, like those in the two thought experiments, are responsive to rainfall on relatively short timescales, typically hours to days, depending on input amount and conduit types (Fig. 105).

A *percolation spring* is not so obviously responsive to rainfall. It can be very slow to react and in Gower cases delays of several days and slow responses over weeks are found. As the name suggests, the throughput flow of water is dominated by percolation, which happens slowly between grains of superficial (drift) deposits and/or permeable rocks, and along cracks in rocks, such as joints or faults or bedding planes.

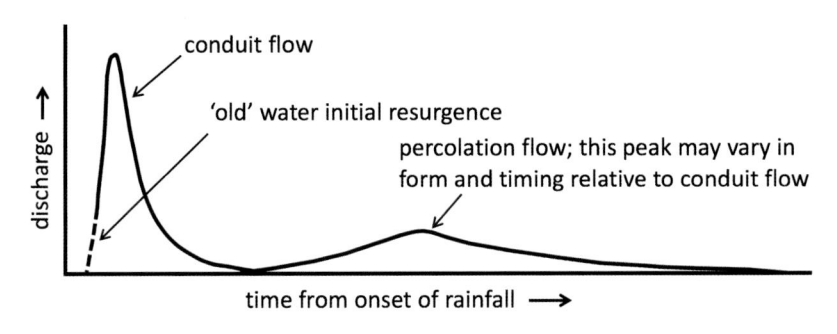

Figure 105. Schematic representation of end-member types of spring behaviours (after Ede 1975). The forms of real discharge peaks can vary considerably. The contrasting extents of rock-surface contact and residence times of conduit versus percolation flows have substantial impact on limestone dissolution and hence water chemistry, as we shall see.

As with conduit flow, the force driving the water to the spring is gravity, but in percolation there are strong resisting forces that slow the movement. The viscosity of water is significant in resisting flow through small spaces, while the very large contact-surface area per small volume of water means that attractive forces become effective. Water is difficult to break at small scales, it is cohesive, and it tends to 'stick' to surfaces according to inter-molecular forces amounting to adhesion and adsorption. We are all too familiar with rising damp, where capillary forces on water in minute pores are sufficient to pull up water against the downwards pull of gravity.

The net effect in a pure percolation system is that the throughput of water is rate-limited. This means that a percolation-dominated aquifer cannot admit or rapidly discharge more water once it is saturated, i.e., 'full', and additional rainwater will tend to pass over its surface, perhaps forming a 'flashy' runoff. By the same token, the percolation spring will have a characteristic maximum discharge when the aquifer is 'full', and discharge will diminish only slowly during protracted drying. We have a classic example of this behaviour, on Cefn Bryn (described below), but before going to the specific cases we should briefly revisit what, in Chapter 4, we considered a special circumstance affecting many springs on Gower.

5.4 Lost accessible depths and misfit springs

The springs and resurgences of streams in Gower mostly prove impenetrable (e.g., Wellhead [14], Trinity Well Ilston [20], Pennard Castle [22], Bishopston Valley, Caswell Valley, Leason [1], and Llanrhidian [3]). Ogof Ffynnon Wyntog is of the same ilk although its tight entrance sumps can sometimes be passed during droughts (Chapter 4). The exit passages are all too small given the evident whole-system maturity and water discharge that they relate to, and the water tends to fountain from conduits that rise from below the water table. We noted that these discharges all occur at around 10 m OD, or slightly above, and we found that the simplest explanation is that original deeper parts of the systems have been 'lost' by becoming stagnant and infilled with sediment, thus causing waters to be forced to surface at the higher 'misfit' levels. This blockage primarily relates to the Ipswichian (MIS 5e) high stand of sea level around 123,000 years ago but it would continue with the Holocene rise of sea level near to its present position, some 7,000 to 5,500 years ago, as valleys and estuaries became filled with alluvium.

In this chapter we will see that considerable variability in both discharge composition and discharge amount relates to systems involving different underground regimes according to drought and flood. This is particularly the case for the substantial Llethrid – Tooth Cave – Wellhead system, which has a deep flooded part within which waters are relatively slow moving, even stagnant. Now we should return to the original main suppliers of Gower's water.

5.5 Gower water supplies

There are several springs and wells that were used for water supply in the vicinity of Wellhead and Parkmill. Wellhead resurgence [14] was first significantly exploited by the Air Ministry when the water was pumped up to supply the Royal Air Force base built and opened on Fairwood Common in 1941, during World War 2. At that time villages and most farms on Gower depended on open wells: "In 1945 there were as many as 38 publicly maintained open wells, and most, if not all, of these were of doubtful purity, and all were inadequate in periods of drought."[7] Parts of north Gower, including Penclawdd, were supplied from Holy Well springs [5, 6 and 9] on Cefn Bryn, although east Gower as far as Pennard and Parkmill had been progressively plumbed into supplies from reservoirs pumped via Swansea. In west Gower, from Rhossili to Reynoldston, water was supplied from the springs at Pitton. Hot Well [7] on Cefn Bryn became plumbed into the Holy Well supply to satisfy new demands due to the American armed forces and artillery installations at Penclawdd and Crofty, in the lead-up to the Normandy landings of June 1944.

In the context of the stressed supply system on Gower, and with a 1944 Parliamentary Act to encourage and support rural water and sewerage schemes (to be approved by the Ministry of Health), a 'Comprehensive Water Scheme' was formulated in 1945. In January 1946 the scheme was discussed at a public enquiry held at Reynoldston Parish Hall. It was estimated to cost £113,000 and involved some 32 miles of new mains with a new 500,000-gallon reservoir on Cefn Bryn. Also included was the building of the Water Treatment and Pumping Station at Wellhead (commonly attributed to Parkmill), where it was to extract a proportion of the resurgence water, treat it, and pump it up to the new reservoir. The scheme ran somewhat over budget (£176,000), and the Wellhead station, costing some £40,000, was officially opened on Wednesday December 8th, 1954. Those facilities were operated initially by Gower Rural District Council, then by West Glamorgan Water Board and then Dŵr Cymru (Welsh Water).

In 1988 a 'Parkmill Aquifer Risk Assessment' was published (Halfacree and Williams 1988) and, while proving the three main local inlets to the Wellhead supply (see below), it also built upon the notion of an exotic component input, referring to specific north-crop locations. The report compared the chemistry of Wellhead water with that from Llygad Llwchwr (SN668178; 229 m OD), which is a substantial resurgent stream and source of the River Loughor, and it found close similarity with aquifer water from the nearby Llwyndewi borehole, supplier of a well-known bottled water. This north-crop site is 7 km northeast of Ammanford, just south of the limestone-founded Carreg Cennen Castle (Fig. 101). The hydrostatic head (pressure) from there would reflect the 217 m height difference relative to Wellhead, more than 30 km away. It seems that this particular site, where the water was supposed to 'go down', became part of the traditional north-crop-supply lore simply because its aquifer was known.

In 1994 daily water extraction and treatment at Wellhead stopped and the plant was only used in periods of high demand or drought. By 1998 the spring was no longer in use, with the site maintained only as a pumping station and for boosting chlorination, as it is now. The reservoir on Cefn Bryn is still used now.

5.6 Wellhead spring water supply

Wellhead spring is the outflow (resurgence) of water that has gone underground via several routes that converge somewhere beneath and east of the (normally) dry valley known as Green Cwm. The whole catchment-to-spring system is referred to here as the Green Cwm System, key features of which are shown in Figures 106, 107 and 108. Wellhead spring discharges 2.5-216×10^3 m^3/day from which the average extraction was ~1.6×10^3 m^3/day (350,000 gals/day). As discussed above, the Green Cwm System is thought to have a flooded and partly infilled deep part that is likely to be largely stagnant downstream beyond Wellhead within the Black Rock Limestone Group and Lower Limestone Shales (British Geological Survey 2011), all beneath alluvium. Upstream there remains a slow-flow phreatic regime where dye tracers are significantly delayed and diluted in drought-flow conditions.

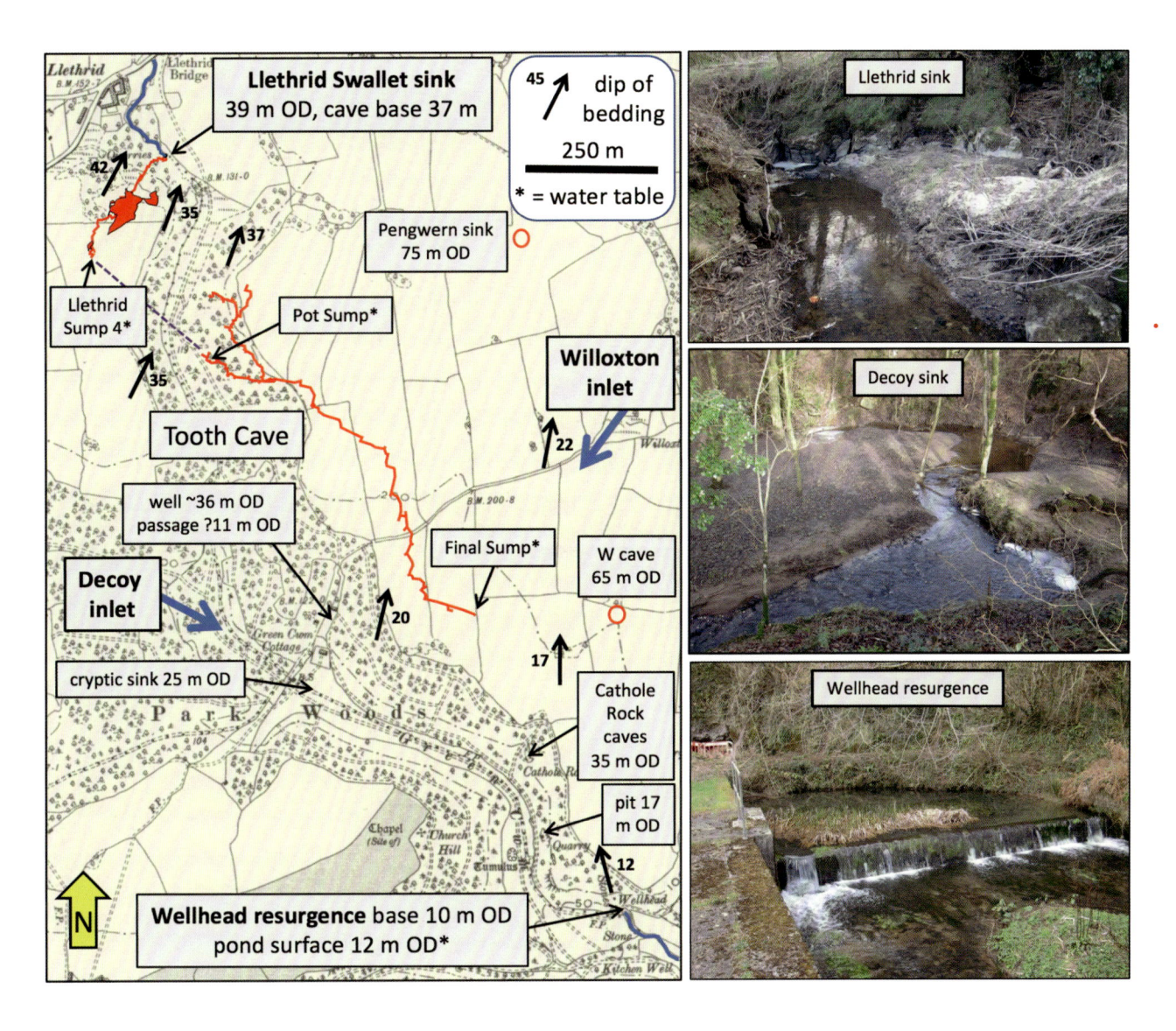

*Figure 106. Known features of the conduit-flow parts of the Green Cwm System feeding to Wellhead resurgence (see also Chapter 4). The photos show fair-weather conditions. Llethrid stream flows south off shales and sandstones (see below) and sinks into the underlying (NE-dipping) limestones at the head of the (normally) dry valley known as Green Cwm. The water course is known in Llethrid Cave as far as Sump 4, which has been proved by dye testing to connect to Tooth Cave (**purple dashed line**). The known cave passages are mostly not directly beneath the valley floor. Additional dye tests prove inlets from Decoy and Willoxton sinks with the confluent waters all resurging at Wellhead. The water course(s) beyond the Final Sump of Tooth Cave remain unknown; a dive by Martyn Farr into the Wellhead resurgence was unable to penetrate more than about 5 m.*

The crow-flight distance between Llethrid sink and Wellhead resurgence is 1.68 km, and the overall fall is 27 m, but within Llethrid cave the water table is no higher than 22 m OD so its slope from there to the spring is ~0.3°. Towards the top of the cwm the water table generally is no more than about ~15 m below the level of the 'dry' valley floor and there is a strong tendency for surface flooding there following protracted heavy rainfall, when the sinks can take no more water. Flooding of the dry-weather-accessible passages in the caves occurs before the surface bypassing and is more frequent; it constitutes a severe hazard for cavers and at one time a real problem in the treatment of the resurgent flood water. (Base map is OS 1:10,560 Sheet Glamorgan 22SE published 1900).

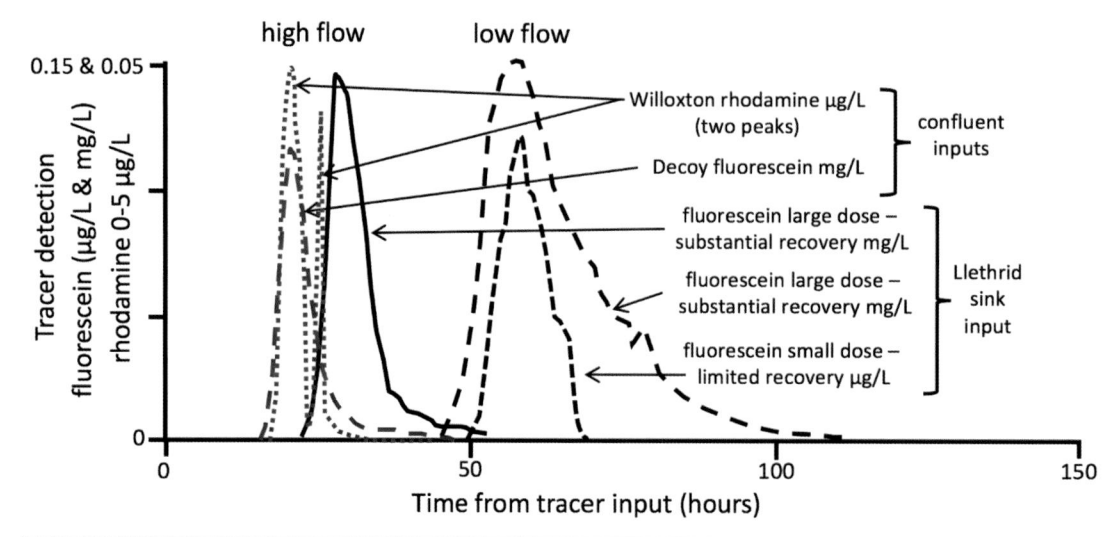

Sink	Notional distance to Wellhead (km)	Previous 5 days rainfall (mm)	Flow at Wellhead (m^{3}s^{-1})	Time to peak (hrs)	Notional velocity (m/hr)
Llethrid	1.68	15.2	0.34	28	61
		4.6	0.15	58	29
Decoy	1.8	53.4	0.54	20	90
Willoxton	1.0	53.4	0.54	20	50

Figure 107. Sink-to-resurgence timings of tracers marking <u>conduit flow</u> in the Green Cwm System[8]. The actual underground distances are unknown, and dispersion is neglected. The main point illustrated is the contrast in conduit through-put times during low- and relatively high-flow conditions. The substantially differing first-arrival and persistence times for Llethrid show flow-path differences; under high-flow conditions water from Llethrid sink mostly bypasses the deeper slow-flow phreatic regime used normally. There was insufficient low-flow input to Decoy and Willoxton to test for a fair-weather slow-flow transfer.

The Llethrid high-flow test was done when Wellhead discharge was almost one third less than the discharge during the Decoy and Willoxton tests, owing to significantly less previous rainfall. This lesser discharge condition indicates that the later peak arrival time for the Llethrid test cannot be taken as a meaningful relative-arrival order. Similarly, the absolute peak height differences are of no consequence, because they relate to different conditions. The graph data are from Halfacree and Williams (1988) and rainfall data are from the Penmaen record.

Tracer added to flood water at Llethrid discharges at Wellhead within 20-30 hours (Baynton 1968-1969; Halfacree and Williams 1988). Tracer-laden flood-water contributions from Decoy and Willoxton sinks also arrive quite quickly at Wellhead (Fig. 107). Because surges in water input tend to push out 'older' water first, increasing discharge at Wellhead occurs earlier than arrival of tracer or turbidity. This gave useful warning of the impending arrival of flood-related turbid water that would need treatment at Wellhead (see below). In any flood conditions there is a net-flow-output compound 'peak', the irregular shape of which depends on the relative volume-rate contributions from the three main inputs and on their respective conduit characteristics, all of which are unpredictable.

In contrast to flood behaviour, tracer dye added during normal low flow at Llethrid sink takes some 50 hours to show at Wellhead. Slow phreatic flow with mixing is indicated by the longer transit time and by considerable dispersion-dilution of the tracer. In any rainfall scenario, there are broad catchments for percolation input and, theoretically, percolation discharge must follow the conduit discharge, possibly days later, and possibly then without local rainfall (Fig. 105). However, as we shall see, rapid runoff and conduit transfer of water dominate, with some percolations through the limestones likely gathering at shallow depths into conduits that feed into the main supplies from Decoy and Willoxton. Distinct peaks attributable solely to percolation discharge are not identified. However, percolation water becomes an important component of the reducing discharge during drought, as signaled by changing chemical composition (Section 5.18).

There are two contrasting rainfall catchments feeding the Green Cwm System (Figs 108 and 109). Roughly 4 km^2 is underlain by limestone that widely has a thin cover of glacial tills of variable permeability. During rainstorms there is little surface runoff except in the west where the artificial Decoy Pond[9] and the stream from it persist at the surface as far as the Decoy sinks. Rainfall on this catchment widely percolates through thin cover and fractured limestone, although some drains quickly into sinkholes, entering the limestone where confluent waters merge into the major conduits. Tracers show that the conduit flows from Llethrid, Decoy and Willoxton are fast (Fig. 107), although there must remain considerable retarded percolation flow as well; this shows clearly in the water chemistry (Section 5.18).

The other catchment, of approximately 3.5 km^2, comprises extensive peat on till, which overlies practically impermeable mudstones (shale) and thin sandstones that form a basin draining to Llethrid. When this ground is fully saturated, or initially when it is particularly dry, rainwater runs off directly into the streams. However, the vegetation, soil and till together also constitute a percolation drainage that can sustain flow to Llethrid even in drought, as in 2018. In that drought of some 89 days without significant rain, the stream entering Llethrid Cave never dried but was sustained by percolation and flow through stream-bed gravels with intermittent showing in slow-flow pools. During a stormy spell 11-14th October 2018 the stream rapidly swelled and then fluctuated in depth as 80 mm of rain fell irregularly. Rainfall versus stream depth measurements show a lag in peak timing of some 6 hours with *rapid* stream rise within 1-2 hours, which is essentially the front of a runoff 'wave'. This is the main reason why the caves in Green Cwm should not be entered if heavy rain is forecast (Chapter 4).

Both catchments are vegetated, with some 25 percent cover of mainly deciduous trees, and with scrub and various grades of 'grassland' ranging from pastures to common-land bog (Fig. 109). Loss back to the atmosphere of rainwater that has fallen is known as evapotranspiration, which comprises evaporation from soil and vegetation surfaces, with loss from plants of water taken up by their roots (transpiration), all influenced according to temperature, wind and air humidity. Such loss is difficult to determine accurately and will be immensely variable in our climate, both through time and in different settings.

In water management generally in non-karst areas, as for example in upland Wales, the 'rule of thirds' is a widely applied rough approximation, in which only one third of rainfall enters aquifers or reservoirs while soils, plants and weather account for a loss to atmosphere of two thirds. However, in marked contrast, the Wellhead catchments are characterised by mainly rapid transfer of water to underground, and, with most water arriving as large amounts falling in short periods, as in storms, evaporative loss is substantially minimised. We must expect more rainfall than the water emergent from the spring, with the difference amounting to the actual evaporative loss. This, of course, only applies if we can exclude an exotic water supply.

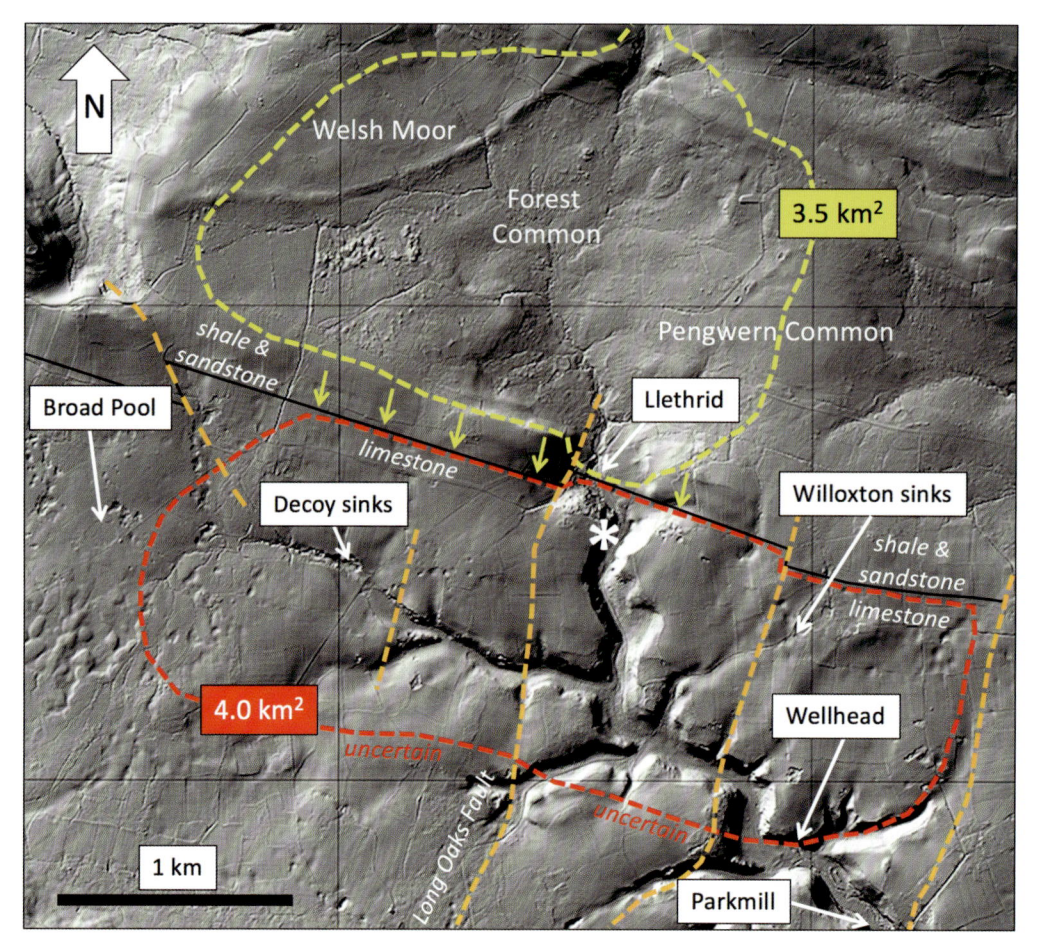

Figure 108. Green Cwm System showing catchment areas feeding water to Wellhead resurgence (see also Fig. 109). The limits are best estimates but 'conservative' where drainage is quite uncertain; the areas are necessarily approximate, but rainfall collected from at least 7-8 km² is probable.

In red *is the 'limestone percolation and conduit-flow' catchment. Surface runoff is limited except to the west of Decoy sinks. Although Broad Pool (pond) floor is sealed owing to a clay content in the till there (Section 5.8), there are numerous dry collapse pits (dolines) and sinks in the vicinity. The limestones beneath this catchment dip just east of north, typically at 15-50 degrees, while the water table slopes south from Llethrid to Wellhead. West and southwest of Wellhead drainage is quite uncertain; except during drought, water rises at Kitchen Well [15] and much must normally flow hidden into gravels in Parkmill valley. Shallow excavations at Parc le Breos Neolithic burial site (Tumulus in Fig. 106) and 200 m south of there have revealed minor flows at the water table. West of Broad Pool water drains north to Llanrhidian [3]. Southeast of Wellhead, in Parkmill, there is a seasonal copious discharge that, together with water from two boreholes there [17], appears to be from a separate system characterized by hardness far greater than Wellhead values (Section 5.18).*

In yellow *is the 'drift percolation and runoff catchment' that drains to Llethrid sink. It is characterized by thick peat and till with localized gravels on relatively impermeable Marros mudstone and sandstone bedrock. Yellow arrows suggest runoff into the limestone catchment. The Pengwern, Forest and Welsh Moor Commons are notoriously boggy and wet, allowing fairly rapid runoff when saturated. Even in drought (e.g., summer of 2018), drift percolation sustains water flows towards Llethrid largely hidden in stream-bed gravels between slow-flow pools. Both catchments are variably vegetated and water loss by evapotranspiration will be significant, although not substantial during heavy rainfall (explained below). The* **white asterisk** *marks the approximate aerial location of the drone that took the pictures in Figure 109. (Image base data contains Natural Resources Wales information © Natural Resources Wales and Database Right).*

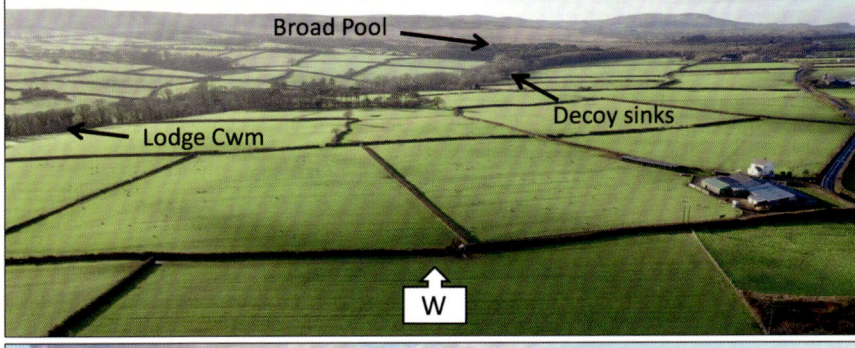

Figure 109. Catchment areas feeding water to Wellhead resurgence (Fig. 108).

*(**Top**) View north over the drift percolation and runoff catchment that drains to Llethrid sink (area within yellow boundary in Fig. 108). The northern limit of the permeable limestones is hidden beneath the foreground trees on lower left and the entire view otherwise is of practically impermeable mudrocks and sandstones overlain by variable thicknesses of glacial till. The commons are mostly heavily vegetated, including tree plantations where excavated ditches enhance rainwater runoff towards the main basin drainage.*

*(**Middle**) View west across till-covered limestones cut by the deeply incised Lodge Cwm dry valley where the rocks are moderately well exposed. The LiDAR image (Fig. 108) shows that much of this cultivated area is perforated by collapse pits (dolines) and sink holes; surface drainage is limited except between Decoy sinks and Broad Pool, although storm-related flood waters occasionally penetrate the otherwise dry valley.*

*(**Bottom**) View broadly east, with features closely similar to the view west. Surface drainage is limited by sink holes, e.g., mid-left and near Willoxton. Tooth Cave and its unknown continuation passages to Wellhead resurgence underlie the fields on the right. (Drone photographs courtesy of Andy Freem).*

5.7 The case for an exotic water supply

We understand that the historical persistence of water flow from both Wellhead [14] and Holy Well [6] underpinned the general belief in a continuous supply from the north-crop limestones (Fig. 101). This view has seemed supported by several observations made before and during exploitation of the Wellhead resurgence. These observations are listed (Table 2) against brief explanations of why, ultimately, they did not require any exotic (north-crop) source. Fuller explanations follow, in several sections.

Original observation	Explanation
Dye-tracer put into Llethrid stream at its sink *during drought* failed to show at Wellhead resurgence, which suggested another input. (this author's italics)	The system has a deep, slow-flow phreatic regime of completely flooded passages. During drought this regime takes the water, slowing and dispersing tracers. More recent sensitive tracing proves the slow-flow connection.
Microscopic examination of sediment discharged at Wellhead resurgence revealed that it must come from a source outside Gower[10].	Glaciers delivered the sediment to the system catchment from outside Gower. Clay-mineral analyses show that what comes out at Wellhead is only what went in at the sinks and also exists trapped in the intervening caves.
Increased discharge without simultaneous local rainfall. This was linked to rain (and snow) falling in the hills to the north of Gower.	Increased discharge during dry conditions that follow a rain-related flow peak is typical of slower percolation drainage following relatively fast conduit throughput (Fig. 105).
Similarity of water chemistry between Wellhead and the north-crop aquifer tapped by Llyndewi borehole suggests the same 'resident groundwater' (Halfacree and Williams 1988).	Chemistry of waters that reside in similar limestones can be expected to be similar in different systems. Chemical, temperature and turbidity analyses prove *local* hybrid conduit and percolation flow behaviour.
During dry weather the hardness is typical of water that has percolated through limestone and from a deep source (mean hardness = 189 mg/L, i.e., 189 ppm) (Halfacree and Williams 1988).	The hardness values, similar to others in Gower springs, are typical of any waters that have had significant percolation and/or residence in limestone. Temperature and chemical analyses provide no indication of any exotic input.

Table 2. Observations and interpretations bearing on the source of Wellhead spring water. Hardness is caused by presence of dissolved inorganic compounds (solute), typically expressed as calcium carbonate but with a small proportion of magnesium carbonate and trace amounts of various other metal compounds. Generally, waters with 0-60 mg/L (milligrams per litre) as calcium carbonate are considered soft, 61-120 mg/L as moderately hard, 121-180 mg/L as hard, and >180 mg/L as very hard.

Further to this tabulation, there are several additional issues that render a northern source with direct hydraulic connection to Gower extremely improbable. Any continuity of 'resident groundwater' in the limestones at Wellhead with those near Llygad Llwchwr – Carreg Cennen (i.e., hydrostatic aquifer connectivity) would entail >30 km in distance and trace beneath >3 km thickness of mudstones, sandstones and coal measures (Fig. 101). Such a connection where hydraulic-head pressure acted over that distance would have to cross numerous structural weaknesses, many faults, each of which could constitute a disruptive permeable pathway, or potential 'leak'. The direct line across the surface crosses approximately 25 steep faults that are substantial enough to have been recorded on geological maps at 1:50,000 scale and many of them would penetrate to >3 km deep. At depth, because of the compressive load (pressure) due to the overlying rocks, the secondary permeability of limestones (fissures due to jointing and bedding) compared with the near surface is reduced so that water flow would occur most readily along steep faults. The notion of a direct hydraulic connection across many such structural barriers to groundwater flow cannot be simply justified.

While the dip of the strata might favour phreatic penetration of water to as much as 400-

500 m below the water table (Worthington 2001), it is not at all clear how or when the supposed connectivity via >3 km depth could ever have been established. Furthermore, passage via >3 km depth would mean that the water would be heated, probably to more than 60°C given that temperature at depth increases here by up to about 23°C per kilometre (Busby et al. 2011; Farr et al. 2020). On Gower there is no thermal signature of such a deep passage.

Taffs Well, the only thermal spring in Wales, discharges at around 21-22°C where the water has taken at least 5000 years to travel some 25 km to a depth of 400 m or more (Farr and Bottrell 2013). The increase of rock temperature with depth means that deep-heated waters that return to the surface via long-lived pathways are not cooled substantially (Gallois 2006), although, of course, admixture of local 'cold' water might result in significant cooling. This effect would cause the spring water temperature to fluctuate according to the mixing with cold water, but it is demonstrated (below) that the outflow temperature of Wellhead water responds sensitively to local climate while there is no sign of any abnormal thermal input. The *minimum* discharge at Wellhead is x29 more than Taffs Well average discharge.

Simply, a north-crop origin for Gower water is physically extremely improbable and direct evidence for it is absent. It therefore only remains to show that the local input into the Green Cwm System directly relates to the output at Wellhead resurgence, i.e., that there is no exotic material addition or dilution, and that the rainfall onto the local catchments is sufficient.

5.8 Input and output: fingerprints in clay

Wellhead spring was notoriously difficult to manage for water supply when there had been storms, because of considerable turbidity (see below). Heavy local rainfall would be followed by an increase in clear water discharge that gave the treatment plant operators, usually Peter Sambrook, warning to get ready with the coagulant alum: aluminium sulphate $Al_2(SO_4)_3$. This chemical was added to the turbid water in order to flocculate (clump-together) and thus settle-out the fine suspended particles before they entered the public water supply. In the light of the original microscope observation that there was a non-Gower sedimentary component in the water (Table 2), a simple mineral-fingerprint test was devised.

Sediment particles smaller than 4 μm in diameter (<0.004 mm) are referred to as 'clay-size'; rather confusingly, such fine material commonly includes clay *minerals*, of which there are many types (here illite, chlorite, kaolinite and smectite). Such small and plate-like particles are readily suspended in flowing water and only settle out slowly to form a sediment. In this test it is this mobility of the clay-sized particles that is exploited. Such fine particles are not separated from one another (not segregated) by transport in suspension, so the constituent-mineral proportions cannot be altered. Thus, the input and output materials will share the same compositional 'fingerprint' if nothing exotic is added. Storm water entering the system with suspended clay-sized material will contain the same suspension at the resurgence. Sediments were sampled from the regional Devensian till, from the inputs at Llethrid and Decoy sinks, from within Tooth Cave where fine sediments had accumulated, and from the Wellhead resurgence pond. The clay-sized fraction was separated from these samples and analysed for the relative abundances of the minerals present (Fig. 110).

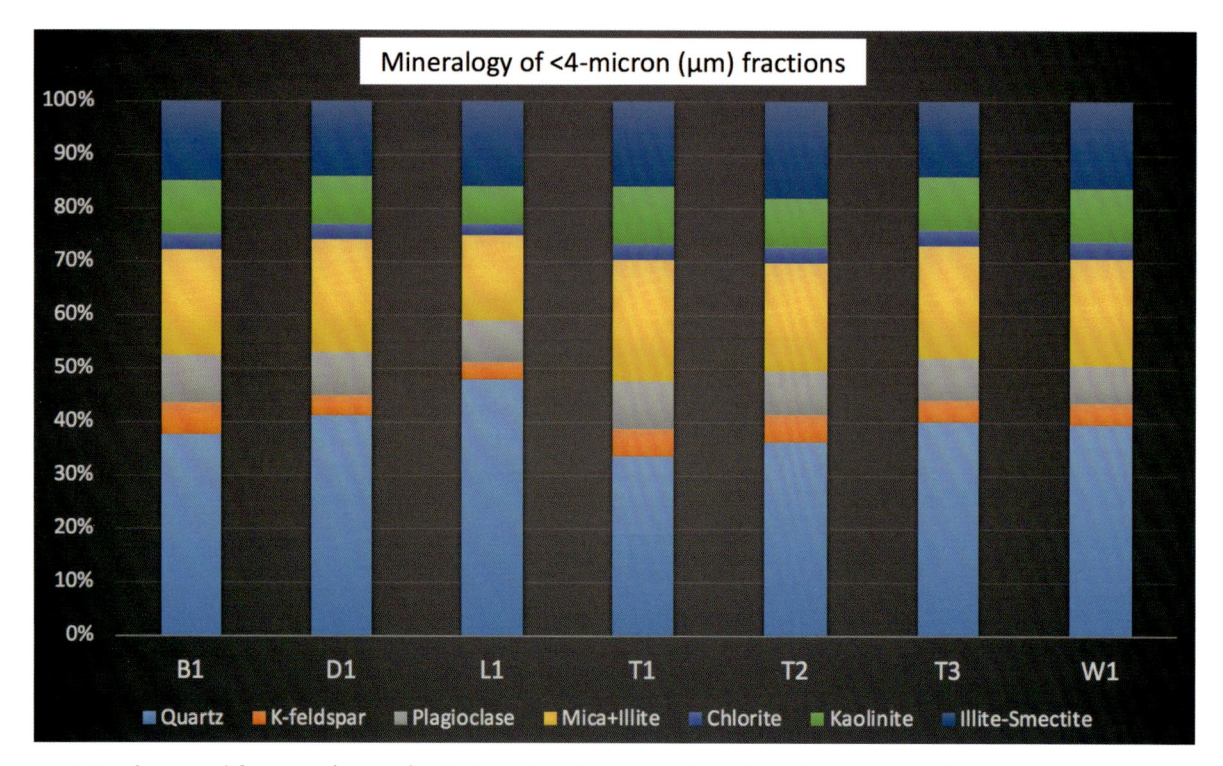

*Figure 110. Clay-sized fraction (<4 μm) relative abundances of the minerals contained in sediment samples from the Green Cwm System, identified by X-Ray Diffraction[11]. The clay fractions constituted 6-9 percent by weight of the samples. Sample **B1** is from glacial till in the catchment of Decoy Sink and represents typical source material. **D1** is from the stream bank at Decoy Sink. **L1** is from Llethrid stream bypass pond, representing main flood input there. **T1-3** are samples from sediment banks within vadose passages in Tooth Cave known to be eroded during storm-related flooding and connected hydrologically to both Llethrid sink and Wellhead resurgence. **W1** is from the Wellhead resurgence pond representing turbid-flood-water discharge.*

The mineral-component proportions of the clay-sized fractions of all of the samples are virtually identical, except for some small variations in the relatively abundant quartz constituent. Thus, the Wellhead discharge reflects only local inputs and has no signature requiring or even suggesting any exotic additional supply. However, the sediment arriving at Wellhead does reflect an origin in the north crop of the South Wales Coalfield, because its source is in glacial deposits transported by ice from there to the catchments on Gower (Chapters 3 and 4). The most obvious component possibly reflecting this source is the quartz, which is relatively scarce in local source rocks.

5.9 Enough water?

Establishing that rainfall penetrating the local catchments is enough to sustain the Wellhead spring discharge is not quite straightforward. We need to consider the long term and should remember that the catchments extents are uncertain, hopefully conservative (Fig. 108), and that the Penmaen record may not capture rain amounts falling farther north (Fig. 104). However, the result is affirmative.

If one assumes that, through time amounting to many years, the volume of water that is stored in the ground above the water table remains constant, then in the same long term the average spring discharge will equate to rainfall amount minus the water lost by evapotranspiration. We can assume that the evaporative loss will be greatest with 'fair-weather' rainfall, but, as shown below, when rainfall exceeds roughly 15-20 mm/day the runoff into the sinks is rapid and the loss is thus drastically reduced. This will apply to the bulk of annual rain, because most falls during poor weather with low potential for evaporative loss. Of course, within any year there will be fluctuations in stored groundwater, which may be drawn down during droughts, e.g., when drift percolation sustains water flow to Llethrid, and is replenished afterwards. Thus, spring discharge in the short term will not correlate directly to infiltrated rainfall and it is only in the long-term that there is balance between infiltrated rain and spring discharge. And, even then, we can only deal in 'ball-park' figures.

Given no exotic addition of water, the Green Cwm System water budget, including average evaporative loss, can be roughly assessed by comparing the long-term daily average rainfall over the catchments with the spring-discharge amounts. Average annual rainfall measured at Penmaen for 1971-2000 is 1224 mm (range 1545-874 mm; Mayes and Powell 2003). The Green Cwm System catchment area is at least 7.5 km^2 (Figs 108 and 109), so the long-term daily average rainfall input amount would be 7.5x10^6 x 1.224/365 = 25.15x10^3 m^3/d, i.e., just over 25,000 cubic metres per day.

Wellhead spring discharges between 2.5 and 216x10^3 m^3/day. Table 3 shows the average daily rainfall and discharge at Wellhead for 1987-1992[12]. The average daily catchment rainfall then was 25.02x10^3 m^3/day, close to the 1971-2000 average. The average daily discharge in the same interval was >19.45x10^3 m^3/day, which is >78% of rainfall, so the loss must on average be <22% of rainfall, although the annual loss varies widely (Table 3). We cannot be certain how much of that loss was actually water lost over the weir, as opposed to apparent evaporative loss, because there is a maximum recordable amount at the weir, and this is exceeded for a few days several times every year.

Year	Penmaen total rainfall mm	Catchment rain daily average	Days with rain	Rainy day average	Wellhead discharge daily average	Discharge % of rain fallen on catchment
1992	1248	25.57x10^3 m^3/day	299	31.21x10^3 m^3/day	>21.6x10^3 m^3/day	>84%
1991	1143	23.49x10^3 m^3/day	280	30.62x10^3 m^3/day	>21.1x10^3 m^3/day	>90%
1990	1142	23.47x10^3 m^3/day	310	27.63x10^3 m^3/day	>14.4x10^3 m^3/day	>61%
1989	1199	24.67x10^3 m^3/day	296	30.42x10^3 m^3/day	>15.5x10^3 m^3/day	>66%
1988	1320	27.05x10^3 m^3/day	306	32.30x10^3 m^3/day	>23.7x10^3 m^3/day	>88%
1987	1260	25.89x10^3 m^3/day	301	31.39x10^3 m^3/day	>20.4x10^3 m^3/day	>79%
1971-2000	1224	25.15x10^3 m^3/day			>19.45x10^3 m^3/day	>78%

Table 3. Rainfall and Wellhead spring discharge during 1987-1992. Simplistically, the disparity between the daily discharge and daily rain (right-hand column) registers the evaporative loss to the system with some (occasional) unmeasured weir overflow.

Uncertainties concerning areas of catchment, catchment rainfall, and unrecorded weir overflow, tell us not to take too much from the data. Evaporative loss must on average be significantly less than 20% of the rainfall (Table 3), which certainly is not beyond expectation here on Gower. Evaporative loss will be least when the rainfall amount is greatest, in heavy showers involving rapid runoff into sinks and in colder months when atmospheric humidity is high and solar energy relatively low. Most importantly, the rainwater so readily goes underground to transfer through the limestones that evaporation must be significantly limited relative to non-karst areas. Consistent with the finding above that the discharge reflects only local inputs, this analysis (Table 3 right-hand column[13]) indicates that there is quite enough rainfall to sustain Wellhead spring without invoking any additional exotic supply.

5.10 Storms, turbidity and Wellhead spring discharge behaviour

When Wellhead water-treatment facility was used to supply water on Gower (1954-1994), rainstorms meant trouble and perhaps overtime work there. Throughout all droughts and during periodic dry and mildly rainy conditions the resurgence water flows clear (Fig. 106). However, intense rainfall amounting to more than about 15-20 mm within 24-hours or less leads to discharge of turbid water; the more intense the rainfall, the more turbid the discharge. This was problematic for the public water supply and needed urgent treatment as it occurred.

Following heavy rainfall, some turbidity originates by erosion of glacial till along ditches and stream banks, before entering the main sink at Llethrid and additional sinks in the Decoy and Willoxton catchments. However, as indicated in the stream passages of Llethrid and Tooth caves (Chapter 4), and consistent with the 'fingerprints in clay' (Section 5.8), much turbidity is derived by erosion of glacial debris along underground passages[14]. During exceptional floods, turbid surface water bypasses the main sinks and reaches down the dry valleys (Figs 111 and 112) to disappear in various locations, thus adding to the input locations for the sediment-laden flows. Moderate surface floods confluent near Green Cwm Cottage tend to disappear in a shallow pond and 'cryptic' sink there (Fig. 106), but not infrequently, e.g., recently with Storm Dennis on 15th-16th February 2020 and with torrential rain on 13th December 2020, the overflow has continued past the Parc le Breos Neolithic burial site (Tumulus in Fig. 106), partly drained into the nearby pit, and passed on through the campsite with water 30 cm deep cutting off the Wellhead water treatment buildings. There are instances of canoeists enjoying the temporary lake.

The discharge at Wellhead resurgence is the eventual outcome of variably independent behaviours in so many system parts we know of, as well as in unknown parts, that the quest to understand it fully is hopeless. Here, before describing the resurgence discharge, we review some historical phenomena to gain insight into the flooding behaviour.

On Thursday 10th November 2005 Matt Carroll recorded a flood into Green Cwm 'dry' valley from Llethrid (Fig. 111 Top). Newspapers reported exceptional downpours north of Gower (up to 100 mm) on the previous Tuesday with rivers generally in southwest Wales at their highest levels since the 1960s. Penmaen weather station had received (only) 40 mm of rain on the previous Saturday. At about 15.00 on the Thursday flood water that had been briefly held back by dams of debris and vegetation in the woods just downstream of Llethrid sink burst into the meadow by Tooth Cave entrance with a broad wave front estimated at 225 mm (9") high. This swamped timber operations and swept large logs down the valley. The slightly turbid water reached about 1 m deep on the valley floor and it subsided fairly rapidly after 45 minutes.

*Figure 111. (**Top**) Green Cwm flood 250 m down-valley from Llethrid sink. At its peak the water level reached the axle of the tractor and was about 1 m deep off the (submerged) track; it subsided within 45 minutes. The rainfall record shows the totals for the 24-hour periods up to 09.00 on the day of the flood. Exceptional rain had been recorded north of here two days earlier (only 27.5 mm at Penmaen), but little fell immediately before this occurrence. (Photograph courtesy of Matt Carroll; only the dog really enjoyed it).*

*(**Bottom**) Green Cwm viewed down-valley 1.2 km from Llethrid sink, from the vicinity of Green Cwm Cottage, on December 13th, 2020. The flood had doubled to this width in less than 30 minutes and then halved this width over the next 4 hours. It had rained heavily during the previous week, and torrentially before the flood peak. (Photo courtesy of Julia and Steve Robson, at that time cut off in their cottage).*

On Sunday 13th December 2020, flood waters penetrated the entire dry valley to Wellhead. Photographs and videos taken by Julia and Steve Robson of Green Cwm Cottage showed that the flood there, more than 1 km down the valley (Fig. 111 Bottom), dramatically increased to a peak within 30 minutes and then took some 4 hours to reduce to a vigorous surface stream. The flooding in Parkmill was seriously deep.

Worse than this, continuous intense rainfall (63 mm at Penmaen) during passage of the remainder of Hurricane Charley, on Bank Holiday Monday 25th August 1986, caused surface bypass flooding in both Green Cwm and Ilston Cwm so as to swamp Parkmill village[15]; intensely turbid water was drawn into the treatment plant during the evening and the spring water still had not cleared fully by the Thursday.

Three main points arise from these events. Firstly, the surface bypass floods follow protracted considerable rainfall so the catchments must have been saturated and the subterranean conduits practically full. Secondly, the surface floods have the general form of an asymmetrical wave (sharply waxing, slowly waning) with a front rapidly deepening on a scale of minutes and a short-lived peak with following diminshment on a scale of a few hours. Thirdly, the rainfall amount is highly variable on a local scale, so whereas the Penmaen record is a valuable indicator (and extensively used; see below) it is not necessarily directly representative of the catchments discussed.

Underground, in Llethrid cave on 18th October 1957, Paul Hartwright and Inett Homes had entered in wet conditions at 10.00 and were in Great Hall by 11.00. At about 12.30 they realised that the water was rising rapidly and that they were trapped. The level peaked at around 15.00 and then fell. They started out at 18.30 under bad flood conditions, although the water clearly was subsiding, and they emerged through a waterfall at the entrance at 20.00. Evidently the water flow rapidly increased for three hours and thereafter subsided over more than five hours; here again is the asymmetric surge-wave off the Pengwern catchment.

The turbidity of the water issuing from Wellhead resurgence evidently in part depends on surface-flow erosion; substantial rainfall runoff causes mildly turbid water to enter the sinks. However, considerably increased turbidity at Wellhead arises without any surface bypass flooding, showing that sediment must be added to the flows underground (Fig. 112). Subterranean erosion of glacial debris is evident in the accessible vadose passages in Llethrid and Tooth caves and it probably also occurs in many other of the pre-glacial passages of the Green Cwm System. In the accessible passages of both caves, eroded banks of fine-sediment-bearing glacial debris occur virtually everywhere (Chapter 4); these are the underground flood-bypass routes that lie above the deeper slow-flow phreatic regime. It has to be concluded that much of the turbidity in the Wellhead discharge reflects an underground plumbing system that had been substantially modified by debris infill during the last (Devensian) glaciation.

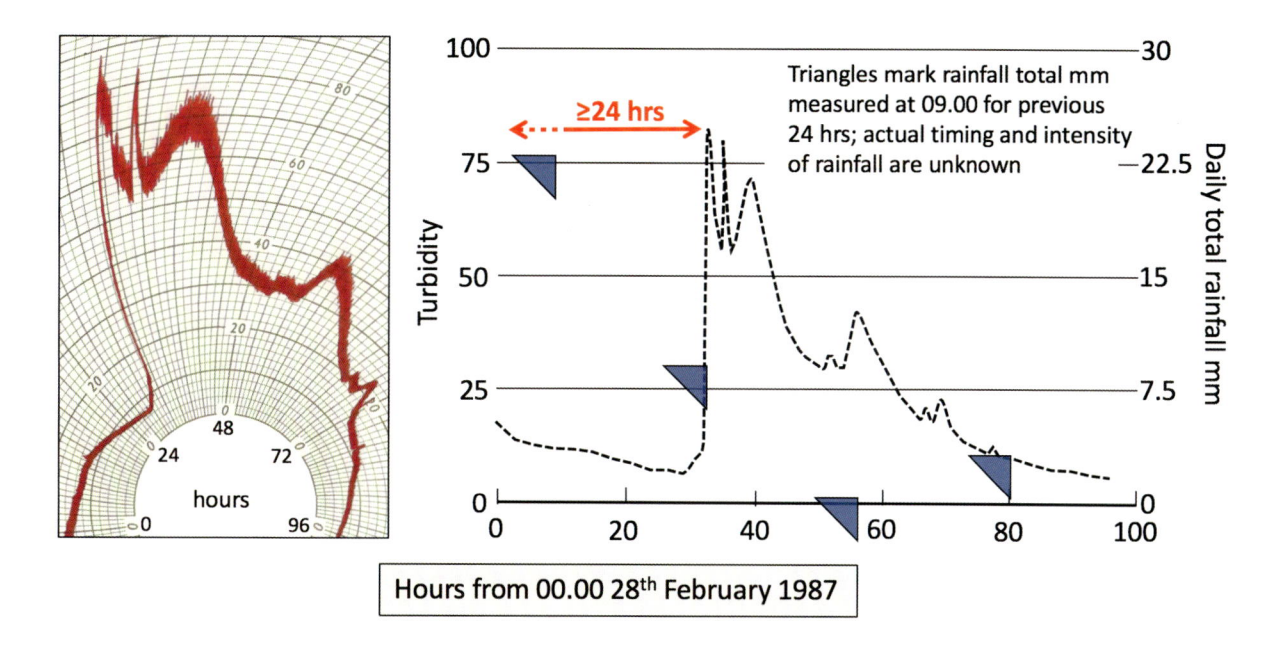

*Figure 112. (Page opposite). Storm-related turbidity in the Green Cwm System. (**Left**) Llethrid stream in flood has built up and bypassed the main sink and here slightly turbid peaty-brown water is entirely swallowed by the new entrance[16] we had excavated though a fault-zone boulder choke (Chapters 3 and 4). Rainfall had been heavy for several hours before this and had just ceased. The stream-water level here rose >100 mm in the 75 minutes of our visit. (**Right**) Typical heavily turbid discharge at Wellhead resurgence showing a low fountain (back of pond) over the constricted exit. Compare with Figure 106.*

*(**Above**) Part of a 7-day circular chart recording of turbidity in 'raw' water extracted from the resurgence pond for treatment at Wellhead; the measurement is explained below. This chart record has been read-off into a time versus turbidity spreadsheet and plotted along with rainfall data (bottom right). The turbidity record is qualitative-relative; there are no absolute values. Rainfall is only recorded (at Penmaen) as totals for 24-hour periods, which masks possible intense downpours. Intense rain within the 24-hour period and daily amounts in excess of 25-30 mm/day tend to cause turbidity that peaks off the recording chart. Clear water spring discharge typically reads around 6 on the turbidity scale. It is the form and timing of the turbidity relative variations that are most informative of the hydrology during heavy rainfall and floods.*

5.11 Enigma variations

Edward Elgar composed fourteen musical variations on the original theme of 'Enigma', commenting that "its dark saying must be left unguessed...". For 'Enigma' here the reader may guess: mysterious or really difficult to interpret turbidity at Wellhead, with lots of variations! There are recurrent features, which we will highlight and try to interpret, but in truth we will never understand all of the various patterns; there are too many variables possibly involved, with some we cannot know or even sensibly imagine. So, what follows is mostly descriptive, perhaps fascinating, and certainly poorly understood. We will first develop a simple model to explain some key features, but then find that it fails eventually when the rainfall pattern gets complicated.

The turbidity records were made at Wellhead by 'nephelometry', which measures the light that is scattered by particles suspended in water when illuminated by a focused light beam; more turbidity causes more light scattering. The records were used to gauge the amount of the coagulant aluminium sulphate $Al_2(SO_4)_3$ that should be added to the 'raw' water to get the suspended particles to clump together so they would settle and be more readily filtered out. A similar measure was applied to the filtered water to prove its acceptable clarity for general supply.

The reader is reminded that the graphs of turbidity essentially show *qualitative* variations through time. Absolute measures of the suspended load, although technically feasible to derive via calibrations using standard suspensions, were not used. Indeed, before the nephelometer was deployed, the Wellhead engineers applied alum according to the visual appearance of the turbidity – the 'eyeball method'. This was something of an art that demanded experience to get it right and done quickly. Big floods would cause the turbidity graphing to go off-scale and occasionally the recording in severe storms would fail owing to associated power cuts.

Below is a representative selection of various types of legible, mostly on-scale, records available for the interval 1985-1994[17]. Here it is always taken that the maximum water input to the Green Cwm System and hence to Wellhead is from the extensive catchment to the north of and leading to Llethrid, and then via Llethrid and Tooth Cave main stream passages (Figs 106 and 108). Decoy sinks have a lesser catchment area but probably a non-trivial input and the Willoxton input is taken as usually the least. None of the patterns of turbidity shown below were formed in association with major floods into the dry valleys (those inevitably went drastically off-scale), so the turbidity has not been complicated by odd input locations beyond the main sinks. The patterns should represent the relatively 'normal' wet-weather hydrology.

Figure 113 captures a common and relatively simple pattern in the turbidity variation at Wellhead, all features of which, however, are somewhat problematical. As in this record, the majority of turbidity events show an extremely rapid onset leading to the maximal increase in turbidity. Clear water discharge always increases ahead of a 'turbidity event' and then the turbidity dramatically increases within minutes leading to a maximum typically within several tens of minutes. Then the turbidity drops off rapidly while high-flow discharge is sustained (Fig. 113 lower panel). We must assume that the increased discharge of clear water occurring before the arrival of turbidity involves 'old' water that is pushed out in response to raised hydraulic pressure communicated down-system ahead of the new flood flow.

The dramatic onset of turbidity seems best explained by flood water entering rapid-bypass vadose passages and flushing loose and dried material in an energetic wave- or surge-front. The Main Stream Passage in Tooth Cave represents such a passage, where steep sediment banks are eroded and tend to collapse, and where organic debris from the surface is lodged in its roof. Some stalactites and protruding limestone edges show corrasion (mechanical erosion) by sediment in suspension (Fig. 88), and fine sand and silt remains patchily on surfaces there, deposited from the waning tail of the last flood (Figs 85, 86 and 87). The drop off of turbidity is taken to represent the general transition to flood water behind the wave- or surge-front, so that the various shoulder forms and flanking peaks somehow register confluent turbid surge-front inputs from Decoy and Willoxton. We know that there is a slow-flow phreatic regime in the Green Cwm System and this is likely to be contaminated by the turbid flood water, so that some of the long recovery to clear water at Wellhead may reflect the outflow from there.

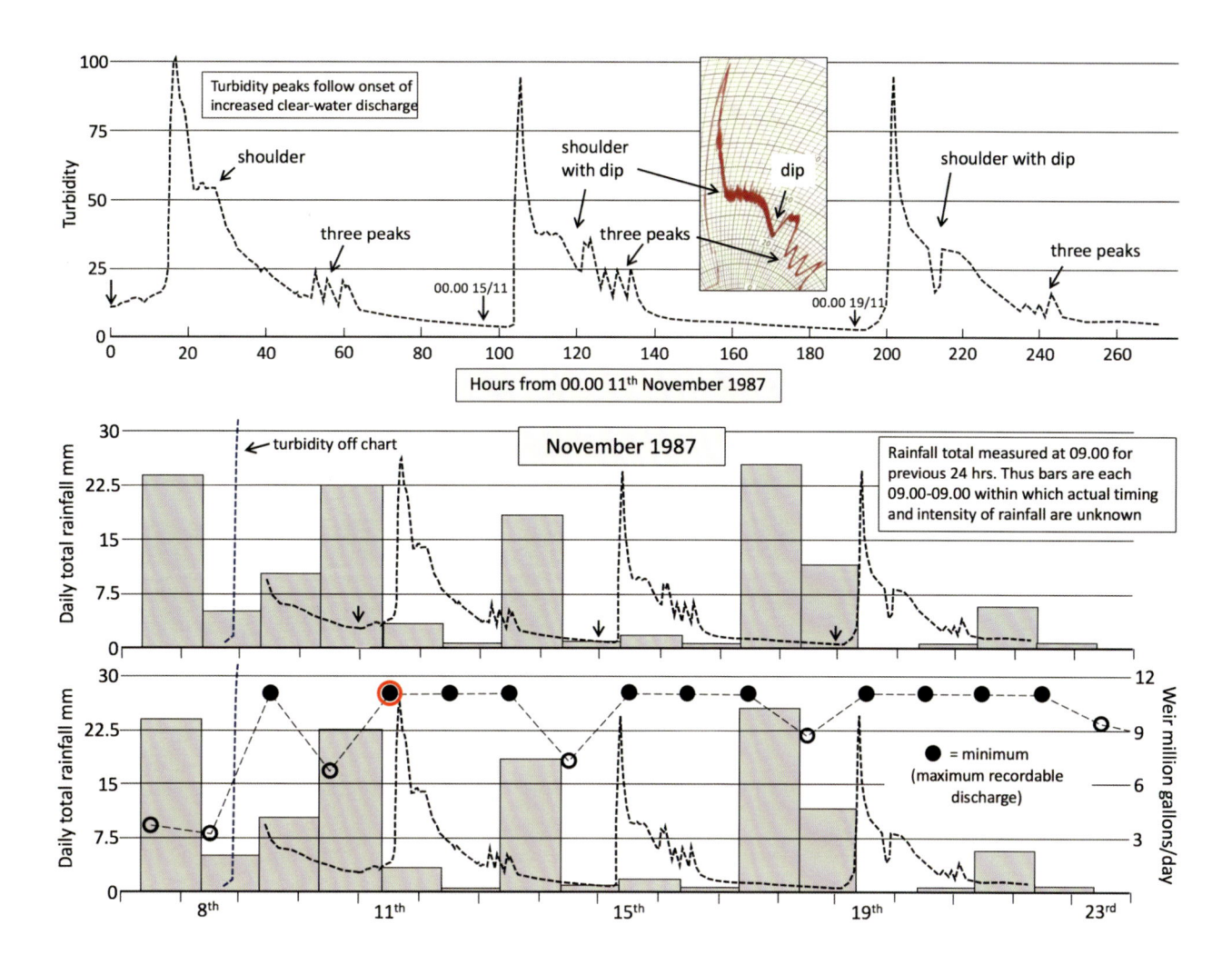

*Figure 113. (**Top**) Remarkable pattern of three turbidity events showing recurrences of key features during a 10-day rainy period in 1987. Each onset to maximum turbidity is extremely rapid with the initial increase occurring within minutes. The drop off from maximum is also abrupt with a rapid decline to a 'shoulder'. The decline through the shoulder is variably irregular, latterly with a 'dip' registering a few hours of relatively rapid decline that is quickly reversed. Such a dip conceivably could represent a clear water 'injection' from the front of a separate confluent turbid input. Most bizarre is the recurrence of the 'three peaks' that each show turbidity increases during about one hour quickly followed by a decline of about two hours duration. Slow recovery to base-level readings of around 6 on the turbidity scale takes about 86 hours from the first peak.*

*(**Middle**) This shows the three events described above with an additional off-chart event beforehand. Also plotted are 24-hour rainfall amounts, recorded at Penmaen, crudely showing the characteristic lag of around 24 hours between peak rainfall and peak turbidity at Wellhead.*

*(**Bottom**) As middle panel but with daily readings of weir discharge[18] added. This reading, lacking actual maxima (unrecorded), nevertheless shows that resurgence-water discharge increases following the first rainfall and remains elevated thereafter, and, importantly, that the discharge remains high while turbidity diminishes. Discharge only diminishes when turbidity returns to near background. As was well known by Wellhead engineers, clear water discharge increases just before arrival of first turbidity, as fortuitously captured on 11th November (**plot circled in red**).*

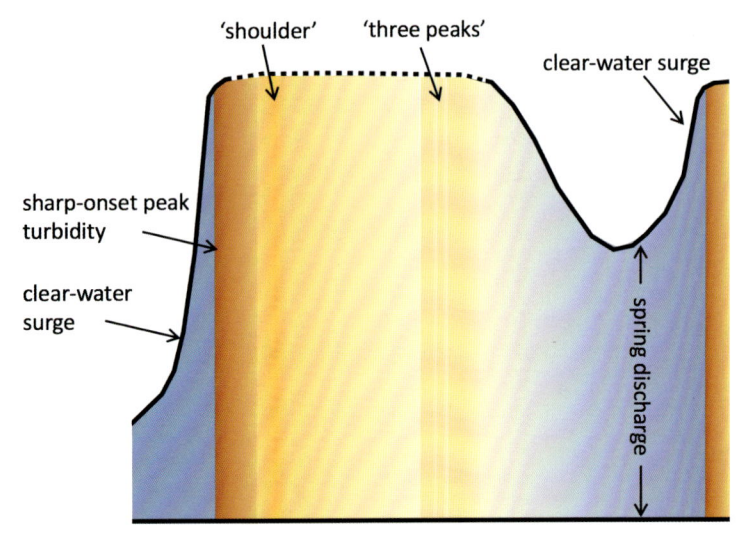

Figure 114. Schematic diagram showing common simple turbidity variation versus resurgence (spring) discharge. One full event and part of the next are illustrated; brown-to-yellow represents turbidity variations similar to 11th-15th November 1987 (Fig. 113).

The key point is that although the actual maximum discharge is never recorded above 10.5 million gallons/day – hence the dotted line – the turbidity events are arrivals all within the duration of high discharge. The sharply peaking turbidity arrivals are most simply reconciled with confluent-input arrivals suggesting original wave- or surge-driven erosion in conduits from the respective catchments. Thus, the first arrival would be from Llethrid, the 'shoulder' from Decoy, perhaps bringing its own clear water front, and the 'three peaks' from Willoxton. This is, however, only a working hypothesis.

The shoulder and off-shoulder patterns are perhaps the most difficult to account for, especially the recurrent 'three peaks' (Figs 114 and 115). These small groups are remarkably similar, most frequently with a large third peak, not always, and with some other more random forms sometimes nearby. They are especially prominent in records from the mid to late 1980s and beg the question of what could cause three slightly asymmetric waves of turbidity to arrive at Wellhead 25 to 40 hours after the main turbidity peak of a single flood event and during elevated discharge (Fig. 113 lower panel). Each of the three turbidity spikes rises more quickly than the fall and lasts for around 2 to 3 hours. With more complicated turbidity patterns (e.g., Fig. 112) the three peaks are less clear, and in some cases they seem completely swamped or clearly are missing.

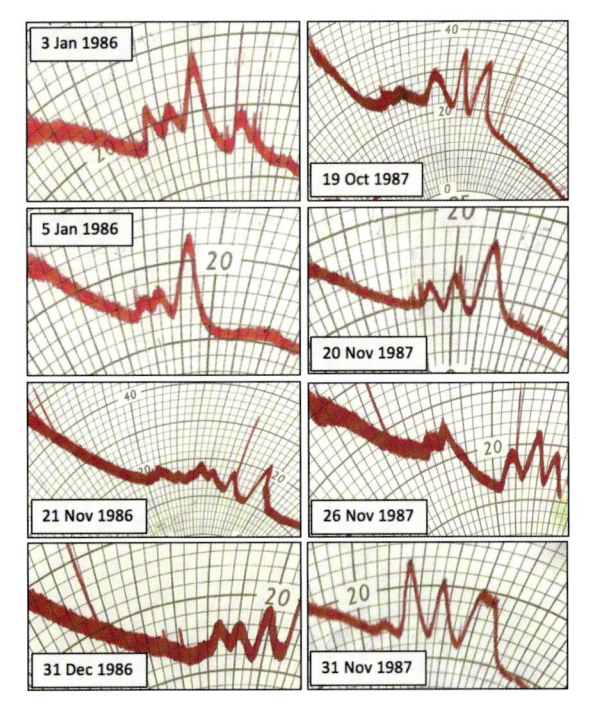

Figure 115. Examples of the mysterious 'three peaks' pattern in turbidity that was common in the mid-late 80s. It occurred during sustained high discharge, on the curve of diminishing turbidity (steep trace to left), and it could be adjacent to apparently random additional peaks to left or right. Sharp-momentary trace excursions were formed by bubbles in the measurement apparatus.

If one attributes the first and largest peak to Llethrid and the variable but recurrent shoulder to Decoy, superimposed upon waning turbidity from Llethrid, then the added three lesser peaks might seem liable to relate to Willoxton input. The three peaks are always evenly spaced in time, which suggests some fixed physical controlling structure, so they might, for example, reflect three input doses from minor branch conduits. Siphoning behaviour is known underground to produce intermittent flows, but there is no obvious way of discriminating what is happening here.

It seems reasonable to speculate that variability in the size, form and position of the peaks in general most likely relates to the rainfall and runoff variability between the catchments. Willoxton has the smallest catchment and it seems fair to anticipate that it will receive less water than the others and hence that its turbidity signal can either be swamped by the others or even never formed and thus missing altogether. This terribly tenuous interpretation may meet its Nemesis in Figures 116 and 117.[19]

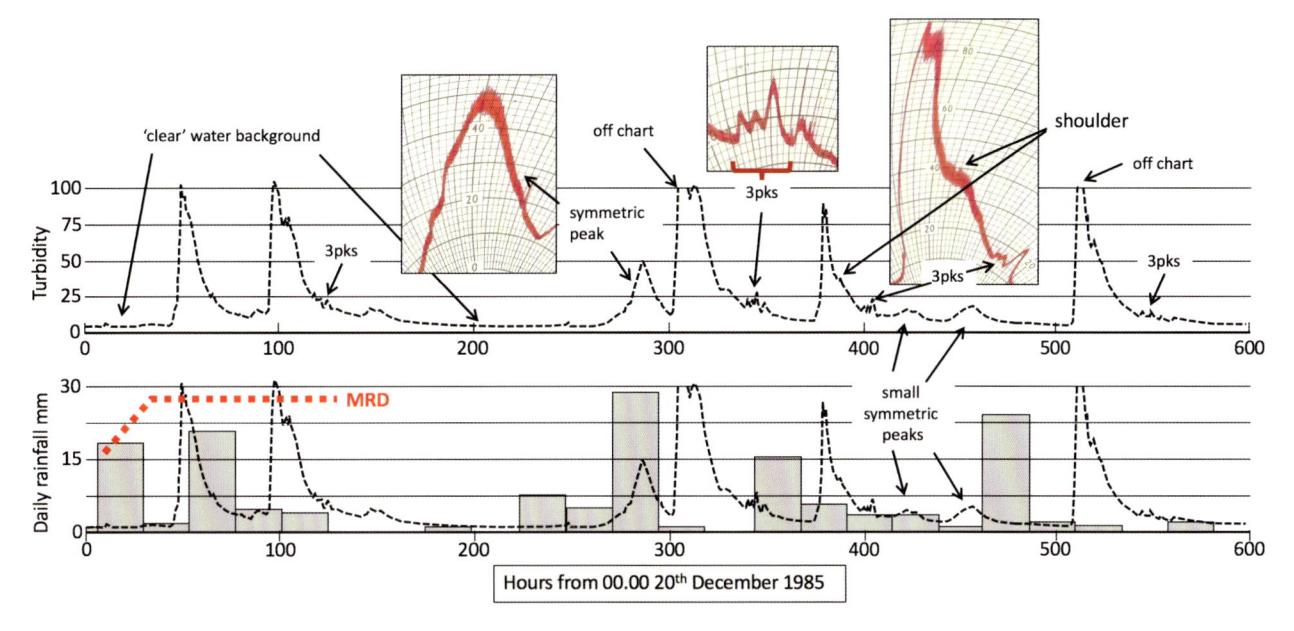

*Figure 116. (**Top**) Enigmatic variable pattern of several turbidity events recorded over 25 days during December 1985 to January 1986. As well as the more usual sharp-onset maximal peaks with some distinct shoulders and some 'three-peak' patterns, there are several symmetrical peaks. The trace of the large symmetrical peak at around 285 hours shows two abrupt increases and a progressive decline. Also shown are irregular maximal peaks. See also Figure 112.*

*(**Bottom**) The 24-hour rainfall record is added, showing the characteristic delay of 20-24 hours between heavy rain and abrupt turbidity onset. Relationship of rainfall to the symmetrical peaks is not clear, although two days of moderate rainfall beforehand may be implicated. Remember that the rainfall record masks any intense downpours. **MRD** is maximum recordable discharge; this attained over 10.5 million gallons/day (48x10³ m³/day) through the first two events, but the record thereafter unfortunately is missing.*

The novel feature of Figure 116 is the occurrence of symmetrical peaks. It is suggested above that the contrasting *sharp-onset* features seem best reconciled with arrivals of waters that picked up their turbidity in some sort of wave- or surge-front that propagated down bypass passages containing erodible sediment. In the symmetrical cases the increase in turbidity is unsteadily progressive for an interval of about 15 to 24 hours, which is difficult to explain. According to the argument that 'steep onset represents wave- or surge-front erosion', one might speculate that the longer-lived increasing turbidity relates to a more progressive increase in underground erosion related to progressively strengthening flow.

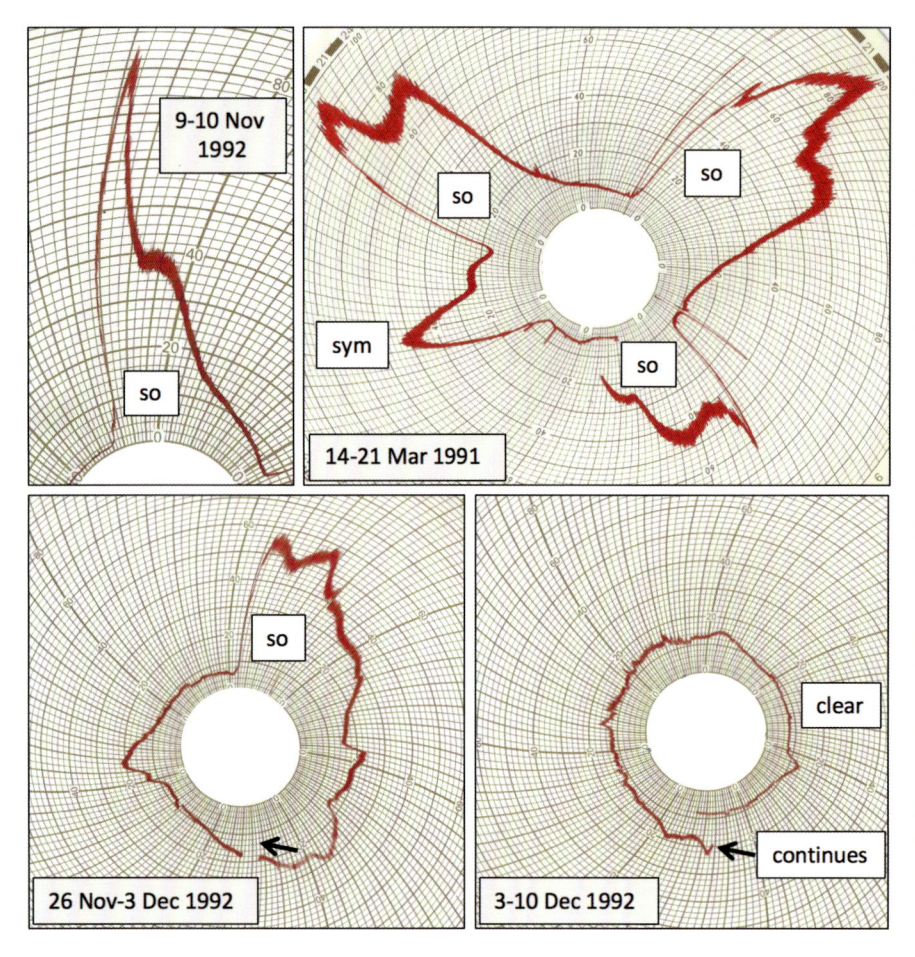

*Figure 117. Enigma variations. (**Top left**) Comparatively simple sharp-onset peak (**so**), shoulder, and progressive smooth decline lacking the 'three peaks'. (**Top right**) One-week record of three sharp-onset peaks and one symmetrical (**sym**) peak. Here it is unclear what constitutes a 'shoulder'. The simplest inference is that the second broad peaks are unusually large 'shoulders', although the third sharp-onset event appears to have a second shoulder where the 'three peaks' commonly occur. The symmetrical event (sym) has a similar shoulder in this position. (**Bottom pair**) Two weeks comprising odd variability leading to a sharp-onset composite event (so) followed by a protracted irregular decline in turbidity to mainly clear water after 11 days.*

Observations of heavy-rainfall-related floods, both above and below ground (Section 5.10), indicate that they typically occur as waves or surges with a steeply increasing discharge front and more progressive decline in the tail. But this is a biased view, because it focuses on the more readily witnessed initial flood behaviour and there is no good record of stream flow variations during more protracted rainfall, possibly relating to several days of rain (e.g., Fig. 116). Stretching the interpretation to its limit then, the non-sharp-onset events and the many variations around the simpler patterns are taken as attributable to persistent but varying conduit flows, some of which may relate to possible variations of rainfall intensity between the catchments. Did it, for example, rain at all on the Willoxton catchment during 9th-10th November 1992 (Fig. 117 top left) and was Willoxton swamped by large events during 14th-21st March 1991 (Fig. 117 top right)?

Overall, this attempted analysis conceivably is itself more turbid than clear; the reader is reminded, from above: "there are too many variables possibly involved, with some we cannot know or even sensibly imagine". So, answers on a postcard please: what on Earth were the 'three peaks' all about, and what happened to them in the 1990s?

5.12 Anomalous spring-water temperatures

Wellhead resurgence was known to have occasional abnormal temperature spikes, the warm ones of which were investigated in this project because of the possibility that they reflected addition of 'exotic' water that had been heated at depth. Two approaches were used. First, a historical record for the interval 1981-1987 was analysed; this utilised the record presented by Halfacree and Williams (1988) and usefully includes the exceptionally hot summer of 1983 and the exceptionally dry time of 1984 (Powell 1984, 1985; Mayes and Powell 2003). Second, spring temperatures with nearby air temperatures were collected at Wellhead [14] and on Cefn Bryn [6], at high resolution of 0.1°C and at 30-minute intervals continuously for one year (Section 5.17). As with other lines of evidence concerning the source of the water, the temperature data show pronounced local influence and there is nothing to indicate any exotic supply from depth.

Wellhead spring-water temperatures normally fluctuate annually between about 8°C and 14°C, with a mean of 11°C (Fig. 118), which is also the mean cave-air temperature and close to the measured groundwater temperature of 11.5°C in Gower at 100 m below the surface (pers. comm. Gareth Farr 2021).

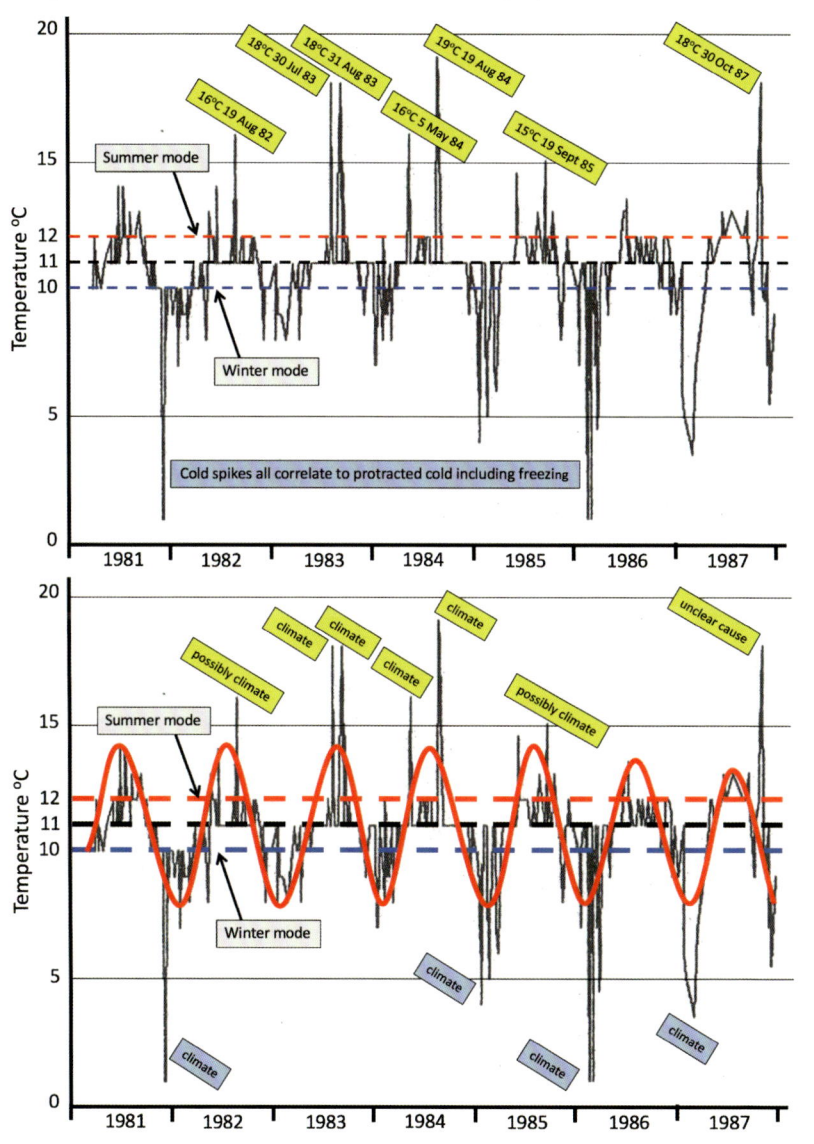

Figure 118. Wellhead spring-water temperatures recorded during 1981-1987, with the graph taken from Halfacree and Williams (1988).

(Top) Seven anomalous high peaks are shown with water temperature and date. Each peak involves a progressive increase and decrease in temperature to and from the maximum, typically of 2-3 days duration; no change is instantaneous, unlike most arrivals of turbidity. The several low-temperature anomalies relate to cold including freezing climate recorded nearby at Penmaen.

(Bottom) The data are interpreted to show the annual variability about the mean of 11 °C and attributions concerning the anomalies. 'Climate' here means prevailing conditions of temperature, rainfall and sunshine recorded at Penmaen and taken to represent the conditions across the catchments that feed water to Wellhead. Figures 119 and 120 show how those 'climate' data can be related to the spring-water temperature.

Temperatures at or above 15°C occurred only seven times during 1981-1987 and not at all during the 12-month monitoring. The four high temperature anomalies of 1983 and 1984 were almost certainly due to exceptional weather affecting the local climate (Figs 119 and 120) and those of 1982 and 1985 conceivably also were. In the latter cases the weather was not especially warm although above average sunshine and no rain preceded the peaks. For the 16°C of May 1984 the weather had been particularly sunny and relatively warm, significantly exceeding normal for late April into May and twice reaching more than 20°C, with notable drought. The 18°C anomaly of 1987 has no obvious cause in the Penmaen data; its stepwise build-up and decline shows it was a real temperature surge. The previous temperatures that year were below the historic mean maximum, but there was a preceding sunny spell with a corresponding lack of rain. Conversely, one of the climatic episodes that looked most likely to form a high-temperature spike, on 13th July 1983 (Fig. 120), had no such registration at Wellhead. Cold-water excursions clearly relate to ambient temperatures.

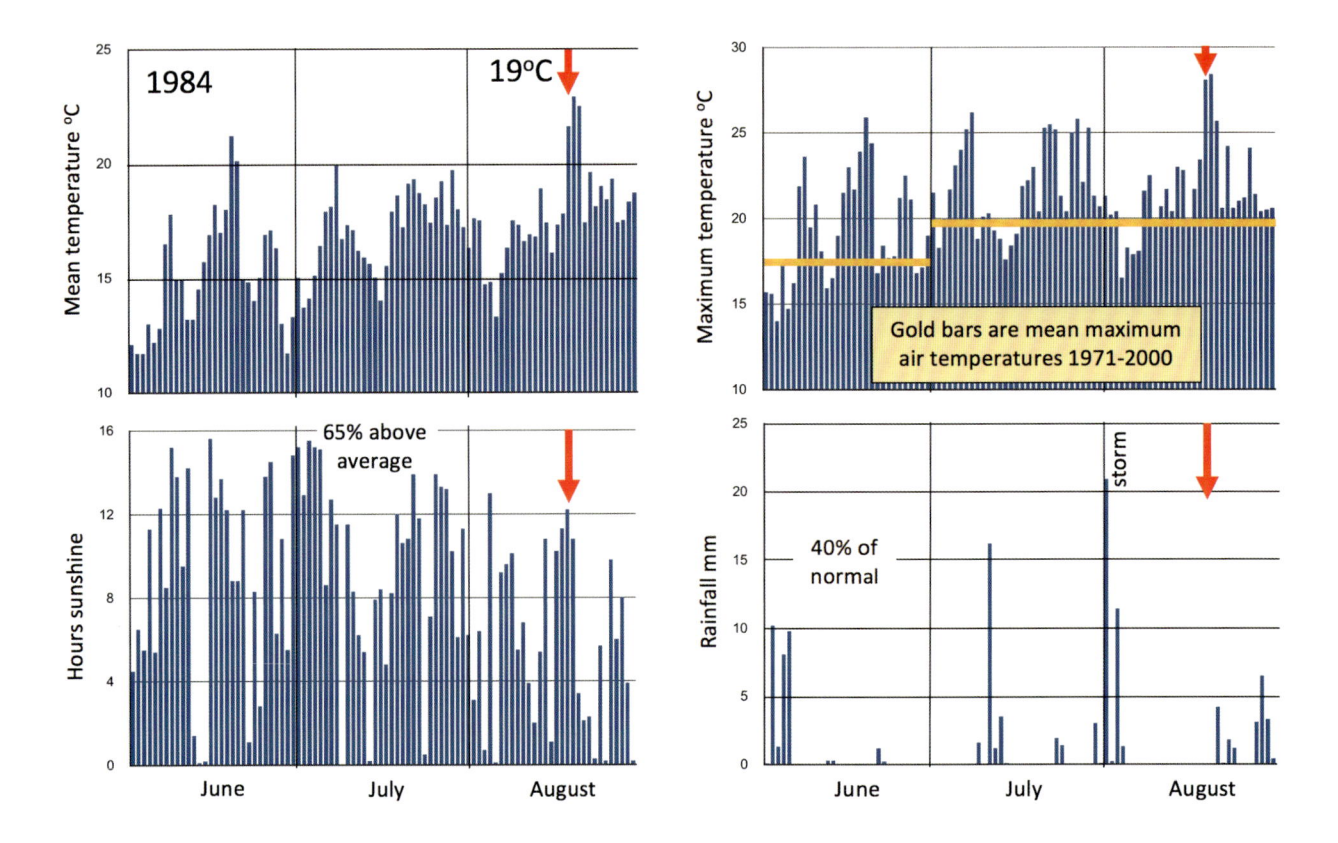

*Figure 119. Summer 1984 daily climate record at Penmaen weather station, which is close to the catchments leading to Wellhead resurgence. The **red arrows** locate the time of the peak water temperature of 19°C measured at Wellhead (Fig. 118). The weather had been warm to hot with notable drought; Broad Pool (SS509910) dried up completely. Of especial significance, night-time temperatures were high in August, with "exceptionally high air temperatures affecting sea temperatures, which reached 20.8°C" (Mayes and Powell 2003). This pattern fairly clearly indicates 'climate' as the likely cause of this exceptional peak.*

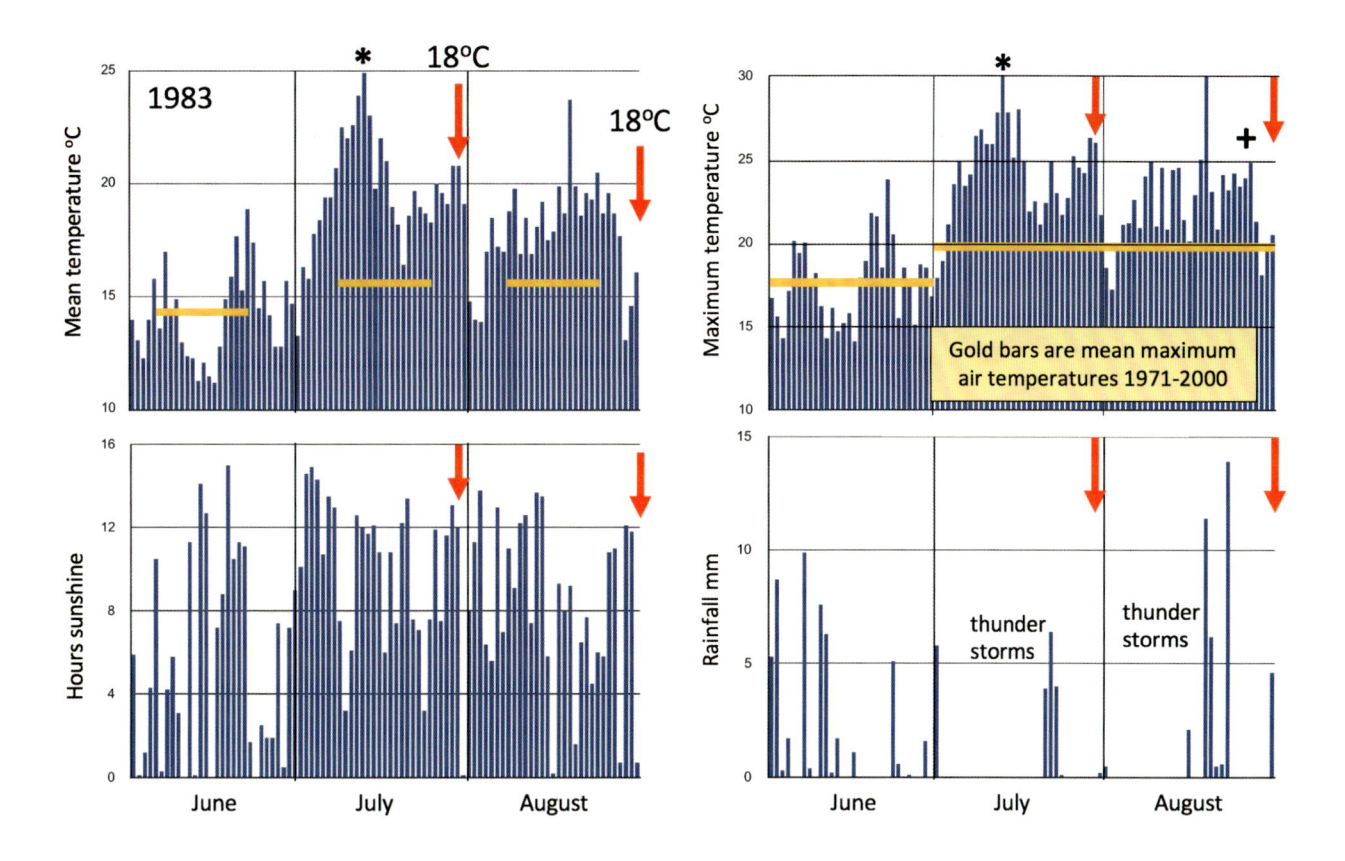

Figure 120. Summer 1983 daily climate record at Penmaen weather station. **Red arrows** *show the times of the two peak water temperatures of 18 °C measured at Wellhead. The July peak involved a sharp rise and slower (~2-day) decline, whereas the August case involved a 2-day rise and a ~4-5-day decline of elevated temperatures (Fig. 118). In both instances the weather had been exceptionally hot with very little rainfall. July was the "hottest in this area this century" and "the second sunniest", while rainfall was only "23% of normal". August was the "warmest, sunniest and driest for 7 years" with the 19th August temperature of 30.4 °C being the warmest day for 41 years. The plus symbol (+) marks when the sea temperature reached 20.6 °C. Strangely, perhaps, the broad high-temperature interval that climaxed on 13th July (**marked by ***), during drought and with both the hottest July day for 50 years and the warmest summer night for 28 years, had no recognizable effect at Wellhead resurgence.*

In summary, most, but not all, of the water temperature anomalies above and below the normal annual variation of 1981-1987 (Fig. 118) can be reconciled with the local climate as recorded nearby at Penmaen. Seemingly the most influential factors behind the higher water temperatures are general preceding warmth, with optimal sunshine and overnight warmth, and lack of rainfall. The inferences are not entirely robust; it is not clear how spring-water temperatures can reflect partial equilibration with near-surface materials or with the relatively stable underground temperatures. The local 'climate' versus spring-temperature coincidences are fairly convincing, but the sites and mechanisms of heat exchange are not clearly known.

The temperature peaks are of several days duration with variable rise and fall rates; some clearly display the lag between peak weather conditions and resurgence time. Variability in this 'response' time is most logically related to flow-rate variation, but given the uncertainties concerning the underground plumbing and rain-input distribution through time, it seems unwarranted to try to interpret any patterns.

Further, since the high temperatures tend to follow drought and consequently occur in low-flow conditions of what is presumed to be phreatic transfer to Wellhead, one has to wonder about the nature of equilibration with ambient temperatures underground during such slow flow. However, high-resolution air versus spring temperature monitoring sheds some light on this issue (Section 5.17). Now we must jump to the other well-known source of water on Gower; although close nearby (Fig. 104), it could barely be more different to Wellhead.

5.13 Holy Well springs on Cefn Bryn

In bygone days Holy Well [6] was known to most people living on Gower. Its extensive surrounding fence keeping livestock out, and the small building frequently visited for chlorination of the water, were prominent on the otherwise naked, common-land hill beside the main road over 'the Bryn'. With adjacent springs[20] it supplied substantial villages on north Gower and, as if by magic, high on the hill, the flow of water persisted through even the worst of droughts. Holy Well's elevated situation lacking any obvious drainage basin seemed to many to require an exotic supply of water.

Figure 121 shows the rainfall catchment and the underlying geology of the site. The catchment boundary is inferred from field mapping and is moderately conservative; within it, rainwater that penetrates the ground (is not evaporated) can be expected to flow towards Holy Well or East Well under gravity. A key feature of the catchment area is that it is persistently *relatively* dry in comparison with the ground immediately downslope to the northeast; it is extensively fairly flat, has different vegetation and at the site of a broad collapse depression (doline) has steep-walled pits that expose some 2 m of well-draining sandy till (Figs 121 and 122).

The published geological map and cross section (British Geological Survey 2002) show that the catchment is underlain by Carboniferous Lower Limestone Shales. Being readily eroded, these strata are poorly exposed on Gower but comprise bedded limestones with some intervening thin friable mudstones (shales).

Significantly at this site (1) these limestones thicken to the southwest and terminate beneath the overthrust Old Red Sandstone strata, and (2) they are not connected to those limestones farther north that extend under the Loughor Estuary and continue up to the north crop (Fig. 101). Thus, the limestones beneath the rainfall catchment constitute a perched aquifer, physically isolated above (not connected to) the other spring-bearing limestones of Gower. This setting implies no simple hydraulic continuity with any limestones farther north.

Figure 121. (Page opposite).

*(**Top**) Rainfall catchment of 0.35 km² likely to supply water to Holy Well and East Well. West Well mostly collects from small springs and runoff northeast of the catchment. The northeastern catchment boundary is roughly coincident with the mapped contact of well-draining Anglian till to the southwest and the clay-rich Devensian till to the northeast (Section 3.7). (Base map © Natural Resources Wales).*

*(**Bottom**) The underlying solid geology shows that most of the catchment has Lower Limestone Shales (LSh) beneath the thin drift cover. The map and inset cross-section, from British Geological Survey (2002), show that Old Red Sandstone (ORS) forms an anticline northeast of the limestone (LSh) that underlies the catchment area. This limestone thus constitutes a structurally isolated unit not connected to the main north-dipping limestones that are part of the main Coalfield Syncline (Fig. 101). It forms a perched aquifer that is truncated to the southwest by the Cefn Bryn thrust fault.*

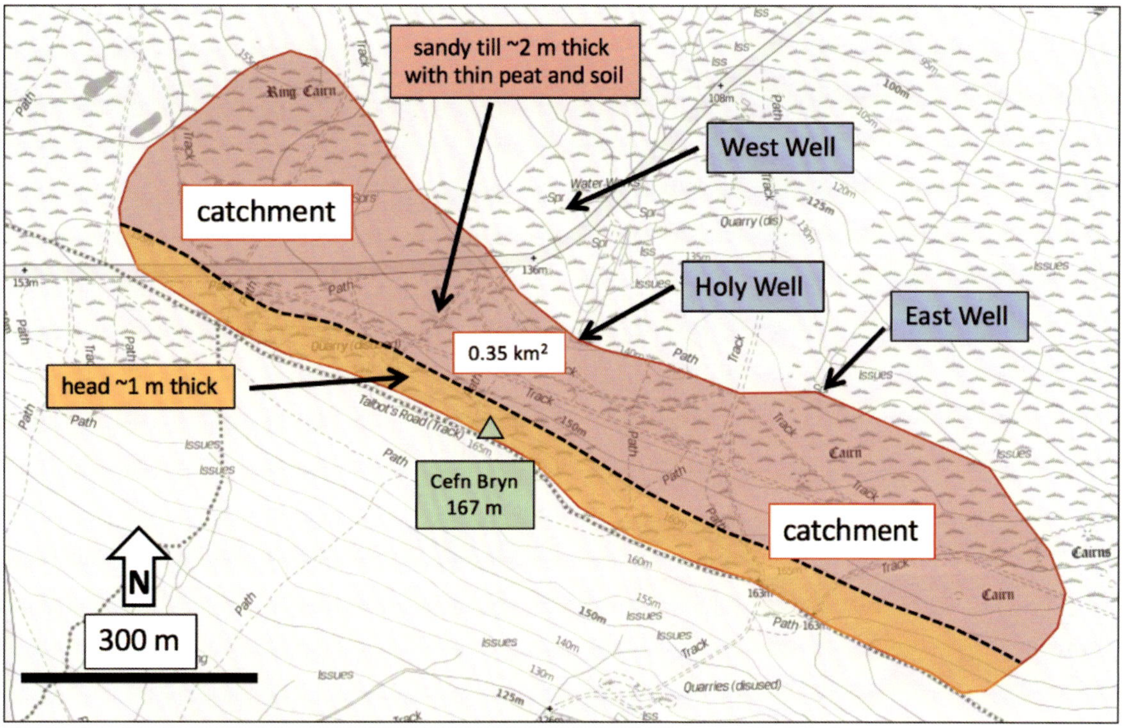
sandy till ~2 m thick
with thin peat and soil
catchment
West Well
Holy Well
East Well
0.35 km²
head ~1 m thick
Cefn Bryn
167 m
catchment
N
300 m

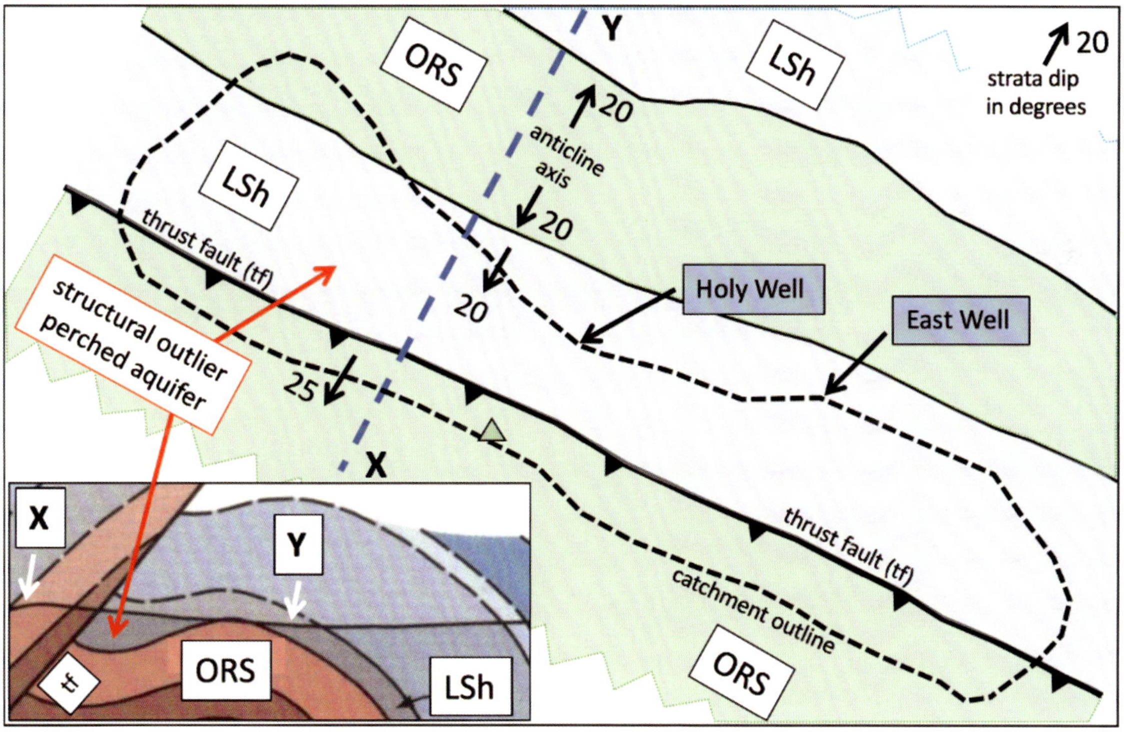
Y
ORS
LSh
20
strata dip
in degrees
20
anticline
axis
LSh
20
thrust fault (tf)
20
structural outlier
perched aquifer
Holy Well
East Well
25
X
X
Y
thrust fault (tf)
catchment outline
ORS
tf
ORS
LSh

*Figure 122. (**Top**) Views to south, showing the broad shallow depressions that result from erosion of till at the springs, and the relatively flat catchment above them. The gently sloping area is thin peat on sandy till overlying limestone and the steeper slope behind marks the boundary with over-thrust, more resistant Old Red Sandstone with thin head cover to the ridge crest. (**Middle left**) View to west at SS500897; a pit within a shallow doline exposes some 2 m of sandy till (**middle right**; scale 10 cm). The collapse is into Lower Limestone Shale, 140 m south of East Well [9]. (**Bottom**) Quarry exposure (SS494898) of bedded Old Red Sandstones and thin peaty head deposits covering the sandstones (camera case 12.5 cm). These form the steeper slope along the southern edge of the catchment, the foot of which marks the position of the fault on which these beds were thrust upwards over and onto the younger Limestone Shales (Fig. 121 cross section).*

5.14 Quick and dirty evaluation of till properties

The catchment above the springs on Cefn Bryn is noticeably better drained than lower down, where peat and clay-rich mud churned by water-seeking livestock can render walking there at least unpleasant. The catchment till is clay poor, of Anglian age (Chapter 3), and constitutes a thin permeable aquifer. Holy Well and East Well springs emanate at the contact of the Anglian till with clay-rich Devensian till, which is widespread to the north and lines the floor of Broad Pool (pond), preventing it from draining into the underlying limestone there. See representative sample B1 in Figure 110. A simple test was devised to demonstrate the contrast in fines content responsible for the permeability contrast (Fig. 123).

Figure 123. Jar samples (45 mm deep) of till treated simply to show contrasts in fines content and, indirectly, porosity and permeability. (Left) Catchment till 140 m south of East Well. (Right) Impervious till 50 m from Broad Pool; this is sample B1 in Figure 110. Sub-soil (B horizon) samples were collected from very poorly sorted till containing distributed granules to sparse boulders. 50 g of sub-4 mm sediment was mixed with 150 ml of water and shaken. The photos were taken after settling for 2 days. The sample from near East Well (left), in the spring catchment area, cleared water in its upper part within 2 minutes of shaking, showing mainly silt and fine sand and lacking a clay/mud fraction. The sample from near Broad Pool (right) did not clear its upper layer of suspended micron-scale / colloidal material, while slowly settling significant mud upon graded silt and sand. In keeping with its well-drained appearance, the catchment till is friable and permeable, capable of storing some water, whereas the lower slopes of Cefn Bryn have a significant clay component and would not drain freely.

This quick test readily shows why the catchment till is well drained in contrast to the rest; it is because it lacks fines that would fill porosity and prevent water flow through it. The spring line most probably registers water emergence at a permeability barrier.

5.15 Rough calculation of Cefn Bryn perched aquifer potential

As a heads-up for the reader, this rather involved section finds that Gower rainfall is adequate to sustain the Cefn Bryn springs without any need for an exotic additional supply of water. Further, it is shown how the springs persist through droughts.

In the following paragraphs, all measures are intentionally simple and conservative[21], and hence approximate, possibly accurate to ±10-15%. The 'catchment' is where rainfall is expected to percolate down to drain out ultimately at the two main percolation-type springs. The springs overlie limestone and rise through and onto till, which, with thin peat, is a superficial cover at least 2 m thick across the catchment. The spring line is at 140 m OD (Fig. 124) with the water table in Lower Limestone Shales (dotted line in LSh) slightly sloping to have a hydraulic gradient of less than 1:100 towards the springs. The cross-sectional area of LSh above the water table (pale blue) is 750 m² and the catchment length is 1,320 m, so the rock volume above the water table is 990,000 m³. This aquifer rock volume will tend to have low ability to hold water (the unconfined storage coefficient), possibly as little as 1-2%, so maximal water storage in the rock will be of the order of 990,000 x 0.015 = 14,850 m³.

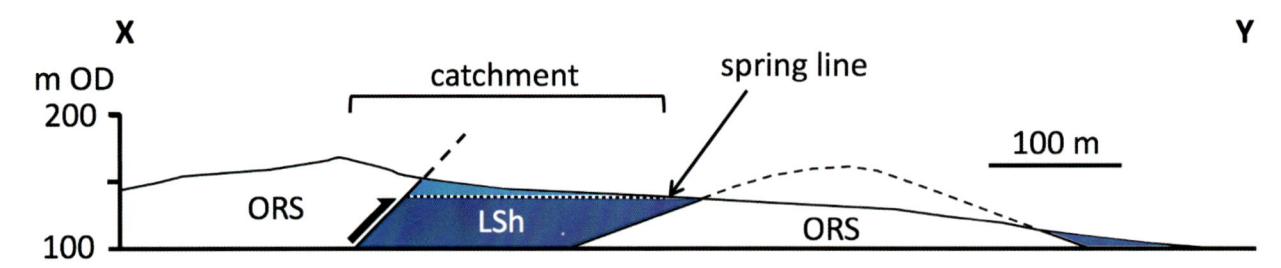

Figure 124. Simplified geological cross section along X-Y of Figure 121. The dotted line represents the water table, which continues beneath the overthrust Old Red Sandstone (ORS). The spring line is not along the contact with the relatively impermeable ORS but appears to lie where permeable Anglian till gives way to impermeable Devensian clay-rich till.

Volumes of the superficial drift layers, taken as 2 m thick, are 540,000 m³ dominantly of permeable till (Section 5.14) with thin soil and peat, plus 84,000 m³ of head overlying the ORS. Conservatively assuming potential water storage to be 5% and 20% respectively, superficial deposit maximal water storage amounts to 43,800 m³. The till and thin peat will be of moderately low transmissivity (percolation rate) and in torrential rain surface runoff will occur once the peat is saturated, although slopes widely are low (~5°) and gullies are few and fully vegetated. Vegetation is predominantly of common-land grasses, bracken and low scrub; trees are absent except at the springs (Fig. 122).

Assuming water storage to be at a constant maximum (aquifer full), annual spring discharge will equal annual rainfall minus runoff and loss due to evapotranspiration. The low slopes and few gullies suggest runoff is minimal. The topography directs most runoff towards the springs where it contributes to the measured discharge, so actual water loss by runoff is considered insignificant and thus is ignored.

Measured mean annual rainfall is 1224 mm (average for 1971-2000; range 1545-874 mm) and measured mean evaporation is 491 mm (average for 1980-2000) (data from Mayes and Powell (2003), measured nearby at Penmaen). Conservatively assuming evaporation plus transpiration is 50% of rainfall (loss by transpiration 121 mm/a), average recharge into the aquifer would equal 612 mm/a. The catchment area is 0.35 km² or 350,000 m², so input would be 350,000 x 0.612 equalling 214,200 m³/a, which is 587 m³/d. Assuming steady storage amount over a long term, this average daily aquifer recharge (input) of 587 m³/d must equal average daily discharge.

For the original water supply to north Gower, Holy Well discharge included flows from East Well and from West Well ("smaller springs in the vicinity"). In terms of relative contributions, Holy Well spring discharges most, with East Well discharging x0.67 of that amount. A minimum net (combined springs) discharge of 318 m³/d was reported in February 1954, and 123 m³/day in October 1958. 180 m³/day and 144 m³/day were measured during the drought in July 2018 (this study). The largest of these measures, at 318 m³/d, amounts to 54% of the average rain recharge of 587 m³/d. Given the generally conservative uncertainty of the data, and having excluded West Well from the catchment, there is *no obvious requirement* for any additional, exotic water supply: the rainfall is enough.

The question then arises as to how the spring flow is maintained through drought periods. The actual varying rate of discharge and hence the persistence of the springs must be governed by overall aquifer transmissivity – effectively the bulk percolation rate. Evidently (historically) the discharged amount never exceeds the rain recharge, but in drought there must be draw-down of aquifer-stored water. Peat generally contributes little to base supply (Bain et al. 2011), but the upper-slope head deposits, soil, till and limestone will all yield stored water by percolation. Thus, during drought, the flow at the springs will be sustained by draw-down according to net transmissivity (percolation rate) affecting the previous rain recharge. Probably transmissivity is least in the till while in the limestone it is likely to be moderated by interbedded impermeable shales that dip into the hill, away from the springs.

Maximal water storage in the limestone (LSh) is estimated at 14,850 m³ and superficial-deposit storage x3 greater, at 43,800 m³ (from above), totaling 58,650 m³. While it is unclear precisely how and how much of this water could be withdrawn to sustain flow during drought, clearly it flows at a reducing rate (see below), but it never fails completely before a recharge.

A study of rainfall distribution through time (Fig. 125) indicates that intervals of more than 90 days (3 months) without rain sufficient to partially replenish groundwater are exceptional. Using the minimum-flow amount recorded for Holy Well (123 m³/day in October 1958), a 90-day drought would withdraw 11,070 m³. As the total storage is conservatively estimated to be x5 greater than this, at 58,650 m³, this would be sufficient to sustain flow during any exceptional 3-month drought. Draw down of stored water certainly sustained discharge continuity during the 89-day drought of 2018; study of the progressive discharge reduction at that time indicates for how long it could have been sustained before drying up. Evidently (see below) it rains too frequently for these springs ever to run dry.

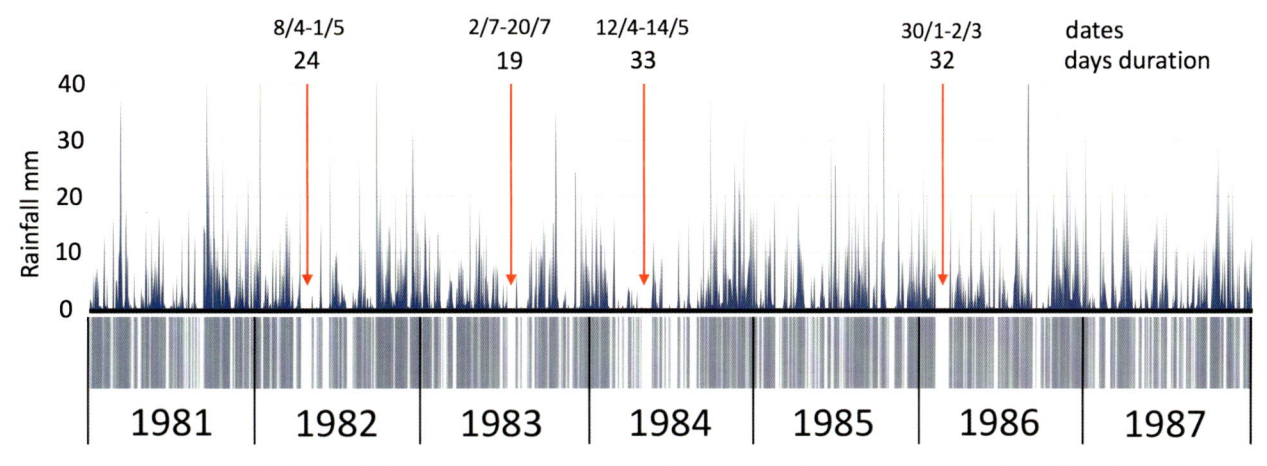

Figure 125. Penmaen daily rainfall record for 1981-1987 relevant to Cefn Bryn springs and Wellhead catchments. Lower panel white stripes are periods with zero or only trace rainfall. 1984 was the driest year. 1983 also had episodes of low rainfall and was particularly hot so that evaporation was maximal. Periods of more than 3 weeks without rain are scarce; the 4 maximum dry episodes are highlighted (19-33 days). The upper panel shows that dry episodes usually are soon followed by significant rain. Intervals of more than 90 days (3 months) without rain sufficient to partially replenish groundwater do not occur, except perhaps in the exceptional year 1984, when, however, the Holy Well springs did not run dry.

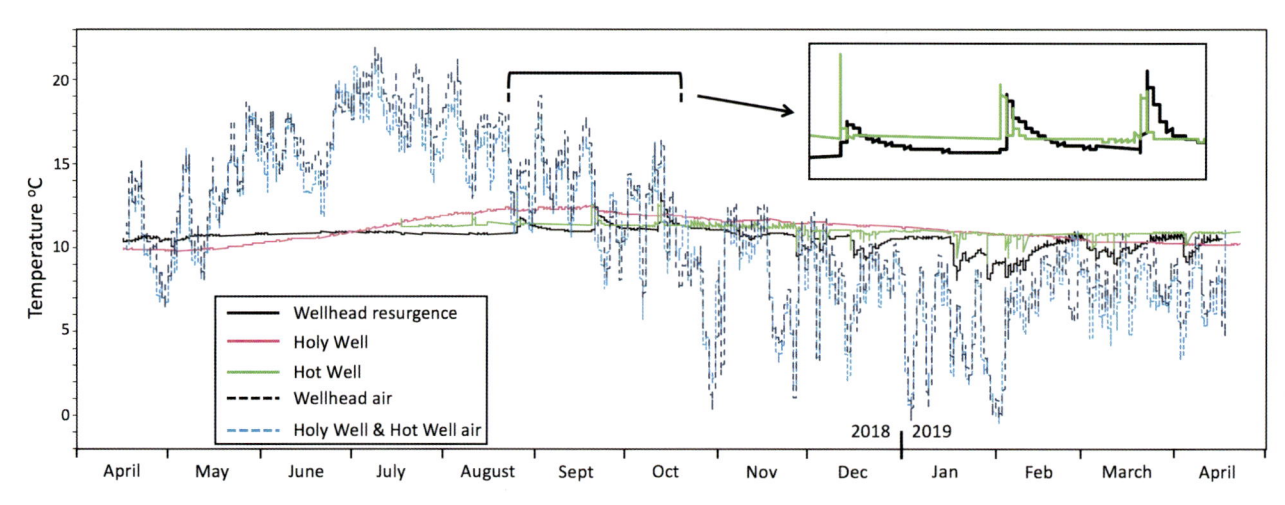

Figure 127. Graphs of complementary spring-water and immediately local air temperatures recorded at Wellhead resurgence [14] and Holy Well spring [6] over 12 months, between 16th April 2018 and 23rd April 2019. Hot Well spring [7] data start on 18th July 2018 and share the same air temperature as Holy Well. The data loggers (type Hobo Pendant UA001-64 bit) recorded temperatures every 30 minutes. Air temperature data are processed via a 50-point moving trendline to make the reading easier. The inset at top right amplifies the discussed August to October Hot Well – Wellhead relations, for clarity.

Figure 127 highlights the distinction between conduit-type and percolation-type springs. While Wellhead resurgence shows spikes in temperature according to the system input and conduit throughput, just as seen with the turbidity record (Section 5.10), Holy Well shows no short-term responsiveness to ambient conditions. Instead, Holy Well shows a broad low-amplitude 'peak-trough' annual cycle that for the warm part lags roughly 2 months behind air temperature. In keeping with its lack of short-term response to intense rainfall (Fig. 126), the Holy Well temperature pattern is typical of a purely percolation-type spring, as is nearby East Well.

Figures 127 and 128 show that in the high-resolution data there is a significant difference in the annual temperature variations between Wellhead and Holy Well. During June through October, when rainfall amounts are relatively low, Wellhead non-peak temperatures vary little around about 11°C, whereas in the same interval Holy Well recovers from winter low to summer maximum; this is evident in the end-of-June crossover (Fig. 127). During the typically wetter period of November through to March, Wellhead temperatures clearly vary 'considerably' (at this resolution) around about 10.6°C, while Holy Well discharge cools steadily, crossing to below Wellhead temperatures in February. A likely explanation for this is that Wellhead is maximally responsive to wet-and-cold season rainfall and that water resurgent at Wellhead typically has mixed with some resident water that equilibrated with rock at fairly steady temperatures. Borehole measurements near Parkmill [17] show the rocks at around 0 m OD are at 11.5°C, which is widely typical. Thus, underground transit to Wellhead may buffer temperatures, holding them down during June through October and then irregularly up when cold rainfall subsequently penetrates the system.

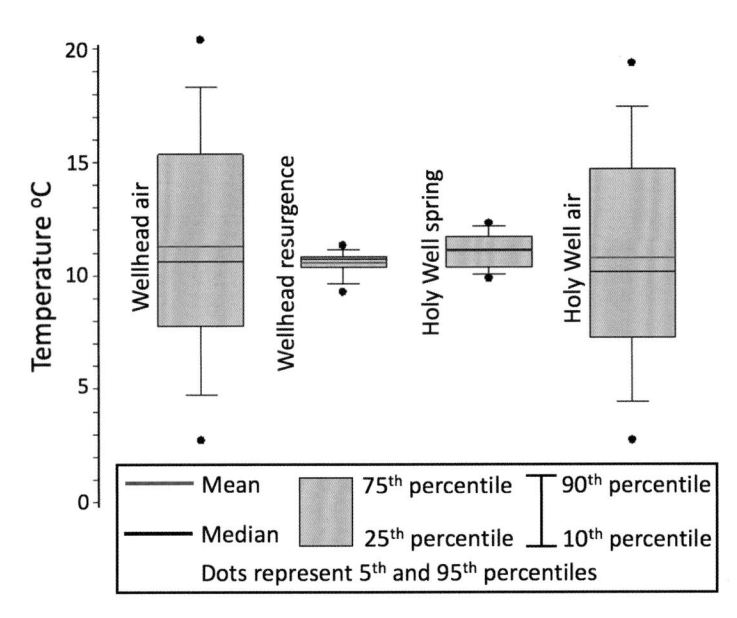

Figure 128. Statistics of the high-resolution temperature record at Wellhead and Holy Well graphically shown in Figure 127. The main feature is the greater variation shown by Holy Well discharge relative to the Wellhead waters. This difference possibly reflects the fact that Holy Well percolated waters gradually follow the climate, whereas Wellhead conduit flows, although more responsive and variable in the short-term, generally have temperatures buffered during transit to the resurgence by mixing with significant volumes of water stored at around 11.5°C, especially during relatively dry periods of low-flow discharge.

5.18 Chemical signatures of spring types

More than 20 springs and many associated sinks in central Gower have been studied in this project (Fig. 104). In the year April 2018 to April 2019 most springs were revisited to measure changes in Total Dissolved Solids (TDS), pH[25] and temperature. The accumulated time-series data are presented here. By early July during the 2018 drought sufficient was known to identify 13 representative waters for full inorganic analysis. Sampling during the drought was not deliberate timing, although it seemed suitable to avoid excessive rainfall dilution of basic characteristics. The unforeseen consequences of sampling what now have to be called *'drought waters'* are intriguing, and those rather unexpected data are presented here too. Before describing the findings, however, it is important to understand the limitations of the data.

'Total Dissolved Solids' (TDS) was measured in the field at the spring sites (using Hanna® Instruments Tester H198301 calibrated at 1382 ppm). Notionally the measure is of ions that are solids dissolved in the water, e.g., Ca^{2+} and CO_3^{2-} from calcium carbonate in limestone. Although the field read-out is as TDS, the measure is actually of the electrical conductivity (EC) of the water, automatically corrected for temperature and multiplied by a factor to give TDS. The readings, however, do not simply reflect true totals of dissolved solids, because in compositionally different solutions the differing ionic content will have differing influence on the electrical conductivity. The 'conversion factor' to get from EC to TDS varies according to overall solution composition (typically in the range x0.5 – x0.64 for spring waters), but, thankfully, we do not need to go there.

Field TDS measurements are not accurate in terms of actual composition, and indeed the inaccuracy varies with actual composition, in some instances a lot. However, provided testing locations and procedures are used consistently at each measurement site, field TDS is a *precise* measure; it is perfectly suited to repeated (and cheap) measurement to track changes though time. In contrast, *accurate* measures of actual composition are derived by laboratory analysis, of which more later.

What do we expect? Relative to most other Carboniferous Limestone terrains of the UK, Gower peninsular karst is of limited lateral and vertical extent. Where limestones are exposed with sinks and springs at the surface the waters exist in what are described as 'unconfined aquifers', in which the pressure at the water table is atmospheric pressure and water moves by gravity according to the hydraulic gradient (slope of the water table) and the hydraulic conductivity (permeability) of the rock. In 'confined aquifers' water is trapped by overlying low permeability material and is under pressure always greater than atmospheric, commonly at depth and often with longer residence times measured in hundreds or thousands of years.

The unconfined Gower aquifers respond rapidly to surface infiltration, so flow rates are high and water residence times are correspondingly relatively short. In general, the deeper the flow path through the aquifer the longer is the travel time and thus there is more opportunity for the water chemistry to equilibrate with the surrounding rock (aquifer). This may be especially true for reaches of the caverns lost below sea level (Section 5.4); really deep stagnant reaches might be considered 'confined'. On Gower conduit-flow residence times tend to be limited, in the order of days, especially during the common rainy intervals, whereas percolation residence times are probably on the order of months, perhaps years.[26] We can expect that for a system with a substantial conduit-flow component variably 'flushed' by rainwater, there will be a compositional trend reflecting variable dilution. In addition, because drought conditions are associated with significantly diminished conduit-flow as well as with reducing percolation-flow rates, we can expect the samples in this study to define 'mixing lines' [27] of varying compositions relative to non-drought waters, as we shall see. The extent of the compositional differences found here, however, was not expected.

Central Gower springs are of two main types: percolation type and hybrid percolation-plus-conduit type. Holy Well [6] on Cefn Bryn is a virtually perfect end-member percolation spring. Our friend the Wellhead resurgence [14] is a classic hybrid type but its 'working parts' are extremely variable in effect through time so the composition and discharge are rather unpredictable. Put another way, in fieldwork it soon became possible to recognize percolation springs by their near-constant TDS. On the other hand, it became clear that the hybrid types would vary to some extent in composition and discharge according to the rainfall, with several drying in the drought. In torrential rainfall Wellhead would act virtually as a pure conduit-type spring, fed by rapid 'flash' runoff into sinks and underground bypass passages. Alternatively, during drought, the sinks would receive percolation flow draining from the catchment superficial deposits (mainly till, soil, peat, and stream gravels) and underground transfer to Wellhead would be relatively slow. Figure 129 summarises the compositional variability through the year and shows that the spring types can be distinguished on the criterion of TDS constancy versus variability. Figure 130 shows the rainwater dilution effect at Wellhead resurgence.

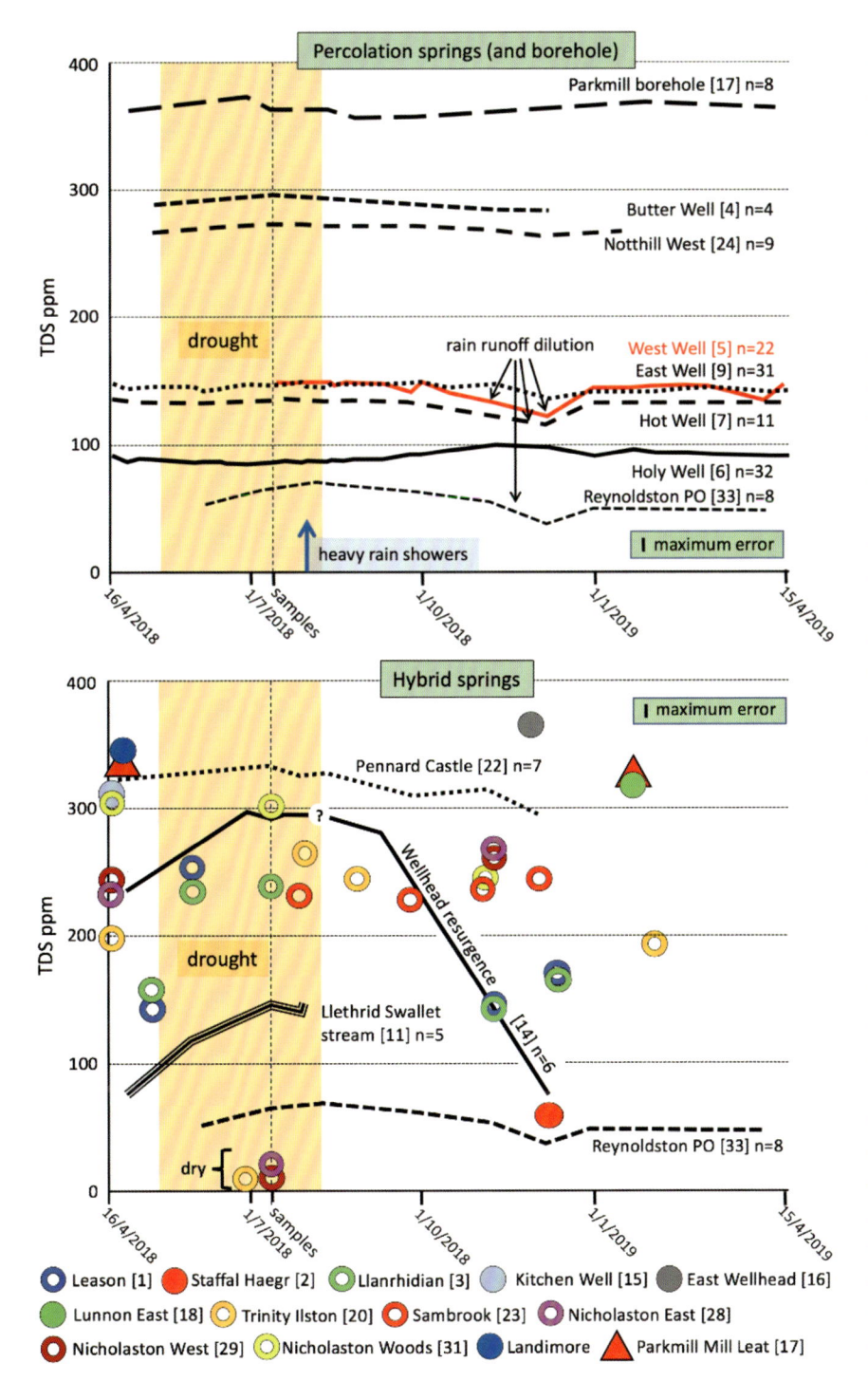

Figure 129. Types of springs distinguished by Total Dissolved Solids (TDS) variation; number in [] gives location (Fig. 104), n *is the number of measurements. The time of collection of samples for detailed analysis is shown within the coloured drought interval.*

*(**Top**) All of these percolation-type springs show practically constant TDS for all or most of the year, as well as only very slow variation in discharge rate (not shown). Reynoldston Post Office spring is included although it is prone to rapid fluctuation in discharge as it gathers surface runoff. Some dilution by runoff is also shown by West Well and Hot Well.*

*(**Bottom**) Various hybrid percolation-plus-conduit type springs. These typically show considerable TDS variation and/or rapidly variable discharge. Pennard Castle and Sambrook springs show rather little compositional change, but their discharges fluctuate widely. Wellhead, Leason, Llanrhidian and Trinity Ilston have sufficient data to show that the TDS content increased during the drought (Trinity dried briefly) and diminished afterwards. Llethrid stream, the main input feeding to Wellhead, also increased TDS during the drought. Staffal Haegr only discharges following torrential rain, when other springs also show runoff dilution.*

Regarding specific percolation springs, at first it seemed peculiar that Holy Well [6] has consistently lower TDS than the springs adjacent on either side, viz. East Well [9] and West Well [5], which are compositionally alike. Although not clear from the ground (Figs 121 and 122), the TDS suggests that Holy Well has affinity with Reynoldston PO spring and it is conceivable that relative to East Well and West Well the Holy Well catchment involves more till and less of the perched limestone aquifer that they all share.

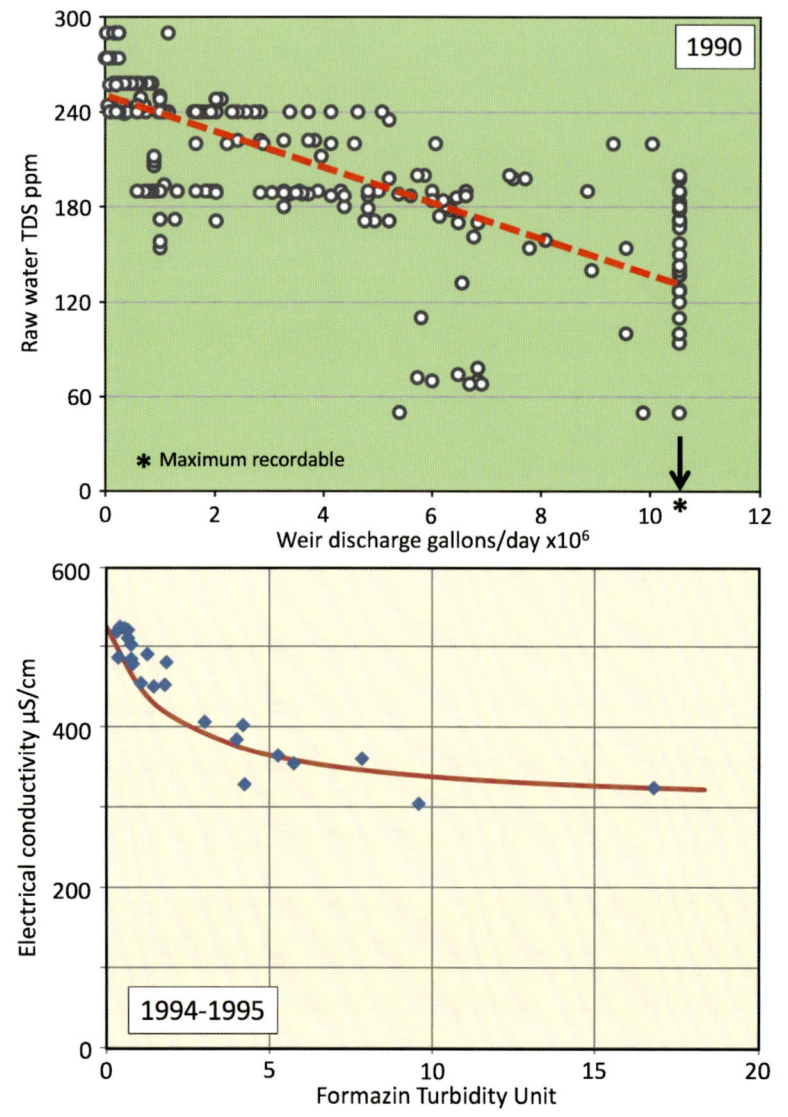

*Figure 130. (**Top**) Wellhead raw-water Total Dissolved Solids (TDS) versus weir discharge as recorded for 1990; the red trendline is the linear best fit to the plots. The discharge does not include the water extracted, at about 0.35x10^6 gal/day, and amounts exceeding 10.5x10^6 gal/day (48x10^3 m^3/day) are unknown.*

*(**Bottom**) Wellhead electrical conductivity (EC) versus Formazin Turbidity Unit (FTU) as recorded for 1994-5 (**diamonds**). (Formazin is a compound used to produce known standard amounts of turbidity). The red curve, a fair fit to the data, is the EC calculated by theoretically relating EC to dilution and factoring for the relation of dilution to turbidity (dilution = 0.5 x turbidity). EC diminishes slowly (asymptotically) below about 400 µS/cm and levels off at around 300 µS/cm.*

The main point indicated by both plots is that, irrespective of the scatter, low-discharge drought conditions are reflected in relatively high TDS values ≥240 ppm and EC around 500 µS/cm, while heavy rainfall with related high discharge and consequent elevated turbidity results in dilution. The discharge for 1990 was less on average than in adjacent years and apparent evaporative loss was greater; also, it rained more often (Table 3). The very general point about dilution applies to any year, as evident during and after the great drought of 2018 (Fig. 129).

The true composition of naturally occurring water has to be determined by detailed analysis, which, in these days of monitoring the environment and relating changes to 'baseline' data, is rather sophisticated. Done properly, water samples are collected at the same time as various field measurements are made, so that laboratory results can be interpreted safely with controls on accuracy. We know that in karst waters dissolved CO_2 fundamentally affects carbonate solubility; bicarbonate (HCO_3^-) is especially vulnerable to CO_2 loss to the atmosphere. Done properly, bicarbonate should be determined during sampling.

In the amateur scope of this project, the samples were collected and frozen at home alongside peas and fish fingers, with the bicarbonate measured later on defrosted samples.

Now, jumping ahead, water geochemists consider that an accurate determination of the cations and anions originally in a solution at equilibrium should balance the positive and negative charges; ionic balance[28] is widely regarded as a reflection of 'no-loss'. Happily, despite the non-ideal sampling, the ionic imbalances of the Gower samples are, with one

Location	pH	Ca^{2+}	Mg^{2+}	Na$^+$	K$^+$	HCO$_3^-$	Cl$^-$	SO$_4^{2-}$	NO$_3^-$	TDS	Ionic balance %
Holy Well	8.20	60.90	9.97	72.00	3.87	30.07	102.09	140.20	11.59	431	4.5
East Well	8.20	37.00	12.18	41.70	3.21	43.88	67.95	29.96	3.21	239	0.1
West Well	8.70	22.70	10.36	36.00	2.92	41.45	58.23	12.97	5.36	190	-0.9
Wellhead resurgence	8.20	17.60	37.06	35.70	3.85	88.58	64.30	21.04	24.63	293	-6.3
Parkmill Heritage borehole	7.70	19.60	65.95	45.60	4.70	152.78	83.24	35.89	36.39	444	-7.7
Llethrid stream	8.20	19.30	14.77	47.70	6.01	62.58	78.53	21.31	4.53	255	-6.2
Reynoldston PO	8.20	26.80	13.20	75.30	1.14	22.75	122.70	7.84	1.73	271	5.7
Notthill West	7.70	65.70	36.81	81.50	8.55	78.02	145.88	55.78	57.09	529	-3.9
Pennard Castle	7.70	41.30	69.95	67.90	8.80	72.33	117.08	48.60	50.60	477	5.7
Butter Well	7.90	57.20	17.50	27.20	5.06	75.58	48.48	36.51	13.66	281	-1.8
Llanrhidian Church	8.70	53.50	32.57	45.90	3.51	68.26	84.55	36.20	29.96	354	0.2
Hot Well	8.30	17.30	11.27	30.10	3.12	35.76	50.09	12.96	3.75	164	-1.0
Nicholaston Woods	7.90	22.50	23.40	38.50	2.70	82.08	76.82	23.97	26.84	297	-15.3

Table 4. Geochemical analyses of Gower 'drought waters' determined at the British Geological Survey Inorganic Laboratory at Keyworth, Nottingham[29]. The data are used with permission copyright BGS, UKRI. Apart from pH and ionic balance all values are in mg/L. Ionic balance at 0% is widely taken as optimal, suggesting aqueous equilibrium and no loss, but imbalance values between -10 and +10 are fairly typical and generally accepted. These results are taken as robust and, except for Nicholaston Woods, adequately accurate.

exception, significantly less than +/-10% (Table 4), which is generally taken as quite acceptable (e.g., Banks et al. 2007, figure 5).

Figure 131 shows the Gower drought waters data as normally plotted, using the diagrams devised by Piper (1944), and also the compositional distribution statistics of the major constituents. The first impression from the plots is 'quite a scatter', but this can be systematically unravelled. Rather obviously (bottom panel), some samples have elevated Na-Cl content, reflecting sea-water aerosol contamination in rainfall on Gower. Also, the high nitrate (NO$_3^-$) compositions register some agricultural and/or man-made pollution (in excess of the limit for drinking water at 50 mg/L NO$_3^-$) with short-path recharge and hence little attenuation in the soils and till. K$^+$ is low, perhaps suggesting pollution is not from a NPK (nitrate-phosphate-potash) fertilizer.

Figure 134 shows the compositions of central Gower drought waters found in this project relative to other aquifer terrains in the UK, including other Welsh systems. Waters recovered from Carboniferous limestones elsewhere, and those from Devonian sandstones, show clustering towards the left in the Piper Plots, which is typical of non-drought waters substantially equilibrated with the calcium carbonate lithologies or cements of their host rocks. Central Gower drought waters are strikingly different.

Evidence from Wellhead shows that normal-to-heavy rainfall drives compositions along mixing lines towards the Calcium and Bicarbonate types, essentially reflecting dilution of the Mg signature. The finding that during the drought some hybrid spring waters became more concentrated in dissolved constituents is most simply interpreted as reflecting an increased contribution of longer-residence percolation water relative to conduit throughput. Percolation waters sampled as drips in caves are known to be relatively rich in dissolved constituents (e.g., Ede 1975) and their increasing influence during drought would inevitably change ionic ratios and hence cause plots to shift.

Whereas the compositional trends towards Magnesium type and Chloride type can be explained by dolomite and maritime Na-Cl contamination respectively, some of the differences of the central Gower waters are likely to reflect impact of the drought on the relative contributions of percolation flow versus conduit flow. Detailed analysis of sinks feeding Leason spring (Chambers 1983) indicated that between drought and storm conditions the conduit-flow component at the spring could vary from 3% to 58%.

While the relative proportions are likely to vary for the various systems studied here, it seems most probable that the Gower 'drought waters' of this study are so distinctive because the longer-residence percolation flows came to dominate the spring-water compositions. This, however, remains hypothetical and more focused research would be needed with samples repeatedly collected during periods of high rainfall.

*Figure 134. (Page opposite). Geochemical plots of central Gower 'drought waters' relative to other regional aquifer and baseline data, mostly from borehole samples. The fields of plots for the central Gower data are shown as **heavy dashed lines** in the other plots.*

*Central Gower data from this study with the field of Welsh data (Fig. 133) outlined in **red**. Data from Bishopston borehole (SS578896) are shown representing non-drought water in Marros mudstones and till immediately overlying the Carboniferous limestones; these waters represent that which enters the main sinks into the Gower limestones. Northern England plot from Abesser et al. (2005), Derbyshire plot from Abesser and Smedley (2008), South Wales and Herefordshire plot from Moreau et al. (2004). Bishopston data, 2002-2011 n=9, are Natural Resources Wales information (© Natural Resources Wales and database right. All rights reserved).*

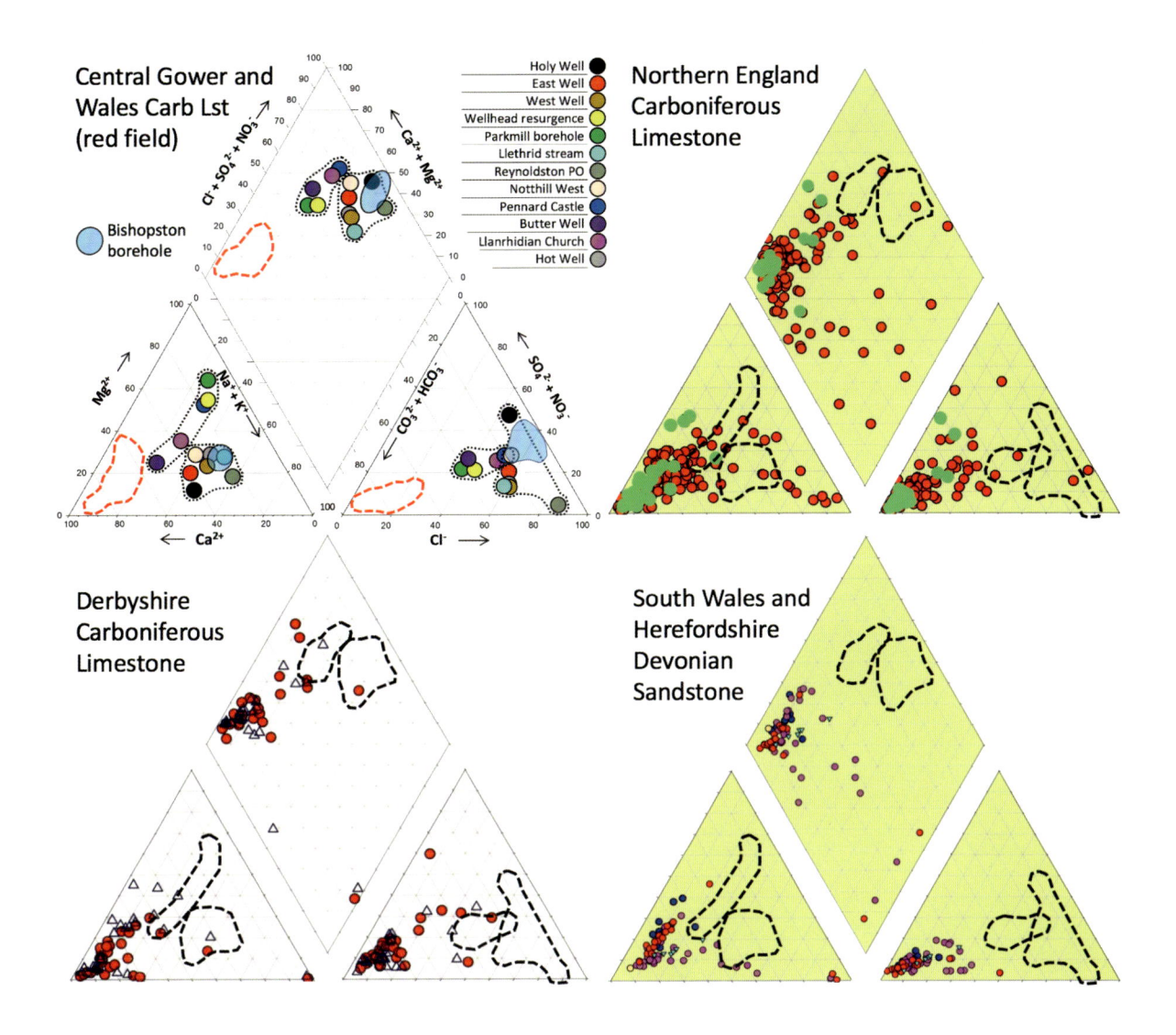

5.19 Tufa and other oddities

The appended gazetteer provides locality information on all of the springs and sinks studied, with photographs of most of them (Fig. 104). Here are some less run-of-the-mill observations.

Notthill West [24] is one of two closely similar springs on the south flank of the southeast continuation of Cefn Bryn; interest in Notthill East was lost in a bramble thicket. The springs are classically of percolation type, with fairly constant composition and unremarkable geochemistry amidst the other ones that are also sourced in Lower Limestone Shales close to underlying Old Red Sandstone, but on the north flank of Cefn Bryn. The distinctive feature of these two springs is that they deposit rather attractive nodules of tufa (porous calcium carbonate; Fig. 135) along their small streams, which tumble no more than 10 metres down to a footbridge 125 m away. The discharge measures on average 271 ppm TDS (Fig. 129) while at the bridge it has 'lost' about 14 ppm (11-18 ppm measured). Clearly the nodules are growing by slowly gaining a fairly consistent small amount of calcium carbonate $CaCO_3$ as the water exsolves CO_2. The interest lies in the uniqueness of this in-stream nodule occurrence on Gower. Conceivably the tumbling waters facilitate the gas exsolution, although microbial mediation could also be occurring.[30]

217

*Figure 135. (**Top and middle left**) Tufa nodules formed in the stream below Notthill West percolation spring (camera case is 12.5 cm). (**Bottom left**) Butter Well as it normally appears, clear and tranquil. (**Right**) Staffal Haegr full-on reflecting flood overflow, in stark contrast to Butter Well.*

Butter Well [4], also known as Saint Illtyd's Well, in Llanrhidian (Fig. 104), is renowned for an occurrence in the 12th century when milk was supposed to have flowed from the spring for several hours. Apparently, "curds" were seen in the gravel bed and "a certain fatty substance" floating at the edge. The spring, at 13 m OD, is striking for being completely unperturbed by storm rainfall that causes the adjacent Llanrhidian [3] (21 m OD) and Staffal Haegr [2] (25 m OD) springs to gush vigorously with turbid flood waters. Butter Well flows clear and is the only Calcium-type spring water of the batch analysed (Fig. 132); behaviourally it is similar to another spring farther west, 150 m beyond Staffal Haegr at 7 m OD ('Rises' at SS49119256), which steadily supplied drinking water that was pure enough to bottle and sell (for a while). The lack of response to flooding, and the relatively high calcium, strongly suggest that Butter Well has a relatively deep phreatic feed to the surface from below the water table. The mysterious 'milk' most probably was related to a subterranean collapse involving calcium-carbonate speleothem (i.e., stalagmite, stalactite or 'flow-stone'), which would liberate a white crush-rock flour and probably small fragments with some of low-density and sufficiently porous and/or buoyed by exsolved gas to float. The debris could well have been flushed away within several hours.

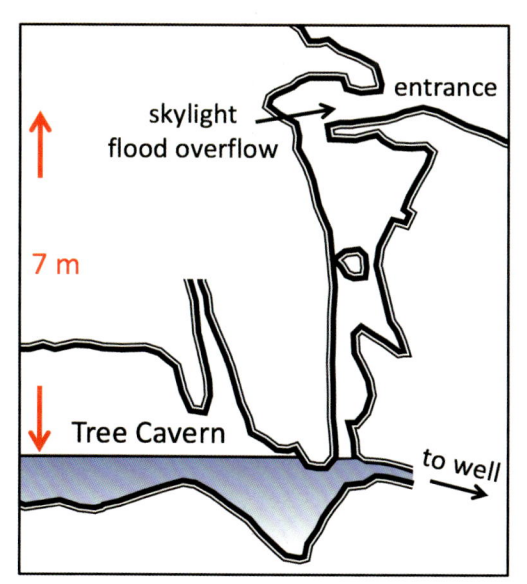

Staffal Haegr [2], probably a corruption of the Welsh 'ystafell hagr' meaning ugly, ill-favoured or unseemly chamber or room, is the name originally applied to a woollen mill that took water conveyed some 450 m by a leat from Llanrhidian Church spring [3]. The property has its own well (7 m OD) that feeds a pond, but here Staffal Haegr is applied to a rather peculiar intermittent spring at 25 m OD that appears either full-on or off; in the former state the discharge has carved a spectacular channel towards the coast (Fig. 134). During heavy rainfall, drainage from the north flank of Cefn Bryn enters several sinks with established connections (see below and gazetteer) to the coastal Llanrhidian and Leason springs. One large western sink (New Park SS492914) of the Oldwalls sink cluster [O] (Fig. 104) only takes significant water when the others to the east are overwhelmed, and it has seemed that this is when Staffal Haegr becomes active. However, this spring is not simply fed. Exploration by cavers led by Maurice Clague Taylor, in 1952, showed that the intermittent spring is a 'skylight' some 7 m above a passage with a persistent stream (Tree Cavern) that feeds the nearby well and pond (Fig. 136). Thus, Staffal Haegr is a flood overflow spring; it could hardly differ more from the consistent Butter Well and its nearby companion of similar ilk fed directly from beneath the water table.

Figure 136. Vertical section through Tree Cavern showing the flood overflow that is Staffal Haegr intermittent spring, redrawn from Taylor (1991).

Water from Holy Well springs and drainages farther west on the north flank of Cefn Bryn (Fig. 104) pass relatively steeply (average 2°-2.5°) through the northern crop of limestones to emerge not far above sea level on the north coast. Moormills sinks [8] and one of the Llwyn-y-bwch sinks [L] (Fig. 104) respectively show the effects of periodic-accidental and deliberate blockage.

There are three drains in the bottom of the 6-7 m deep Moormills pit; they are tight and repel normal humans. Occasionally, during or after heavy rain with considerable runoff from Cefn Bryn, the drains block with vegetation and the pit becomes an impressive lake (Fig. 137), which can persist for weeks. Dye testing (Baynton 1968-1969; Chambers 1973) shows that this water normally takes about 24 hours to discharge at the predominantly conduit-type Llanrhidian Church spring [3]. West of the Oldwalls sinks cluster [O] (Fig. 104), one of the several Llwyn-y-bwch (Grove of the buck) sinks has been deliberately blocked to form an ornamental lake (Fig. 137); the overflow from this continues as before, to Leason spring [1].

Chambers (1973, 1983) studied the geochemical behaviour of these north Gower sink-to-spring systems in immense detail, sampling the waters at one or two-hourly intervals for up to six weeks. His work proved that occasional sampling would miss much natural variability and also that, despite the relatively short underground routes, discharge compositions would be hard to predict. A focus of the early studies (e.g., Baynton 1968-1969; Chambers 1973, 1983; Ede 1975) concerned the dissolved carbonate load in relation to rainfall and thereby the rate of karstification of the Gower limestones. Chambers (1973) found that the total hardness concentration varies directly, inversely or not at all with precipitation. He suggested that the rather counterintuitive (and less common) direct correlation of increasing rainfall with increasing dissolved material content could reflect the fact that turbulent water in conduits is more aggressive (dissolves more) than in the laminar flow involved in percolation, or that it could result from higher water pressures allowing more CO_2 to exist in solution, thus increasing the aggressiveness. Regrettably, fifty years after these enthusiastic pioneers meticulously studied Gower springs, we are still not really much wiser here.

*Figure 137. Sinks that normally drain water to the north coast of Gower (Fig. 104). (**Top left**) Moormills sinks normally take two streams that include discharges from the north Cefn Bryn percolation springs discussed above. The water issues at Llanrhidian Church spring [3]. (**Top right**) Occasionally after heavy rain a lake persists at Moormills owing to natural blockage of its small drains. (**Bottom**) Llwyn-y-bwch sinks [L] take Cefn Bryn runoff that is proved to discharge at Leason [1]. The original drain in this sink has been blocked to form a lake for memorable moments at the Oldwalls wedding venue (daughter and son-in-law for scale).*

Although Gower limestones form unconfined aquifers, it would be wrong to consider the water tables as uniform surfaces. Baynton (1968-1969) captured an intriguing case from the excavation in 1921 of the well at Green Cwm Cottage [12] (in Green Cwm; Fig. 104). According to one person involved in sinking the well, it was dug to 25 m below ground level where, at 11 m OD, water almost 4 m deep was found flowing slowly in a narrow passage from which there was a strong draught. The well then "immediately" filled to nearly 22 m OD and then drained again back to 11 m OD. Some years later the well was dry and then partly filled with rubbish (when we used it to practice cave-ladder techniques); now it is filled to the top with stones. The inflow of water during digging, as described, is a relatively common occurrence where aquifers possess unconnected fracture systems variously filled with water. It is probable that such a feature was intercepted in digging and discharged its water into the well making the water level rise only to fall again as it drained to the lower level. Interestingly, the water level at 11 m OD in 1921 was very close to the level of the Wellhead resurgence at 10 m OD (before the weir was built there) suggesting a very low hydraulic gradient in this direction which, if connected, would suggest a high transmissivity typical of an open conduit system. Such a low hydraulic gradient means minimal water flow (as observed when the well was sunk), and therefore fair opportunity for the water to equilibrate with the limestone aquifer.

And finally, it is perhaps worth a brief encounter with springs and two boreholes [17] drawing water in and around Parkmill (Fig. 104). One borehole, within the Heritage Centre[31] and penetrating to 32 m below sea level (-32 m OD), has been monitored for composition and, because of its constancy, is plotted with the percolation springs (Fig. 129). Springs that emanate beneath, in and above the leat at the Heritage Centre, close to the borehole, have similar near-constant compositions. They all tend to plot with 'main series' waters that indicate more equilibration with dolomite than others (Fig. 132). The second borehole, 160 m to the north towards Lunnon (from Llwyn onn – Ash grove), penetrates down to Limestone Shales and found copious fresh water at 50 m below sea level (-50 m OD).

Of interest is that these Parkmill waters, along with Pennard Castle [22] and others east of Wellhead [16 and 18], but not including Trinity Ilston [20], all have field TDS in excess of 300 ppm (Fig. 129), setting them apart from others; they all show some NO_3^- contamination so they are not ancient. Trinity Ilston is a variable mainly conduit-type spring, fed during drought mainly through valley-floor gravels and periodically drying, and otherwise running as the Killy Willy stream, which partially sinks [19] near Ilston Church to resurge 200 m upstream from Trinity Well. The 'TDS 300-plus' waters are spatially distinct from others and here are assigned to a 'Lunnon domain', apparently not actively connected with the Green Cwm System. The boreholes, with fresh water to 50 m below sea level (-50 m OD), show that there is no saline intrusion despite the fact that high spring tides reach close to the Parkmill Heritage Centre.

5.20 Sinks and springs: what have we learned?

The springs on Gower are supplied by local rainfall: it *is* all our own water. It was reasonable previously to invoke an exotic, north-crop source, not least because it actually does seem mysterious that the springs persist in acute drought. Setting this apparent mystery along with the knowledge that the limestone hills to the north across the South Wales Coalfield are manifestly 'full' of water, giving rise to substantial persistent rivers, e.g., Loughor and Tawe, hydraulic connection via the syncline is very appealing. It is not obvious that water flows continuously during drought, albeit reduced, unseen within valley-floor gravels towards Llethrid sink and on underground to Wellhead. And the existence of a perched aquifer of Lower Limestone Shales hidden on the flank of Cefn Bryn above the Holy Well springs is similarly not obvious. In both cases, however, we have found that conservative assessments of the relevant catchments, with application of rainfall records fortunately made nearby, show our own rainwater to be more than just adequate.

Though not good for farmers or gardeners, the epic 89-day drought of 2018 on Gower was fortunate for this project in allowing measurements of the waning discharge from Holy Well. This showed that a drought of some 140 days would be required for those springs to run dry; evidently 20-week droughts have never occurred in living memory, and hopefully they never will!

We have learned that Gower has two main types of spring, represented by the two main springs studied. Percolation springs show fairly constant compositions and slow responses to extreme variations of rainfall and temperature, typically with a considerable lag and smoothing-out effect. Holy Well constitutes a classic of this type.

Hybrid springs, on the other hand, are essentially variable in most respects; Wellhead resurgence is an excellent example. The hybrid is of conduit flow with percolation flow, the former of which can vary dramatically according to fair or foul weather, as can both during protracted drought. In rainstorms runoff into sinks is dominant, with underground rapid bypass via vadose cave conduits and resurgence via undersized conduits. The water flow is essentially surge- or wave-like, with rapid increase and relatively protracted decrease, although in complex rainfall patterns the rise and fall of water levels everywhere is unpredictable. Storm runoff causes erosion so that suspended sediment is added both along streams and in underground flooding passages.

In contrast, during dry episodes, the water flow can be almost imperceptible. Where rapid runoff once occurred, there is then a transition to percolation dominance; vegetation, soils and glacial till slowly release stored water that drains to the sinks, while underground the conduit flow is via slow-flow flooded (phreatic) passages beneath the water table. The Green Cwm System draining to Wellhead resurgence is a classic of the Gower hybrid type. Its water chemistry accords with rainfall, which tends to dilute the dissolved constituents, but its various catchments and complicated underground plumbing render even partial predictive understanding beyond possibility. A possible exception is in the low-flow involvement of dolomite (see below).

Also apparent is that central Gower springs near the peninsula coast reflect existence of former parts that are now below sea level and at least partially infilled and stagnant. It is suggested that the common occurrence of discharges from undersized and impenetrable conduits at or just above 10 m OD relates originally to the interglacial (Ipswichian) high stand of sea level around 123,000 years ago. Before then springs are likely to have been located close to the contact with low-permeability Triassic rocks, generally now at some 25 m below sea level (-25 m OD) and hence some distance beyond the present shoreline. Similarly, the latest post-glacial rise of sea level that led to alluvial filling of valleys, as in the Loughor Estuary and between Three Cliffs Bay and Parkmill, will have reinforced the infilling of lost former caverns.

This study has attempted to do some justice to the meticulously kept records of the daily monitoring that was undertaken for years at Wellhead Water Treatment and Pumping Station. In the quest to understand the water supply we learned that Wellhead resurgence water temperature was responsive to local climate moderated by the prevailing underground temperature of around 11-11.5°C. We found that the early engineers correctly identified a 'non-Gower' exotic component in the sediment discharged at Wellhead, but, by clay-mineral fingerprinting, we were able to prove that the exoticism resulted from the last (Devensian) glaciation. The local catchments are plastered in till and gravels transported by glaciers from the north crop of the Coalfield, and it is this 'exotic' material that went into the sinks and came out at the resurgence, with some picked up from underground where it had penetrated during the glaciation (Chapter 4).

The Wellhead turbidity records proved irresistible to try to analyse, especially knowing that there are daily rainfall records to match. We learned that turbid heavy-rainfall water surging into the Green Cwm System would increase the discharge of pre-existing clear water, so Wellhead engineers knew what was coming. Turbidity typically onset sharply and peaked to maximum very rapidly and then, overall, would wane more slowly, but the records are complicated by 'trains' of peaks. In brief, if there was a single heavy downpour evenly distributed over the local catchments, then there would be turbidity peaks that could fairly sensibly be attributed first to Llethrid, being the greatest input in terms of volume, then to Decoy and lastly to Willoxton. It is clear that spikes in turbidity arrived successively during sustained high discharge of water, such that they must record the arrivals of different conduit flows into the main surge. It is tempting then to try to construct a model plumbing system and order, but, realistically, there are too many unknowns and unknown unknowns for such an analysis to be warranted. Perhaps the remarkably consistent subsidiary 'three peaks' of turbidity that existed in the 1980s and thereafter disappeared will constitute the next enigma to fathom, now that the water-source issue is resolved.

The chemistry of the Gower spring waters analysed in this study shows interesting contrasts with other karst regions, reflecting not only the limited scale of the peninsula outcrop and its wild coastal setting, but also the impact of drought. While other limestone regions of Wales and England typically are dominated by Ca-HCO$_3$ type springs and wells, the Gower spring waters analysed here are mostly Ca-deficient (Fig. 134) and classify as Mixed type. This in part reflects the drought conditions at the time of sampling, since normal-to-heavy rainfall evidently leads to mixing-line trends towards more 'normal' Calcium type. In the Green Cwm System and at Parkmill the drought caused deep-aquifer waters in dolomitic rocks to dominate discharge and thus caused a trend towards Magnesium-type composition. Such rocks are largely bypassed in floods when ordinary limestones take most water.

Several hybrid springs showed systematically increased dissolved constituents during the drought and decreases afterwards (field TDS, Fig. 129); some dried up during the drought. The increases during the drought are taken to reflect increasing contributions from relatively enriched percolation waters, which otherwise tend to be overwhelmed by more dilute conduit flows. Ede (1975) noted that dripping waters in Gower caves, representing percolation, are more concentrated than underground conduit waters.

Springs compositionally distinct from the 'main series' limestone waters (Fig. 132 observation 5) involve percolation via Lower Limestone Shales (≤150 m thick, alternating calcareous shales and thin impure limestones) and proximity to the underlying Old Red Sandstone. Reynoldston PO spring involves only Old Red Sandstone strata and till but is unlike any composition from the Devonian sandstones of South Wales and Herefordshire (Fig. 134). Permeable Anglian till is involved in several cases, perhaps locally containing significant Triassic debris as reflected in elevated sulphate (SO$_4$) at Holy Well; industrial pollution is not to be expected there. The relatively limited residence time of water in the ground on Gower means that **there is little time for natural attenuation of man-made pollutants before return to the environment in streams and rivers** (e.g., fuel spillages or nitrogenous compounds from fertilizers or sewage).

There is a distinct signature of Na-Cl reflecting the seaside setting and susceptibility to maritime aerosols owing to prevailing south-westerly and westerly winds. Whereas Cl$^-$ concentrations are generally low (<50 ppm) elsewhere, those on Gower are especially high, averaging 85 ppm and reaching 146 ppm (Fig. 131); relative to other areas, the waters trend towards Chloride-type compositions (Fig. 134). Although there are distinct tidal influences on some low-lying coastal karst areas of South Wales, e.g., around Ogmore Vale (Robins and Davies 2015), there is no evidence of coastal saline intrusion where it might be expected, e.g., at Parkmill where fresh water is found to 50 m below sea level (-50 m OD). The most marked Na-Cl contamination is in springs high on Gower where there is no hydraulic connection to the coastal limestones. On the other hand, fresh-water springs at beaches along the shore are common, locally producing quick sands, and flow into the marshes and farther offshore is probable but unquantifiable.

Sinks and Springs Notes

[1] For any rock unit that continues at depth underground, geologists refer to its occurrence at the surface as its outcrop. In this instance coal-bearing layers of tough sandstone and relatively soft mudstone 'crop out' respectively to form ridges and hollows that impart a pronounced grain in the landscape. The rocks are widely blanketed by a thin superficial layer of glacial deposits but are exposed where this is missing.

[2] Map redrawn from the simplified 1:625,000 Bedrock Map of Wales in Howells (2007). Sections, redrawn so vertical equals horizonal scale, are from the more detailed British Geological Survey 1:50,000 Geology Series Sheets 247 Swansea (published 2011) and 230 Ammanford (published 1977).

[3] It was so secret that I never recorded its whereabouts. Now, some 55+ years later, I cannot find it.

[4] There are also important springs in western Gower, for example at and above Pitton (Pitton resurgence and Talgarth's Well; SS429880 and 427881), at Delvid (SS424927) and Lagadranta (SS425928) (from Llygad Rhandirau, meaning source of the stream on the sharelands; Philip Stephens pers. comm.), at Raven's Well (SS430911) and at Burry Head (SS456903), respectively draining catchments on Rhossili Down, Llanmadoc Hill, Harding's Down and between Harding's Down and Ryer's Down. The fascinating spring of Ogof Ffynnon Wyntog (SS432863), by The Knave, has been discussed in Chapter 4. The spring in Horton (SS474857) is not for drinking! In eastern Gower a substantial spring in Caswell Bay (SS593877) issues water drained from south Clyne Common, while, in the late 1800s, water from a spring on the west flank of that bay (SS589875) was pumped by a windmill near the top of Redley Cliff to add to a local supply. Subjected to inconvenient winds, commonly either too light or too strong, the windmill was badly damaged by a gale in the winter of 1887/8 and had ceased to be used by 1900 (detail from Wendy Cope at: https://sites.google.com/site/ahistoryofmumbles/the-windmill-at-caswell-the-water-supply). The Devil's Spring (aka Westcliff Spring; SS5472187424) beneath Southgate is one of several along that stretch of coast and reputedly polluted.

[5] This remarkable record of temperatures, rainfall and sunshine was collected within the study area (SS531888) by John S Powell. With impressive speed it was made completely and freely available via the Met Office National Meteorological Archive. It proved invaluable.

[6] Some authors use 'resurgence' for a spring where a sinking stream is known to reappear, but on Gower that is not straightforward to apply. Wellhead [14] is the only case where 'resurgence' is sometimes used, as here; its water is not just from the known sinking streams.

[7] From notes prepared by Henry Davies, Engineer and Surveyor, for the opening ceremony at 'Parkmill Pumping Station and Water Treatment Plant', 8th December 1954.

[8] Various fluorescent dye additives, mainly fluorescein or rhodamine, were used to prove water-flow linkages. Small quantities of fluorescein would rather alarmingly turn huge amounts of water a lurid green. Normally, activated charcoal in small permeable bags was placed in streams to capture the diluted

dye and detect unseen flows. As youths we discovered that the urinals in a certain pub discharged directly into the nearby stream, which had turned a fantastic green!

[9] Decoy Pond (SS515909), now drained, originally was created for 'harvesting' of ducks that were enticed to the water by decoys and then driven to their demise along curved channels away from the pond. It is visible in Figure 108, half-way between Broad Pool and Decoy sinks.

[10] Baynton (1968-1969) wrote that the initial evaluation of the resurgence for general water supply, undertaken before the Wellhead station was built, stated "microscopic examination of the sediment revealed it must come from a source outside Gower, the most likely place being an area above Ammanford". Frustratingly, despite extensive searches, no record has been found detailing the microscopic evidence behind this. However, it is most likely to be in the presence of abundant quartz grains (see text).

[11] Analyses were kindly and expertly undertaken by James Utley at the University of Liverpool. Details available from the author.

[12] Complete daily records are only available for this interval; hand-written, they record 'weather' and 'gallons surplus', which is water leaving the resurgence pond over the weir to enter the river. The actual discharge is this surplus plus the extraction amount (~1.6×10^3 m^3/day); it is a minimum because during flood conditions weir measurements are limited so that any excess above 48×10^3 m^3/day is not recorded. Obtaining each figure for annual discharge required adding together 365 sets of 6- or 7-digit numbers; tedious to say the least!

[13] The 'more than' symbol (>) reflects the fact that some discharge water is lost over the weir and unmeasured. This, however, would never be sufficient to alter the case that the annual rainfall is quite sufficient for the evaporative loss plus discharge without need for any supplementary supply.

[14] Baynton (1968-1969) first noted the subterranean sediment source of the turbidity seen at Wellhead.

[15] High tide was at 10.39 (British Summer Time) and not tabled as especially high, but the storm was notable for its low barometric pressure and, since high spring tides reach into Parkmill, the exceptional flooding may have resulted from the rain runoff being impounded by storm-surge sea water. Nevertheless, tide is not needed for damaging floods. On 5th June 1931 storm water rushing down Ilston Cwm took out the walls of Parkmill school and police station (in Stonemill, the eastern part of Parkmill; SS550891).

[16] At this time the dig was in progress. A stone embankment and steel gate were yet to be built to restore the overflowing stream to its original natural course into the dry valley. That restoration is now complete. In extreme floods the bypassing water here is up to 2 m deep.

[17] The author is indebted to Ian Murphy and Jason Hyatt of Dŵr Cymru (Welsh Water) for both allowing and facilitating actual access to these and other records remaining at Wellhead, and to Peter Sambrook for aiding in their appreciation; his handwriting in them is pervasive!

[18] Weir discharge is a measure of water that overflows from the resurgence pond (Figs 106 and 112) and continues as the river in Parkmill valley. The measurement recorded is limited to values up to about 10.5 million gallons/day (48×10^3 m^3/day), above which the level could not be sensibly measured.

[19] Some further consideration of the behaviours accorded to the three inputs to the Green Cwm System is provided as an appendix following the Springs Gazetteer at the end of this book.

[20] Strictly, Holy Well is the name of the most productive spring. Its discharge was collected in a 37,000-gallon storage chamber into which were piped discharges from East Well [9], West Well [5] and minor

nearby 'issues'. The collection and treatment site, leased from the Duke of Beaufort in 1914, was known generally as Holy Well and measures of discharge referred to Holy Well were actually of the combined flows.

[21] ORS sandstones and conglomerates can hold water, mainly in near surface fissures and weathered zones (Moreau et al. 2004). For this exercise the ORS *rock* is ignored, but not the overlying head; its inclusion would add slightly to the overall storage capacity, which could only improve the aquifer potential.

[22] Channel sites at East Well and near Holy Well captured the entire discharge. A plastic-coated tablecloth secured so as to line the channels allowed water to be collected in timed intervals and accurately measured. Plotted results are the average of at least 10 measurements. This could not have been done without Peter Sambrook, the most accomplished measurer of Gower water and expert table-cloth manager.

[23] Figures derived from the raw data.

[24] Hot Well was named for mists supposed to derive from heated water; data do not support this thermal origin.

[25] pH was typically alkaline, mainly in the range 7.5-8.9; the full data are not relevant here.

[26] Groundwater hydrologists commonly use pollution as an indicator of maximum age. Agricultural pollution signifies an age of no more than a few centuries; pollutant CFCs (chlorofluorocarbon compounds) did not exist until the first half of the 20th century.

[27] Mixing lines are features of graphs depicting chemical composition. They reflect the relative contributions of the 'end-members' that become mixed. Pure water mixed with water containing 100 units of solute 'x' can only (ignoring trace quantities) range in content from 1 to 99 units of 'x', depending on relative contributions to the mixture. The 2-part mixtures must define a straight mixing line, in this case between 0 and 100 'x'. In natural waters we often consider compositions containing various solutes (see below), but the principle remains the same; a mixture of complex composition A with complex composition B must fall on a straight mixing line joining the A and B end-members. Graphs depicting 3-part mixtures must have a composition within the triangle formed by joining the three component compositions...

[28] Ionic balance concerns charges that are chemically equivalent; hence the measured concentration of each ion has to normalised by dividing by its atomic weight and multiplying by its valency (ionic charge). Hence, e.g., Ca^{2+} at 60.9 mg/L (from table above) is divided by atomic weight 40.08 and multiplied by valency 2 = 3.039 milliequivalents/L (meq/L). Ionic balance then, in meq/L, is: (Σcations-Σanions) * 100 / (Σcations+Σanions).   (Σ means total).

[29] The author is immensely grateful to Elliott Hamilton and Michael Watts for conducting these analyses.

[30] Tufa, now severely damaged, lines the 'cascade' beneath the spring of Ogof Ffynnon Wyntog (SS432863) and probably also formed by CO_2 loss to atmosphere.

[31] Roy Church generously allowed frequent free access to the springs and borehole in Parkmill Heritage Centre.

A gazetteer of the more photogenic springs is provided in the Appendix Section 7.3 on page 266. Further analysis of Wellhead data is provided by Geoff Williams, in the Appendix Section 7.4 on page 274.

Broad Pool, a pond on cavernous limestone by virtue of a sealing layer of clay-rich Devensian glacial till. In the summer of 1984, it dried up completely.

Depth versus temperature measurement of a Gower borehole, with Gareth Farr.

Always first to the spring.

6 STORMS, DUNES AND ROCKSLIDES

6.1 Introduction

This chapter concerns *historical* changes of the Gower coast. It focuses on rather striking phenomena and features that many people will be aware of but perhaps may not have fully appreciated. Historical changes have mostly occurred only gradually, in a time span similar to our own lives, but some have been dramatic. Hopefully what follows will illustrate this. The choice of topics is necessarily selective and while humans have modified Gower in the ~11,650 years of post-glacial times (Holocene Epoch), for example with cairns, forts and field systems, archaeological features are addressed in other publications and are mostly neglected here. We will be concerned primarily with natural changes and how some of these have affected people.

Peninsular Gower openly faces the Atlantic Ocean, with prevailing strong winds from the southwest or west and with far-travelled swell waves that build up in the shallowing water on approaching the coast. The entire peninsula feels the influence of the sea, with salty aerosols driven inland from waves breaking along the shores, both rocky and sandy, in just 'normal' gales (Fig. 138).[1] Gower spring waters register salty rainfall (Section 5.18). Big storms that these days are given names mostly initiate as huge energetic phenomena along the western Atlantic coast and their considerable remnants can arrive here, often with serious consequences. We tend to think in terms of climate change now causing increasingly frequent extremes of heat, drought, wind and flood, but there have been some pretty awful episodes in the past, as we shall see.

*Figure 138. Atlantic storms impacting westernmost Gower (vicinity of SS4087). (**Top, previous page**) Cliffs between Fall Bay and Mewslade Bay reach 50 m OD and at high tide take the full impacts of large breaking swell waves. (**Bottom, previous page**) The head of Worms Head is over 45 m OD and takes the brunt of swell waves during a south-westerly gale. (**Above**) Waves break more than 2 km from the shore in Rhossili Bay during a fresh to strong westerly gale.*

As the Late Devensian ice sheets melted and receded globally, rising sea level (Fig. 13) flooded the broad valley of the ancestral River Severn to form what would become the Bristol Channel. By 8,000 years ago the rise rate slowed and from 6,000 years ago sea level was only some 5-7 m lower than it is now (Fig. 139) and rising only very slowly. Big storms from the Atlantic drive deep-water sands in the Bristol Channel landwards (Pye and Blott 2009) and the net effect of a series of them is that sand accumulates in sand waves, banks and bars within the depth range of normal wind-generated waves and hence into the intertidal range. The post-glacial advancing sea thus drove sands reworked from the glacial outwash west and south of Gower towards the land. By 6,000 years ago the south coast would have appeared similar to today, except for extensive moundy marsh land where the moraine of the Tawe – Nedd ice remained exposed (Fig. 139).

Tides would have been of the same considerable range then as now and tidally exposed sand banks would have been similar too. Sand banks and barrier beaches would have fronted the Loughor Estuary and sand driven ashore would have built precursors to the modern dunes at the back of the bays exposed to the west and southwest. Valleys that previously were flushed and partly filled by glacial outwash debris, such as at Caswell, Pwll-du, Three Cliffs, Penrice, Llanmadoc (Burry) and Loughor, became sediment traps and thus gradually filled with sand, silt and mud (alluvium) that ultimately formed the sedge wetlands and salt marshes of today.

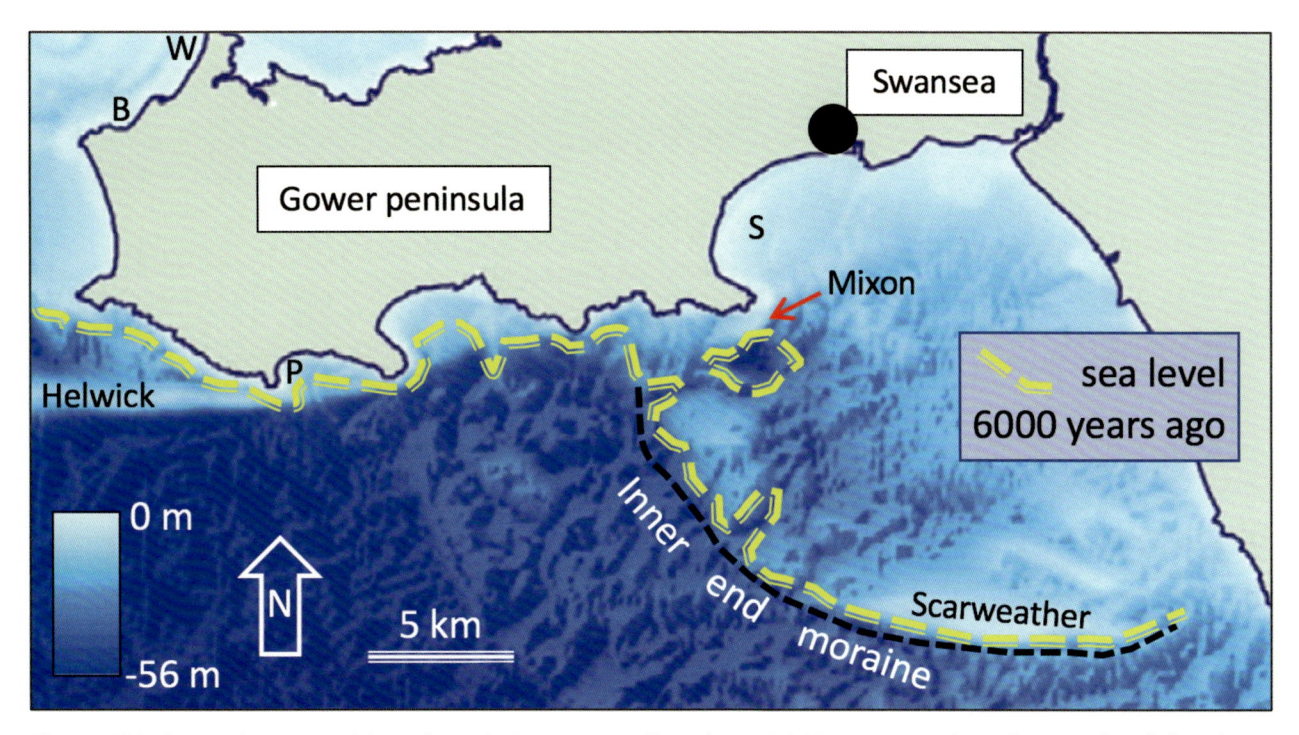

Figure 139. Approximate position of south Gower coastline about 6,000 years ago, based on sea level then being some 5-7 m below OD, according to data from inner and outer reaches of the Bristol Channel (Shennan et al. 2006). The base map is from Gibbard et al. (2017). The moraine is the Late Devensian recessional feature due to the Tawe – Nedd glacier (Sections 3.6 and 3.17); it may have become somewhat reduced in the last 6,000 years owing to marine erosion. Neither Helwick nor Scarweather sand banks would have existed as such until the sea depth and tidal regime became more like they are today. A lagoon may have existed off what is now Mumbles Head, but cricket and tea on the Mixon sand bank would have to wait most of the full 6,000 years for the tidal feature to form there (web site: A history of Mumbles - Cricket, tea and tragedy on the Mixon Sands by Carol Powell).

*Peat layers dating from about 5,500 to 3,790 years ago are today exposed between the tidal limits at Port-Eynon (**P**) and material from layers at Broughton (**B**) is ~4,000 years old (Philp 2018 and pers. comm.). A cockle midden was briefly exposed as a layer within peat close to submerged tree remains at Whiteford (**W**) (Fig. 140), perhaps related to Bronze Age occupation (4,100 to 2,750 years ago) and certainly significantly older than the Saxon (410-1065 AD) sea-food middens known within the dunes nearby. Peat deposits with submerged roots and trunks of oak, hazel and birch, and relics of Bronze Age and Iron Age occupation, are well described from Swansea Bay (**S**) (Sherman 2011).*

So-called 'fossil forest' remains of tree stumps and root systems, commonly in growth positions and with long sections of trunk lying nearby, are widely exposed with peat layers where beach sands have been stripped during storms, for example at Whiteford, Broughton, Port-Eynon and Swansea Bay. Artefacts of human activity, especially ancient trackways, that became exposed in medieval times may have been interpreted then as evidence of villages lost to the sea, which, through time, became folk lore handed down through successive coast-dwelling communities. Perhaps more telling of the environment was the temporary exposure, before loss by erosion, of a cockle[2] midden within the peat layers at Whiteford (Fig. 140). This confirms existence of just the right intertidal sand grain size and water depth needed by healthy cockles, as gathered by the author to this day on the intertidal shore of Whiteford and Broughton.

*Figure 140. Holocene remains at Whiteford Sands (vicinity of SS443960). (**Left**) Growth position roots of a deciduous tree, probably oak, which lived for more than 40 years according to growth rings (inset). (**Right**) Ancient cockle midden within peat. The large size of the cockles (ten pence coin for scale) is the same as those that can be collected near here today in non-commercial quantities. Abundant evidence nearby of Bronze Age inhabitation of the area favours such an age (4,100 to 2,750 years ago) for this midden near the top of the peat layer.*

Considering the more recent history of the medieval period and afterwards, the effects of the slowly rising sea level seem apparent in successive losses of coastal land and some infrastructure of Gower, especially during the so called 'Little Ice Age' from about 1300 to 1850 AD. Some epic storms destroyed previously habitable spaces at scales that suggest involvement of exceptionally raised seas resulting from intense low barometric pressures, known as storm surges.

A great storm and associated surge coincident with exceptional high tides on January 30th, 1607, devastated land and took many lives as it funnelled far up the Bristol Channel (Bryant and Haslett 2002, 2007; Horsburgh and Horritt 2006). Its impact on Gower was 'moderate' but at Tears Point the sea flipped over large slabs of limestone (Fig. 34) and it apparently breached the barrier dunes at Llangennith. The event was investigated regionally in detail and at first interpreted by Bryant and Haslett (2002) as possibly resulting from one or more tsunami waves. However, the long duration at high level of the associated floods, together with the fact that a coincident intense low-pressure system moved east away from Wales and caused similar severe problems on the North Sea coast, confirmed that the event was caused by an exceptional storm surge combined with the high tide (Horsburgh and Horritt 2006). The latter observation constituted "slaying of a beautiful hypothesis by an ugly fact" (TH Huxley 1870), as no tsunami could so affect the North Sea coast as well as the Bristol Channel.

Apparently according to a more or less 100-year recurrence interval, a great storm in 1703 demolished the mansion in the vicinity of the Port-Eynon Salthouse that had stood there since the early 16th century, as well as wrecking the fleet of skiffs that were used to export locally mined ochre and were moored behind Sedgers Bank (Tucker 1951). Sedgers Bank (SS471843) then was more extensive, but sea level rise and storms have reduced it now to a very narrow and diminishing tidal islet just 60 m long. Similarly, the Old Rectory at Oxwich was destroyed by 'tidal action' in 1805; the chunk of land where it once stood now is mostly missing (see 1790 painting at https://artuk.org/discover/artworks/the-old-rectory-oxwich-224880).

Because the tidal range around Gower is large, for spring tides about 6 m in the Loughor Estuary and 10 m near Mumbles, vast tracts of sand are exposed repeatedly during low tides. Strong westerly and south-westerly winds then readily blow sands onto the shore and, as clearly shown by Figure 141, repeated exceptional gales may drive it far inland.

6.2 Driven out by sand

Through the 'Middle Ages' until about 1250 AD the climate of Britain had been benign, a time referred to as the Medieval Climate Optimum or Warm Period. Thereafter conditions deteriorated, leading towards the so-called 'Little Ice Age', and from about 1300 AD episodes of cold and of strong winds became troublesome (Brown 2015). Irrespective of the ongoing political turmoil and battling across Wales, which was bad enough, life in rural Gower throughout this time probably was mostly grim. Frequent storms arriving from the Atlantic and the incumbent dynasty of harsh Lords of the House of Braose rendered life hard for many.

The de Braose stronghold of Pennard Castle was built with stone towards the end of the 13th century, on the site of a previous less substantial ring fort. A group of timber and daub dwellings and the stone-built St Mary's church, in existence by 1291 (Latimer-Davies 1928 in Lees and Sell 1983), lay close by constituting the original village of Penarth (Pennard; Fig. 141).

Fable has it that one of the nasty men of the de Braose dynasty[3] really annoyed some frolicking fairies so that these ethereal spirits in revenge brought a storm that overnight besanded the castle. It is more probable, however, that trouble with blown sand developed late during the 13th century and continued at least into the 16th century. Sandy waste lands were documented in 1317, land rents were reduced in 1478, and the besanding caused abandonment of the Pennard church and associated lands by 1525.

Similar to Pennard, an early church uncovered in Penmaen Burrows evidently was abandoned early in the 14th century. This church was claimed to lie within a buried village named Stedworlango (Morgan 1894), but a tythe map of Penmaen and interpretation that 'worlango' derives from 'furlong' suggested to Seyler (1920) that the name better fitted a site nearer to Penmaen House (Fig. 141) and that there was no evidence for a village setting of the besanded church. Also, the original Nicholaston church, built in 13th century, was literally removed piecemeal from Nicholaston Burrows to the present site in the 14th (where it was later rebuilt). The stem of the font in Nicholaston church is described as part of a stalagmite utilized in the original building where, nearby, quarrying had broken into a limestone cave. The font stem is carved from tufa that could have been found in the cave; lacking concentric growth structure, it is not strictly stalagmite. Below modern Rhossili, in the 'Warren', the old church and village there suffered both besanding by 14th century storms and erosion of the coast by the sea (Toft 1985, 1988), as continues today.

Those of us who have been here for 50 years or more will know that much sand has been lost from some dune fields (e.g., Fig. 141) and also that some new dunes have been 'quick' to grow, e.g., at Hills Tor (Fig. 142). We will consider the de-sanding of Gower beaches later, but for now it is relevant to know that the climate has improved since the end of the 'Little Ice Age' in the mid 19th century, at least in the sense that seriously damaging storm-related besanding has ceased.

Ophrys apifera, the bee orchid, another secret hidden in the dunes, once found never forgotten.

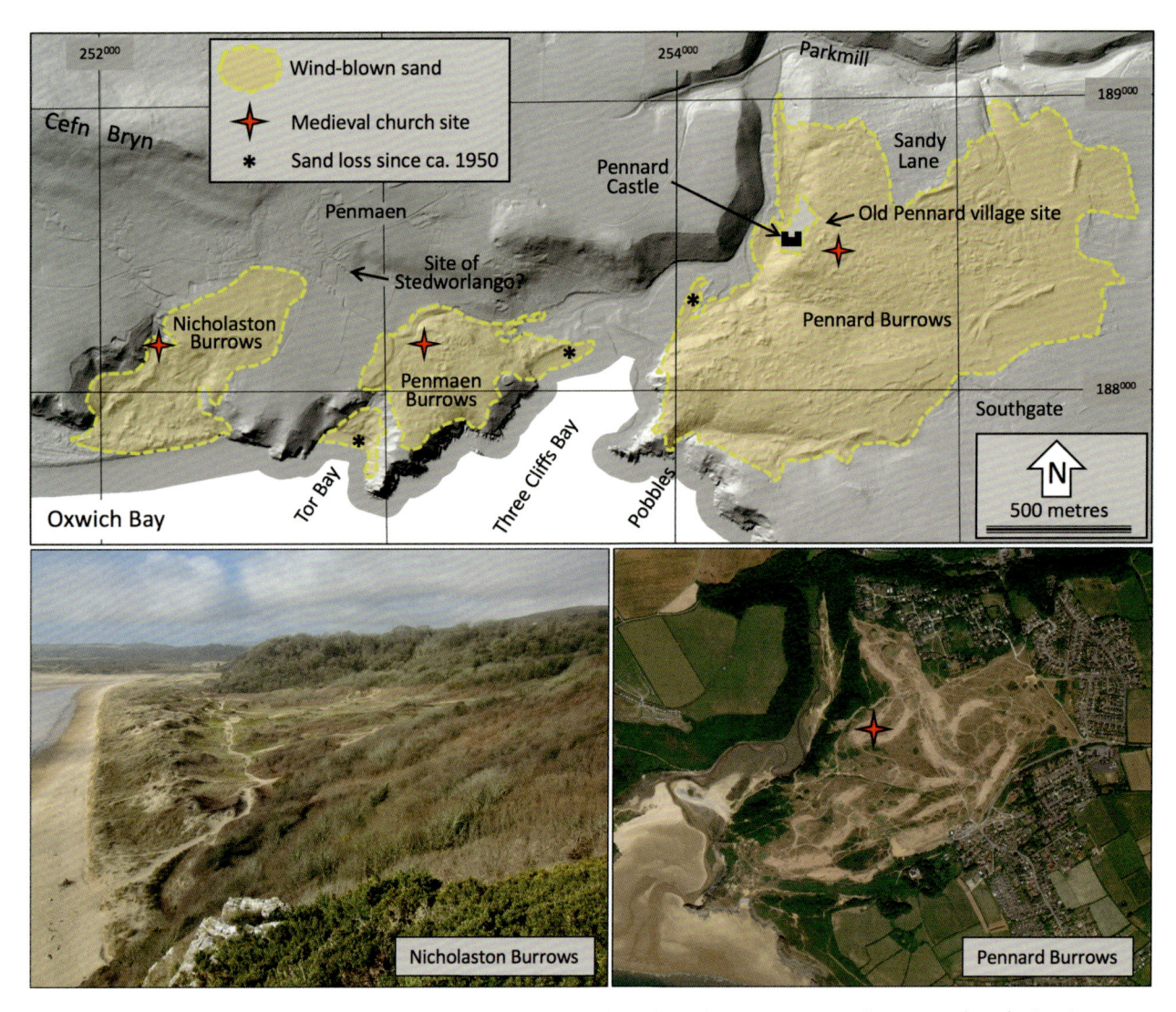

*Figure 141. Blown sand that caused abandonment of Medieval settlements on south Gower. (**Top**) The dunes distributions, forming the extensive 'burrows' across land widely over 60 m OD, clearly shows the prevailing wind direction of the time, from the southwest. The blown sands mainly advanced from Oxwich Bay, initially stripped during low tides, up the Nicholaston, Tor Bay and Pobbles slades cut into the cliffs (Section 2.10), and up the valley of the Pennard Pill behind Three Cliffs Bay. The sand limits drawn are slightly modified from the British Geological Survey 1:50,000 Superficial Deposits Map Sheet 247 (BGS 2011), superimposed on a shaded relief LiDAR image (base data © Natural Resources Wales). Asterisks mark sites where the mismatch between the mapped limit and ground topography (shaded relief) is a reflection of sand loss since the map revision of around 1950. (**Bottom left**) View towards west. Nicholaston Burrows rise from Oxwich Bay to the level where the photo was taken at 55 m OD and spread northeast onto the platform above. (**Bottom right**) Google Earth view of Pennard Burrows (and golf course) widely over 65 m OD and extending beyond, beneath the housing estates to the northeast. The site of the medieval besanded church is shown.*

6.3 Comings and goings of dunes

It is widely believed that the major dune systems of Gower took on their main distribution and forms during the stormy besanding period of medieval times onwards from the 12th century. However, there inevitably would have been a long history of dune development ahead of the rising sea levels, with the landwards beach migration that followed the last glaciation. Of course, the early-formed post-glacial dunes would be reworked so that what we see now is mostly their latest stage of development. Here we demonstrate the comings and goings with particular reference to Whiteford Burrows; in the following section we consider evidence for dunes long gone.

Whiteford Burrows in the main have probably been stable for thousands of years. A barrier sand bank and some dunes are likely to have formed nearby as rising sea level reached over the peat and 'fossil forest' (Fig. 140) that Bronze Age people had inhabited before some 2,750 years ago. Additionally, a recessional moraine underlies Whitford Burrows (Fig. 143), at least in part, so it would be wrong to attribute the growth of the entire promontory to the medieval besanding episode. Medieval sea-food middens in the middle of the promontory are not significantly buried and the dunes and slacks there reflect long-term stability. However, LiDAR images now available show that they are susceptible to significant marginal erosion and rebuilding (Figs 142 and 143).

Figure 142. Whiteford Burrows. (continued next page...)

ALL OUR OWN WATER

*Figure 142. (**Top, previous page**) Whiteford Burrows viewed towards south-southwest (see Fig. 143). This middle section of the dune-capped barrier of the Loughor Estuary (visible on the left) has been eroded back as much as 230 m in the past 50 years and sand is also lost from the tidal beach to the west. Recent dune erosion has exposed an in-situ mound of glacial moraine with boulders, cobbles and pebbles set in a clayey matrix (Fig. 143; SS446962), which is flanked by younger peat layers. Extensive boulder fields extending north from here to beyond the Whiteford lighthouse are marine-reworked remains of a recessional moraine dominated by Coalfield erratics (Section 3.14). Drone photo courtesy of Andy Freem.*

*(**Bottom, previous page**) Whiteford Burrows viewed towards northeast from Hills Tor. In the past 50 years sand has accumulated so that the dune limit has migrated westwards from the middle of this view to the present position on the left (see Fig. 143).*

*(**Above**) While the southern part of the main dune area extended westwards (middle view) a new pile of blown sand accumulated against the limestone cliff of Hills Tor building from the level of the persons (at about 6-7 m OD; **red circle**) to 30 m OD (**yellow arrow**). This is linked to the beach via a wind gap and has largely been stabilized by the binding due to underground rhizomes of the pioneer plant Marram Grass.*

*Figure 143. (Page opposite). (**Left**) Sand dune erosion and growth at Whiteford since 1963. The pre-1963 line is from the Ordnance Survey map published in 1964 and the 2018 line is from Google Earth. The **red asterisk** is the location of the drone that photographed Figure 142 top image, the middle image was photographed from Hills Tor and the **black asterisk** locates the dune that has formed against Hills Tor, shown in the bottom image. The western and northern dune-field limits shown are the reaches of high spring tides. During low tides Whiteford Sands are exposed to prevailing westerly winds.*

*(**Top right**) Recent new exposure of Whiteford Devensian recessional moraine poking through the eroding dune near Whiteford Point (at **gold star**). This is dominated by coalfield erratics, mostly coarse sandstones with ironstone. Scale is 10 cm.*

*(**Bottom right**, located at **green star**) Extensive storm beach composed of reworked Whitford Moraine dominated by coalfield slabby coarse-grained sandstones, probably mostly Pennant Sandstone. The view is a little north of due east, up the Loughor Estuary. Scale is a Large Munsterlander.*

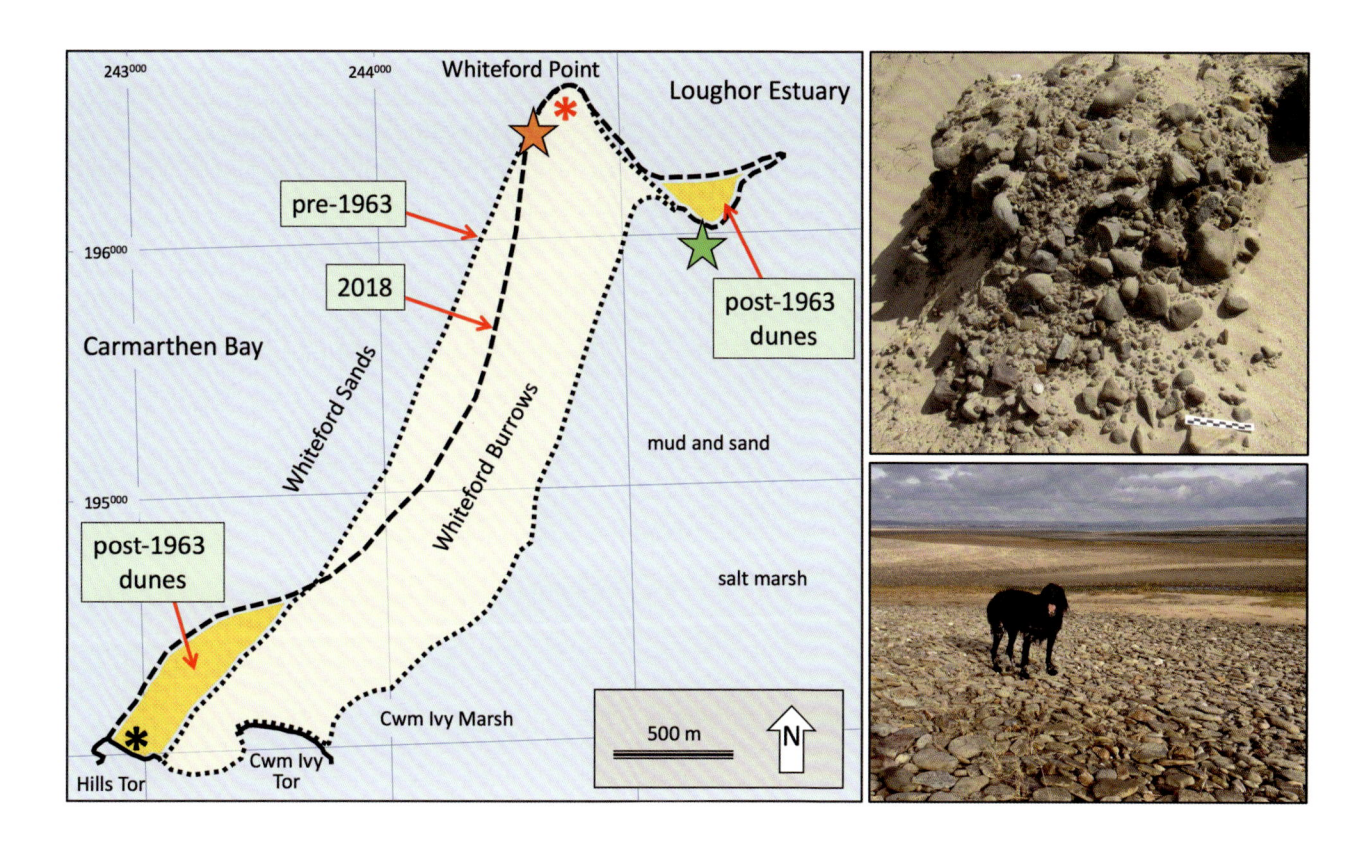

6.4 Transient images of missing sand

In the chapter on landscapes, we considered various styles of limestone dissolution that can contribute to coastal erosion. It was suggested that the beautiful and intriguing microrills at Little Tor (Fig. 17 Top) register dissolution along the contact of the limestone with the sand of a long-lived dune that had been eroded away, but for which there is photographic evidence (Fig. 20). Microrills are well known (e.g., Ford and Williams 2007, p. 324), but, perhaps, the significance of their occurrence is not.

The extraordinary sinuosity of some microrills is difficult to explain unless capillary action is involved in the migration of the water that dissolves the rock. Capillary action can only occur in narrow tubular spaces, including connected pore spaces, so for cases on the Gower coast we have to invoke the former existence of sand that is now missing in front of the microrills. On steep surfaces microrills generally trend downwards strongly influenced by gravity, but on some surfaces the mm-scale grooves seemingly defy gravity with tortuous deflections. Here surface tension and intermolecular forces (liquid adhesion and cohesion) within micro-scale spaces must be involved. Unaltered microrills are always about 0.5-2.0 mm wide, which reflects the physics of the capillary flow and its limitation to operate only at the small scale. Water can migrate along contacts in almost any direction according to runoff or evaporation involving capillary action, and it seems from occurrences of microrills that this is most likely to occur against homogeneous fine-grained limestone, typically micrite (calcareous mudstone), which in some cases has the polished appearance and feel of porcelain.

Microrills are especially well developed on surfaces of micrite layers exposed in low cliffs on the southeast flank of Caswell Bay (Fig. 144). Although this extremely tough rock superficially appears homogeneous, some patterns pick out and deflect along subtle layering. Other examples occur in the north of the bay, near the piped stream outflow from the nearby spring

237

(SS592876). A question logically arises regarding the former existence of the sand that must have been banked against these low cliffs for some significant duration. There are no dunes there now. The cliffs on the southeast side of the bay, close to the microrills, have Patella 123,000-year-old deposits on them and so they have not significantly altered in form. As microrills do not form along rock fractures here, it seems inescapable that sand was once banked here by south-westerly gales, rather as did occur at Little Tor (Fig. 20). No old photographs from the 1800s show such sand, and it is speculated here that it may have been a medieval dune accumulation. If microrills do form by corrosion behind long-lived sand banks, or dunes, that are now missing, they are in effect a key to reconstructing an historic past. But, as we shall see, the agent that forms dunes can also cause their record to be erased.

Figure 144. (Page opposite). Microrills on fine-grained limestone (micrite) at Caswell Bay (SS594873).[4]

(Top left) *Microrills occur on smooth surfaces that commonly have the appearance of being polished. These surfaces occur in patches that stand proud of more normally weathered limestone. Microrills are typically within about 0.5-2.0 mm wide, in this case widening downwards as they become confluent.*

(Top right) *View northwest towards Caswell Bay. All of the microrill occurrences are on the distinctly grey micrite beds thinning from ~2 m to ~1 m away from the camera. Please just look and wonder!*

(Middle) *Although the microrills generally trace downwards, they locally show convergence or divergence, while some straight courses clearly follow pre-existing weaknesses in the micrite and are wider than normal.*

(Bottom left) *The extreme sinuosity of the microrills requires the involvement of capillary forces acting in addition to gravity. Traces connecting across several grooves reflect subtle influence of original sedimentary layering in the micrite (near horizontal or slightly down to left).*

(Bottom right) *Although the micrite superficially appears homogeneous, in places the microrills clearly show deflections along some otherwise cryptic sedimentary bedding or lamination, and they also pick out some fractures at steep angles across the layers.*

Microrills occur at the base of Hills Tor (SS428940), where blown sand tends to come and go banked against the cliff on the seaward side of the 'new dune' (Fig. 142 p. 236). What is interesting here is that the exposed microrills have in places been progressively obliterated by sand blasting (Fig. 145), especially on rock faces fully exposed to the brunt of wind-driven sand at the top of the west-facing beach. Thus, the microrills are erased, technically speaking 'ablated', when the sand behind which they formed is removed; sand deposited from wind and hiding microrills as they form, gives way to erosion by wind-driven driven sand and hence erasure of the microrill record.

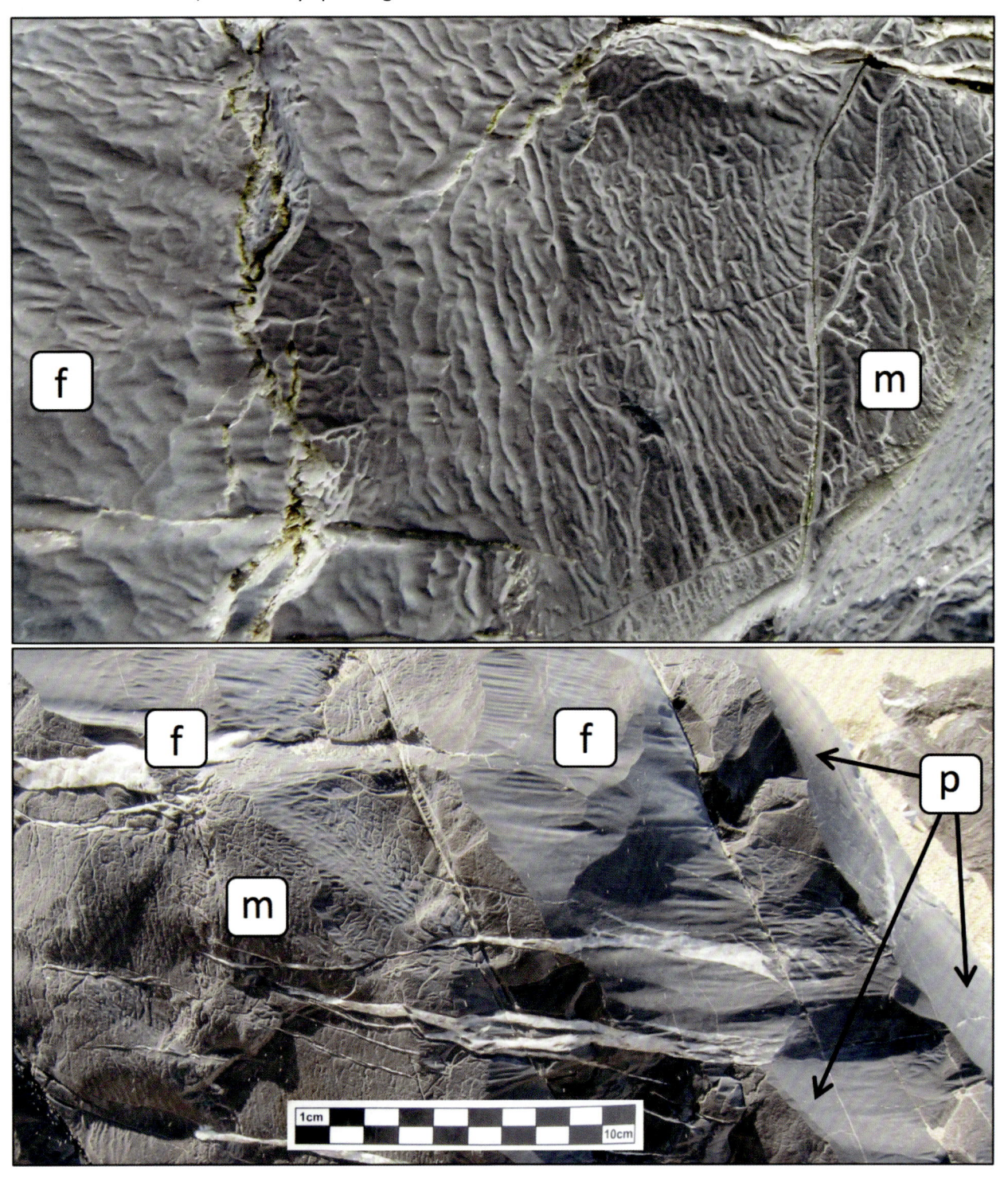

Figure 145. (Page opposite). Microrills eroded by wind ablation at Hills Tor (SS428940). Wind direction right to left.

*(**Top**) Partially sheltered microrills (**m**) on the right become progressively obliterated from the middle towards the left, where horizontally orientated wind-ablated flutes (**f**) gradually wear down and remove the rills. Field of view is roughly 0.7 m wide. (Photo courtesy of Jessica Winder).*

*(**Bottom**) Microrills (**m**) are preserved on the lee side of the limestone outcrop, while adjacent sides clearly are being worn down to produce sharp-edged smooth surfaces with flutes (**f**) parallel to the wind direction, and sides more directly facing the wind are polished flat (**p**) by sand blasting.*

6.5 Beach stripping and revelation by storm

It is quite normal for 'big' storms from the Atlantic, for example those now named by the Met Office, to remove considerable volumes of sand from the beaches exposed to them. Especially when coincident with high tides and involving storm-surge-elevated sea levels, the water that piles against and pounds the shores necessarily has a return underflow that sweeps the seabed sand and suspended sand off into deeper water. This normal process tends to be prevalent in winter and the sand then gradually returns with less energetic weather during late spring and summer.

It is apparent, however, that some Gower beaches over the past decades have lost so much sand that it is no longer possible to run and play freely in the surf without stubbing a toe, or worse. Notably, Port-Eynon, Horton and Langland beaches have progressively lost sand and in places become stony with exposed patches of ancient peat and clay. This long-term loss is of concern regarding impacts on tourism and it is considered in the next section. Here we look at the less serious seasonal beach stripping and focus on some rather pretty, if ephemeral, results.

Many of the sandy coves on south Gower mark geological fractures (faults and joints) that erosion has exploited. Most of the folding and faulting that accounts for the patterns of the rocks today occurred progressively from Late Carboniferous through to Early Permian times, which was about 300 to 290 million year ago. The rock deformation caused faults to form along which open spaces became filled with mineralising fluids, mostly 'saline' water, from which mainly calcite ($CaCO_3$) but also hematite (Fe_2O_3) and galena (PbS) were precipitated. Late Triassic dilation of the faults allowed fine grained red sediments to become washed down into many of them. Here we examine some examples of this mineralisation where storms reveal beautiful fault-rock formations that tend to be covered in fair-weather conditions.

The recent 'big' storm Ciara originated across the Atlantic as an extratropical cyclone and impacted Gower on 8-9 February 2020. It was followed less than a week later by Storm Dennis, on 15-16 February. At Mewslade Bay (SS420871) so much sand was stripped by the two storms that the bare-rock platform beneath was revealed to show beautiful cleanly polished calcite veins amidst broken limestone (Fig. 146). Under normal beach conditions practically no calcite veining is seen, so evidently some 80 cm thickness of sand had been stripped. Half of it had returned by early August.

ALL OUR OWN WATER

*Figure 146. Sand loss at Mewslade Bay due to Storms Ciara and Dennis in February 2020. **(Top left)** View west from above Mewslade Bay, with Tears Point and Worms Head in the distance. Compare this normal swell-wave scene with Figure 138 Top. **(Top right)** Detail of Mewslade beach showing the view-point location of the pictures taken after the two storms. **(Middle)** Views inland and towards the Bristol Channel from the spot located above. Rarely seen, the beach platform fault rocks are exposed showing rather beautiful clean polished surfaces of calcite mineral veins amidst broken limestone chunks. **(Bottom left)** Calcite vein margin not normally visible; pole is about 1 m long. **(Bottom right)** Radiating calcite crystals that grew upwards within a former fault void containing mineralising water (wedged limpet shells for scale).*

Low-tide views down onto the rock platforms from the cliff tops almost anywhere on south Gower show a spectacular pattern of the limestone layers (bedding) cut by numerous fractures, some of which are faults on which the rocks on either side have shifted relative to each other, and some of which are joints, where the rock is simply fractured but not offset. One exceptional mineralised fault several metres wide is sometimes well exposed on the upper beach of one such platform (Fig. 147); lower down it passes into a deeply eroded gully. Here successive growth layers of calcite are variably stained or overgrown by the iron oxide hematite (Fe_2O_3), which is partly altered to the iron hydroxides goethite ($FeO(OH)$) and/or limonite ($FeO(OH) \cdot nH_2O$).

The beach exposure reveals what must be a long history, perhaps amounting to thousands of years, when growth conditions fluctuated so that mineral layers varied in crystal size and colour. Many of the several dozens of growth changes appear to be cyclic alternations of iron mineral content – the red-white alternations – whereas others affect crystal size and layer thickness, some of which probably represent temperature fluctuations. One could go on and on, and so detract from the natural beauty of this occasional revelation. It is an 'admire only' locality, not always fully revealed and never to be pointlessly raped by hammer and chisel.

Figure 147. Spectacular calcite vein-mineral growths formed in an opening fracture flooded with calcium carbonate- and iron-rich mineralising water. Red bands are mostly of original hematite growth; various whites and greys are due to differing styles of calcite crystallisation. (Scale is 10 cm) (continued next page...)

Figure 147 continued. This site is commonly mostly obscured by pebbles, cobbles and boulders, but here has been swept by storm waves. It is practically impossible to collect samples from here as the minerals are naturally extremely friable and the surfaces smoothly abraded. Wet surfaces are especially beautiful to admire.

(**Top left**) *This vein is one of the many that cut Gower limestones broadly trending north-south. The calcite crystals shown grew into fluid-filled space, adding layers towards the right (west). A counterpart mirror-image opposite side of the open space with minerals growing towards the left (east) is not so well developed.* (**Top right and bottom**) *Detail of mineral growths showing abundance of facetted geometric forms that reflect original well-formed calcite crystal faces; the views are vertically down, and the growth direction is towards the top of the images, which on the ground is towards west. Scale is 10 cm. The colour banding reflects compositional and/or temperature variations of the fluids forming the minerals through time, with the reds indicating iron mineralisation, mostly original hematite growth, although some bands may reflect entrapment of fine suspended mineral grains. Red hematite-goethite-limonite-rich sediments occur locally in pockets and elsewhere on south Gower they have been sufficiently abundant to be mined for ochre or reddle (Section 2.4).*

6.6 The beautiful sand monster: Helwick

Old picture postcards can remind us post-war kids of golden days and acres of pure sand at the seaside. Sadly, some favourite Gower beaches of old have become unpleasantly stony, locally with exposed patches of ancient peat and clay. These include Port-Eynon and Horton, which is a single beach, Langland and Limeslade. In these cases, although storms and fair-weather waves seasonally sweep and return the sand, sometimes shifting it from one end of the beach to the other, as between Horton and Port-Eynon, there has been a long-term net loss (e.g., Fig. 148). Each of the affected beaches certainly was and hopefully will remain a jewel in the necklace of sandy bays that contribute to Gower's status as an AONB (designated 1956), and perhaps now more than ever they are important as attractions in the tourist economy. The degradation is certain; simple comparisons of the 1:10,560 maps of several Gower beaches clearly show losses (e.g., Fig. 148), although some others are perhaps less noticeable on the ground.

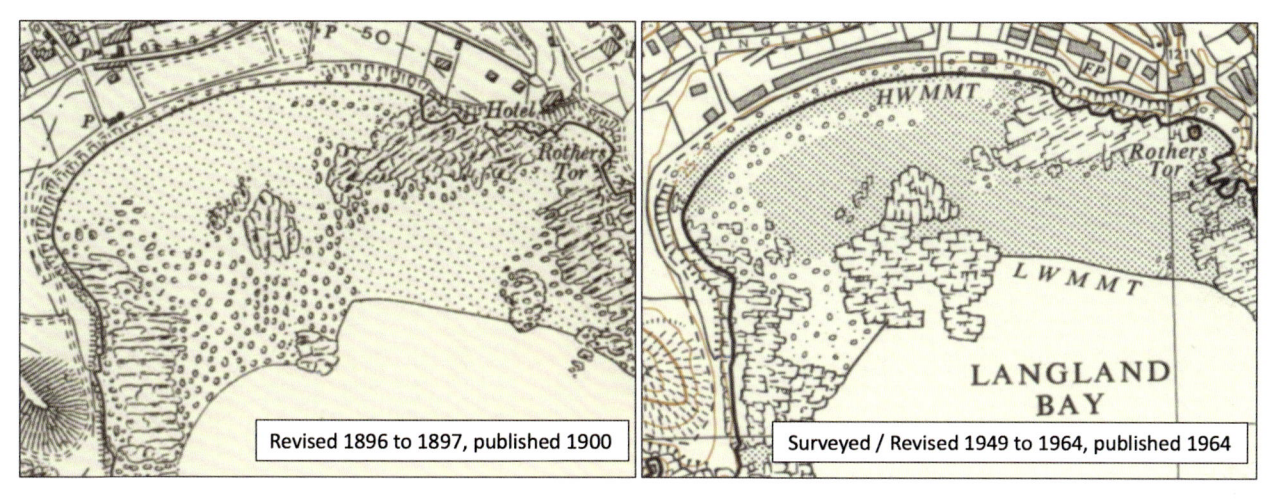

Figure 148. Successive 1:10,560 maps of Langland Bay (SS606873) showing significant beach sand loss over some sixty-plus years. The later map (right) shows the Low Water Mark of Mean Tides (LWMMT) closer to shore than previously (it was 'Ordinary' Tides on the older map: LWMOT) while rock in the middle of the bay is more extensive, both showing that the sand top surface is lower than it was. Similarly, the stony foreshore is more extensive near the high-water mark in the later edition.

Offshore from Port-Eynon Point and continuous westwards out beyond Worms Head is the infamous Helwick sand bank. It is 13.5 km long, up to 2.7 km wide and up to 40 m tall (Schmitt et al. 2007). Although it does not emerge, it is within a few metres of sea level on low tides and it has caused havoc over past centuries for coastal seafarers and local fishermen alike, because huge waves can form and break over it. A lightship station was established there in 1846 and from 1927 until 1977 Lightship 91 Helwick was anchored off the bank to warn mariners. This vessel is now a very familiar and treasured possession of Swansea's maritime museum. Although long known, the true hydrodynamic beauty of the Helwick sand bank has only recently been revealed by detailed seabed mapping, with its active 'life' as an ever-changing wave- and tidal-current-shaped mass only now appreciated (Fig. 149).

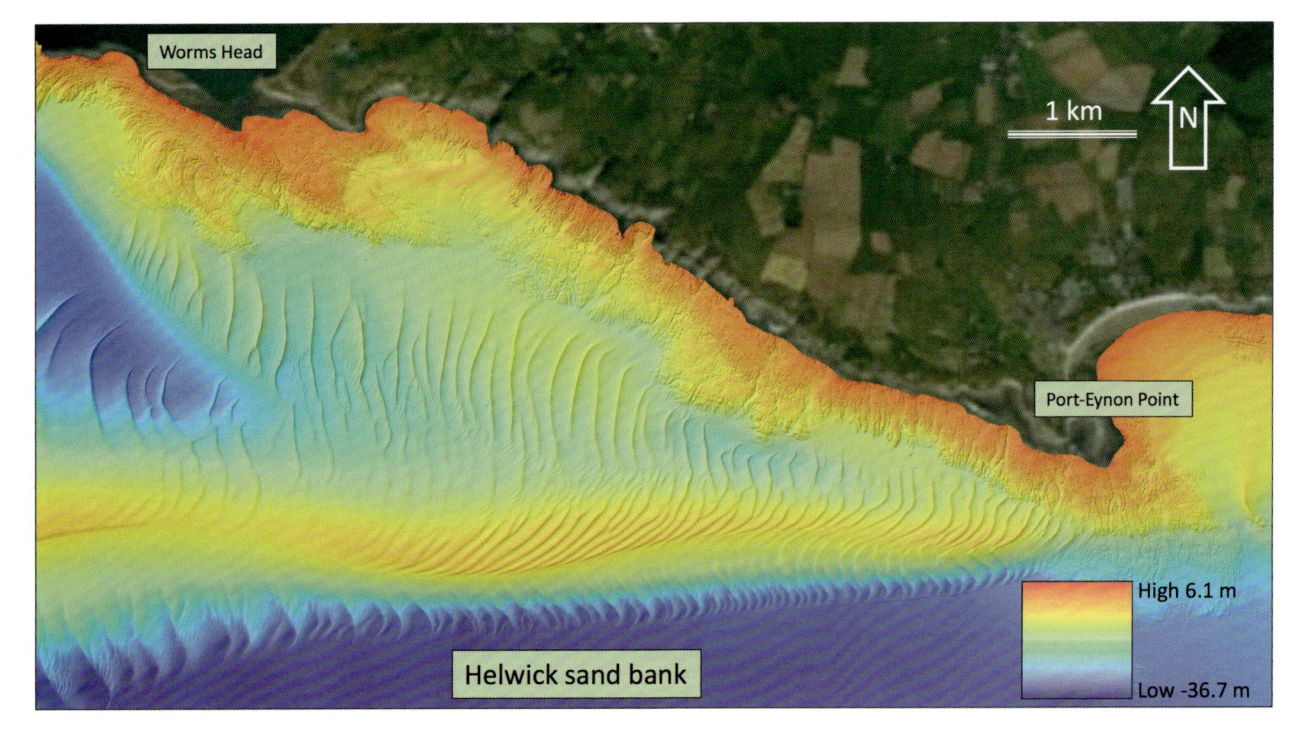

Figure 149. Inner reach of the Helwick banner sand bank, showing the beautiful physical patterns of the main bank and surface sand waves that continuously respond to tides and waves. It is, perhaps, reminiscent of a monstrous centipede. Obviously, the colours and shading are to reveal depths and slopes, but the details of the forms and their spacings hold wide-ranging information on current interactions with Port-Eynon Point and on the associated seabed sand movements. Depths are relative to the level of the lowest astronomical tide: LAT. (Base image multibeam bathymetry, at 2 m x 2 m resolution, from The Admiralty Marine Data Portal, kindly processed and provided by Gareth Carter). The sand-waves on either side of the bank reflect clockwise sand movement, eastwards along the north on the incoming tide and out westwards along the south during the ebb (Schmitt et al. 2007; Schmitt and Mitchell 2014). The sand-waves between the coast and the bank move eastwards at between 20 m and 200 m per year (HR Wallingford 2016).

Cutting to the chase concerning this beautiful sand monster, dredging from its western end to extract mainly building sand, since the 1950s, has led some to believe that dredging is the cause of the long-term sand loss from some Gower beaches. In the year 2000 an application for further dredging of 300,000 tons per year for 15 years was submitted and led to a public inquiry in 2006, with Welsh Government input. The inquiry found in favour of removal of 150,000 tons per year for 7 years, subject to stringent monitoring. The 'precautionary principle' – don't do it unless you know what you're doing – was advocated, despite a huge gulf in terms of plausible physical explanation for a link between supposed cause, dredging, and anticipated effect, beach-sand loss. Nevertheless, it was true that there were no data either way and the somewhat heated exchanges at least served to fuel the necessary investigations.

Helwick is known as a 'banner' sand bank, because it appears tied to a headland, here Port-Eynon Point (Fig. 149). It gains its form and activity primarily from the extreme tidal regime of the Bristol Channel. Another local example is the Nash sand bank, 'tied' to Nash Point, the protruding belly button of South Glamorgan. The dynamics of such banks are complex, but the associated headland in strong tidal currents causes rotating currents (vortices) that reverse between flood and ebb phases, in turn resulting in convergence of moving sand to form linear banks 'tied' to the headland (Lewis et al. 2015).

Helwick bank is generally up to 20-25 m tall, locally 40 m, and sand waves register strong eastward currents on the north side during incoming tides and westward currents on the south side during ebb tides. Although it can look steep in enhanced sea-bed images (Fig. 149), its slopes are gentle, 0.6° on the north flank and 3.5° on the south flank (Schmitt et al. 2007). Its surface sands are continuously reworked and in the medium term of several years its overall form and volume change according to the 'wave climate', which is the timing, frequency, duration and energy of storm events each year.

During 6 years of monitoring of the sand content of the Port-Eynon and Horton (single) beach there proved to be no causative link between dredging from Helwick and beach erosion (Phillips 2008). Further, a study through 1991-2002 (Lewis et al. 2015) found that the volume of the Helwick bank decreased over an 8-year period at a rate x6 greater than the sand taken by dredging in that interval.

The bank evidently alternately gains and loses mass via seabed migrating sand waves and there is no seabed connectivity for sand removed by dredging from Helwick to link to beach-sand volume (HR Wallingford 2016). The Wallingford report found no 'evidence provided' from anywhere around Wales for changes to coastline and the wider physical environment resulting from licenced dredging. But that is not to say there is no effect anywhere, just that there has been no provided evidence of it. Hopefully the new high-resolution seabed imaging, together with the associated understanding of the dynamic nature and large scales of seabed sand movement, will provide the necessary evidence and so help allay previously held natural concerns, or at least provide the data essential for objective evaluations.

The bottom line is that both the Helwick bank and local beaches lose sand in major storms, with redistribution of this material by tidal currents and with Helwick able to recover volume. The reason(s) for the long-term Gower beach sand loss remains enigmatic. Casting off for a moment from evidence-based understanding, the reader is reminded that the storminess of the 'Little Ice Age', which seemingly onset on Gower from about 1300 AD and ended in the mid 1800s, piled a huge amount of sand higher on the coast than previously: really, a lot! It seems just conceivable that the recent historical beach sand loss is part of some return or readjustment to a dynamic equilibrium that existed in the bays before the 'Little Ice Age'. Golden sandy beaches are not fixed entities in a changing climate; we old timers were just born lucky.

6.7 Pwll-du storm beaches

It is widely believed that the prominent mounded banks of limestone pebbles and cobbles in Pwll-du Bay originated as waste from the quarries on the west flank of the bay, but this is not quite so simply the case. Pebble-cobble storm beaches existed at Pwll-du long before the heyday of quarrying activities there in the (late) 17th, 18th and 19th centuries. Right at the back of the bay and impounding the marshy land behind, is a distinct bank mainly of cobbles (Fig. 150, bank 1). This is succeeded seawards by a further bank (2) upon which were built first the Ship Inn (aka Pooldie House, possibly originated in the 17th century) and then the Beaufort Arms (built in the 1830s; Holt 1996), which sustained the many thirsty quarry workers and sailors. These buildings were behind and protected from the sea by banks 3 and 4.

Farther seawards are banks 5, 6 and 7. Bank 7 has existed in similar form for over 100 years to this day (Fig. 150). Close examination of the banks shows that quarry waste fragments, identified by being somewhat more irregularly shaped and less rounded than others, predominantly occur in bank 5 with some in 6 and 7, although the seaward face of 7 is dominated by supremely well-rounded forms. This observation fits with the public houses, intended to service the quarry men and women, being founded on and behind existing stable banks 2 and 3, essentially safe from contemporary storms. Thus, it is bank 5 that marks the main phase of quarry activity and is

ALL OUR OWN WATER

most likely to have formed during one or more storms of the late 18th or early 19th century. Most probably bank 7, the currently active and long-lived storm beach, existed in the 19th century.

Figure 150. *(Top)* Pwll-du storm beach pebble-cobble banks labelled in order of formation. Each bank 1-6 has been breached by the stream, which after heavy rainfall is impressively powerful. Bank 7 is commonly breached by the stream, but in these images, taken more than a century apart, the water passes through it onto the beach. The **inset** at top right is a shaded relief LiDAR image showing that the buildings in the west side of Pwll-du are founded on early-formed banks, labelled 2. (Left photograph is from the British Geological Survey GeoIndex Onshore catalogue, reference P201079, and right-hand image is from Google Earth. The inset uses base data © Natural Resources Wales). *(Middle)* Details of banks 1, 5 and 7, located by the **asterisks** above. Bank 1 is partly covered and infiltrated by blown sand and, seemingly being the first to form at the back of the bay, could be thousands of years old. Bank 5 has noticeably the most irregular and least well-rounded rock fragments, although this would not be apparent from a galloping horse. (continued opposite...)

*Figure 150 continued. Banks 6 and 7 have a greater proportion of well-rounded forms. Various Anglian and Devensian erratics occur sparsely within the banks. The buildings, formerly the Ship Inn and Beaufort Arms public houses, are founded on bank 2, behind bank 3. This situation and the relatively poor state of rounding of bank 5 deposits suggest that it is bank 5 that primarily reflects the active quarrying, which occurred here mainly from the early 18th century, with lesser reflection in banks 6 and 7. (**Bottom**) Results of a protracted North Atlantic storm in late October 2020, originating from Hurricane Epsilon. An exceptionally high bank surmounted the 'normal' active bank due to waves focused into the northeast corner of the bay. At low tides a substantial head of rainwater that was ponded behind the bank allowed pebbles and cobbles to be lifted and swept seawards to form this developing breach on the seaward face. Note the extreme rounding of the stones.*

Pebble-cobble storm beaches are common at and beyond normal high-tide limits in south and west Gower, irrespective of whether or not there has been any significant quarrying. In places they are partly obscured by having been built upon, as at Caswell and Langland, and/or are partly buried beneath dunes, as at Three Cliffs, Oxwich, Port-Eynon, Llangennith and Whiteford, like bank 1 at Pwll-du. In many cases they formed barriers behind which marshes developed.

Pwll-du storm beach banks are striking, however, and perhaps beg some special consideration in this narrowest of the bays. Firstly, Pwll-du Bay must have been a subaerial valley that took very considerable glacial outwash during both the advance and the recession of the Late Devensian ice that at its maximum covered it and extended beyond the present coast (Section 3.6). Outwash from the catchment basin that includes what is now the east part of Fairwood Common and the west side of Clyne Common was funnelled into and must have partly eroded Bishopston Valley. Thus, considerable local and exotic glacial debris will have passed through the valley before it became flooded when sea level rose. Some of the debris inevitably would have filled across the original valley floor[5], with that upstream of the modern beach becoming buried in finer alluvium and marsh peat. So, some of the early banks could include original glacial outwash debris. Certainly, the outer beach platforms on the west side of the bay are littered with cobbles and boulders that arrived as north-crop erratics and the banks themselves contain significant amounts of that erratic debris.

Secondly, and perhaps quite significantly, it is probable that during sea-level rise the bay was partly 'protected' by a barrier of broken cliffs that extended some way across it from the west side. On the west side of Pwll-du Bay (Figs 151 and 152), seawards of the main quarries, is a promontory of huge loose rocks, best known of which is the 'Ring Rock', a prominent near cubic mass that was used as a mooring by boats in the bay. The promontory is locally known as the 'Needles' (Holt 1996), comprising heavily broken limestone that is more or less in situ and towards the land shows solution tubes and calcium carbonate 'flowstone' indicating a former small cave system. Beyond the promontory is Bantam Bay (Fig. 151) and Gravesend.

The quarried rock and Needles promontory are made of Oxwich Head Limestone, which generally dips down eastwards towards the shoreline and is cut by numerous fractures (faults and joints). At first sight, the intense fragmentation in the promontory is enigmatic. Neither limestone dissolution nor faulting can explain the jumble of rocks, while the promontory provided shelter for boats and was not broken by quarrying. We know that ice flowed south over this site during the Last Glacial Maximum (Section 3.6) and that ice probably would have scoured this fragmental promontory away if it had existed as such then. On the other hand, during the glacial recession considerable outwash must have passed the end of this promontory along the valley floor that is now the bottom of the bay infill. The sea floor drops steeply away from the eastern end of the Needles promontory[6], so the original valley side would have been steep when this area would have been susceptible to periglacial freezing.

Figure 151. Pwll-du Bay viewed towards west. The main quarried area, locally called Goonland, is between the storm beach and the dark promontory of huge rocks that extends into the bay; it is now largely obscured by woods (top image) but extends close to the skyline (bottom image). Quarried material would be constrained to tumble or slide on wooden sledges to the beach between the parallel ridges, known as slides. The promontory is a natural jagged jumble of angular blocks locally known as the 'Needles' where, near the grass, there is abundant evidence of small cave passages with flowstone. Prevailing south-westerlies batter the promontory and, given that storm beach debris in Pwll-du Bay is not all from quarry waste, continued erosion of the partial bay barrier could have made a significant contribution to the banks. The quarrying ceased at the beginning of the 20th century, over 50 years before the lower photo was taken. Soon very little of the quarry will be visible. (Black and white photograph from the Richard Burton Archives, Swansea University).

*Figure 152. (**Top left**) Ring Rock and the Needles promontory viewed towards east. The manifest jumble of broken rocks is quite exceptional at sea level and rather enigmatic. (**Top right**) Huge masses of limestone above the quarry have shifted under the force of gravity down on the bedding planes that dip towards the sea, opening rock crevasses where they became detached. The boxed letter **Q** marks the same rock face as in the bottom image. It is not quite clear whether the crevasses are a result of the quarrying or of earlier natural erosional under-cutting. The same problem exists at Oxwich, where the catastrophic rock falls that have occurred there recently from this same situation perhaps indicate what could happen here. (**Bottom**) View of main exposed quarry faces in the same limestone layer, labelled **Q**; the quarry above Q and on the left is obscured by trees (see Fig. 151). The line labelled fault is the approximate position of the surface trace of a fault that has the rocks on the farther side dropped down several metres relative to the near side. On the far side on top of the main quarried Q-rock layer is a jumble of broken rocks and pinnacles closely similar to the 'Needles' promontory. None of these broken rock piles would have withstood being over-ridden by glacial ice and it is suggested that the break-up must have occurred after the ice had melted back north of here (see text).*

The dip of the limestone layers towards the sea aided in the quarrying, because the broken rock would readily slide or tumble down the sloping quarry floor, in the 'slides', to be broken further at the beach and loaded into boats. However, the same dip also allowed huge masses to detach and shift downwards under gravity so that crevasses up to several metres wide opened in the higher slopes (Fig. 152). The main quarry faces are in and above a thick and massive layer of limestone (Q in Figure 152). The upper limestones are the same layer as the 'Needles' and are distinctly broken, as on the promontory, which must have assisted the quarrying. The simplest explanation for the broken nature of the 'Needles' is that the limestones there, and those above the Q layer, are laterally unsupported, essentially unconfined at their edges, and have collapsed according to gravity, thus spreading apart and tilting as the natural fractures opened freely. Any view down onto the beach platforms shows the criss-cross of the closely spaced fractures that affect most of the limestones and, of course, when such layers lie unbounded on the top of ridges and cliffs they will naturally fall apart.

Given that the Needles and the pinnacles above the quarries could not have withstood the passage of ice if they existed broken as such before the glaciation, it logically follows that the breakage of those rocks came after the ice had melted back. At that time the valley was incised to leave the valley sides steepened and liable to break-up when freeze-thaw cycles could have assisted the breakage. In this scenario, one might then suspect that the massive-rock detachments to form the crevasses occurred as the ice that filled the valley melted and removed its support, rather similar to The Bulwark landslide on Llanmadoc Hill, which resulted from loss of support by ice (Section 3.12). As we shall see, the detachment of huge blocks of rocks on inclined bedding planes can occur where there has been no undercutting by quarrying, so conceivably some of the big rockslides are late glacial features.

6.8 Oxwich rockslides: the full story

The steep northeast flank of Oxwich Point (SS507857), now mostly wooded, has long been prone to major rockslides. On Good Friday 1855 hundreds of tons of rock crashed down near St Illtyd's Church (Tucker 1951) and more gradual collapse has long been shown by an open 2-3 m wide rock crevasse by the footpath high in the woods (see below). Almost the entire coastal cliff, which is composed of the Oxwich Head Limestone, has been quarried to varying extents, as evident from the many half-tube remains of blasted drill holes in the rock faces. The tough, thick limestone layers (beds) dip at around 20° northeast, directly down to the shore, and several have thin intervening weak partings of coal, mudstone or clay. The layers are cut by numerous steep fractures, with faults trending subparallel and also oblique to the shore, and with cross-cutting joints that render the rock mass broken into huge angular chunks. This geometry of fractures together with the dip towards the shore and the weak partings between layers make the cliffed slopes especially prone to gravitational collapse. Relatively minor detachments before 2009 produced the huge boulders on the beach oddly known as the 'The Dices', reflecting their block form.

Now, again, before we 'blame' the quarrying, it is worth pointing out that, just as at Pwll-du, it is not clear that the excavations were entirely, or indeed at all, responsible for the rock sliding that has occurred. A closely similar geometry of layers with weak partings has also led to rock sliding with opening of rock crevasses above Llethrid (SS530911) in a place where there was no undercutting by human activity. At Oxwich, cliffed alcoves high in the woods (e.g., SS506856) could be post-glacial collapse scars, but it remains unclear whether quarrying extended up so far. This is a don't know!

The very obvious modern collapse scar (Fig. 153) initiated with block sliding in 2009. This left a straight, near vertical, partly red-stained wall newly exposed above huge fallen blocks on the upper beach. A further collapse occurred there in early January 2019 with a further major increment three weeks later. These later events involved most of the 2009 wall failing with formation of a new scar both farther up slope and more extensively along the slope towards Oxwich. Further lesser falls occurred in 2020.

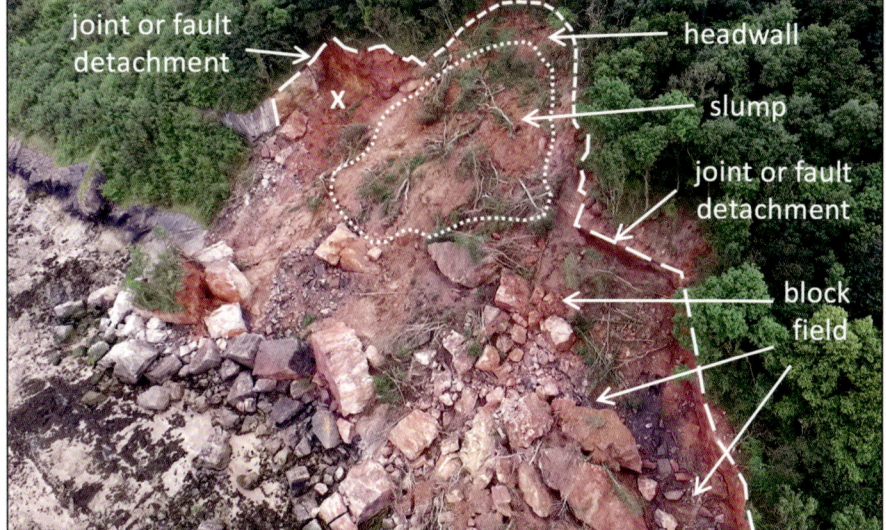

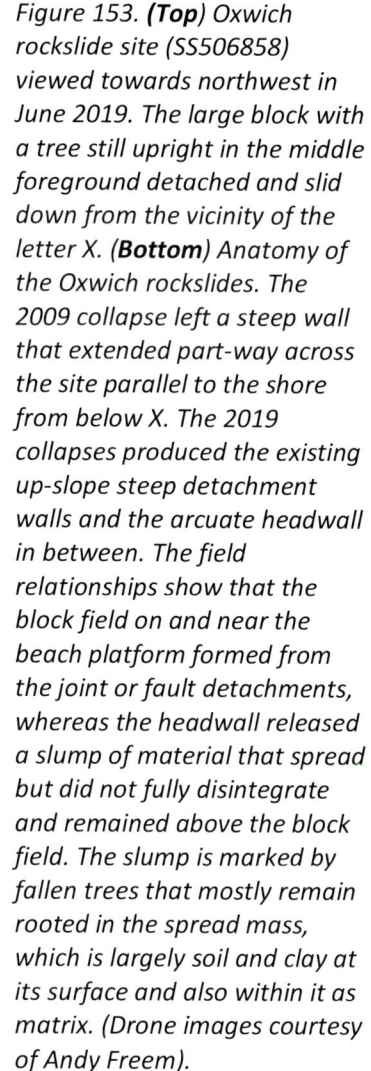

*Figure 153. **(Top)** Oxwich rockslide site (SS506858) viewed towards northwest in June 2019. The large block with a tree still upright in the middle foreground detached and slid down from the vicinity of the letter X. **(Bottom)** Anatomy of the Oxwich rockslides. The 2009 collapse left a steep wall that extended part-way across the site parallel to the shore from below X. The 2019 collapses produced the existing up-slope steep detachment walls and the arcuate headwall in between. The field relationships show that the block field on and near the beach platform formed from the joint or fault detachments, whereas the headwall released a slump of material that spread but did not fully disintegrate and remained above the block field. The slump is marked by fallen trees that mostly remain rooted in the spread mass, which is largely soil and clay at its surface and also within it as matrix. (Drone images courtesy of Andy Freem).*

*Figure 154. **(Left)** Two red mineralised faults are revealed where the 2019 slide-block became detached (X in Figure 153). The wall on the right is a straight continuation from the left-hand fault. **(Right)** Rock crevasse 2-3 m wide that extends for some 100 m along the slope high in the woods above the main collapse area. The left-hand body of rock is essentially unsupported where some 20 m downslope there is a huge cliffed embayment.*

The 2019 collapse evidently followed from an earlier release below it (in 2009) and then involved two contrasting styles of mass movement. Up-slope-progressive failure (retrogressive failure) first released blocks that slid and tumbled down to form the block field, and this in turn released a slump with characteristic arcuate headwall detachment and travel limited by some blocks below. The slump mass evidently is clay-rich; it was somewhat impervious and formed small debris flows, and it may have involved a pre-existing hollow on the limestone surface within which there was Anglian glacial till (Fig. 24). The distinct red colour of many of the rock faces exposed by the collapses is a clear sign of iron oxide and hydroxide mineralisation.

The Oxwich rockslides have delivered to the vicinity of the beach a fascinating record of the early history of the rocks involved. This record may well prove temporary as the fallen blocks become weathered and perhaps eroded by wave action during high tides. Although the features described here are close to the beach platform and the blocks themselves provide some 'shelter' against further falls, *visiting here and certainly lingering here without a watchful observer is not advocated*, for obvious reasons.

The red coloration of many of the block faces indicates iron oxide and iron hydroxide mineralisation (hematite, goethite and limonite) and, of course, this is on fracture surfaces on which the blocks have readily parted (Figs 153 and 154). What follows is an account, derived from examination of the fracture surfaces, of the successive processes recorded.

We know that the limestones and other Carboniferous-age rocks of Gower were fractured and folded in late Carboniferous and early Permian times by a major mountain-building event that occurred to the south in Europe (Section 2.4). The layers were thrown into the positions they occupy now and were cut by faults that channelled hot mineralising fluids towards the surface (Section 6.5). This produced what we may usefully call Phase 1 mineralisation (Fig. 155 Top). Phase 2 is when the mineralised fractures became penetrated by cold waters from above, probably mainly in Triassic times, involving infiltration of fine sediment that accumulated in open spaces (Fig. 155 Bottom). Phase 3 was when the fractures seemingly were reactivated, perhaps only slightly, so that the fracture fills became broken up with the debris falling into further open spaces (Fig. 156 Top). Phase 4 most probably occurred during the Pleistocene glaciations, as surface debris including thermally shattered fragment fell into the open spaces (Fig. 156 Bottom).

Missing from this order it is also clear that at some time(s) the limestones were beneath a water table, because the rocks are penetrated by several phreatic conduits that range from a few centimetres to a metre or so wide (Fig. 157). This dissolving of the limestone probably initiated in Phase 2 when we know the rocks were flooded. The Phase 2 Triassic age is also indicated because nearby there are phreatic tubes with fine sediment infills closely akin to the late Trias pothole that formed on Worms Head (Section 2.4, figure 10).

Figure 155. *(Top)* Mineralisation along an early fracture, probably formed during Late Carboniferous to Early Permian times. Limestone with crinoid fossils forms the fracture wall and has been corroded to etch-out the crinoid pieces (*circled*). Calcite crystals with dog-tooth forms grew onto the limestone and were then coated in a crust of iron oxide and iron hydroxide, forming the hematite-goethite-limonite crust, all from fluids that permeated the fracture. The **inset** shows the characteristic dog-tooth calcite crystal form, which can be recognised both beneath and poking through the crust. This is the hot-fluid (hydrothermal) mineralisation that accompanied folding and faulting of the Carboniferous rocks, here regarded as Phase 1.

(Middle) Partly filling open spaces against the mineral growths, laminated calcareous silt and mud has accumulated, washed in by cold water that penetrated from the contemporary surface. Labels **li** mark soft-state liquefaction features where sedimentary lamination has become contorted while the deposits remained wet. Labels **f** mark a small fault that cuts no higher than the top label, which, like the liquefaction, indicates disturbance during sedimentation. Labels **c** mark broad V-shaped channels into which the pink layer thickens; this indicates an episode of water flow and erosion. Thickness variations of other layers in general reflect disturbance or differential compaction. This cold-water sedimentation constitutes Phase 2.

(Bottom) In places on other fracture faces the dog-tooth calcite lacks the iron-mineral crust and is directly draped by laminated silt and mud. This rather beautiful occurrence suggests that the hydrothermal plumbing system of Phase 1 was far from simple.

*Figure 156. (**Top**) Heterogeneous fracture infill. This fragile deposit remained adhering to a fracture face shortly after a major collapse. The black, brown and orange fragments are pieces of the early hydrothermal hematite-goethite-limonite mineralisation. The green fragments indicate that the copper carbonate malachite ($Cu_2CO_3(OH)_2$) was precipitated along with the iron oxide and hydroxide. The cream fragments and coatings are of the sedimentary infill, showing that this infill mixture formed after the sedimentary lamination that itself followed the hydrothermal mineralisation. It represents Phase 3 taken to record some fault reactivation, which could well have happened during Paleogene or early Neogene crustal deformation (Fig. 21).*

*(**Bottom**) Also adhering to some fracture faces was a fragile deposit that includes numerous angular fragments of mudstone, chert and limestone, along with a finer grained version of the deposits shown above. The angular fragments suggest freeze-thaw (thermoclastic) fragmentation and the material appears to represent a fracture infill formed during Pleistocene, Ice Age, times, representing Phase 4.*

The rock dissolution that is recorded on several of the fallen blocks at Oxwich (Fig. 157) is also apparent in the nearby cliffs where remnants of laminated sediment infills are poorly preserved but closely analogous to the occurrence on Worms Head, which is of Late Triassic origin. The Oxwich rockslides for the moment demonstrate a fascinating geological history in some detail, but this may not be so well preserved for long.

Figure 157. Phreatic tubes in fallen blocks of the Oxwich rockslides.

(**Top**) *Phreatic tube formed by dissolving of the limestone beneath a water table and now exposed on a fracture surface. The scale bar in the centre is 10 centimetres.* **S** *marks a location where there is encrusting calcium carbonate 'flowstone' (speleothem) including small stalagmites, indicating that the early-formed tube must later have drained for the crust to form.*

(**Bottom**) *Partial section of a phreatic tube (forming the 'window' onto the dark debris behind). The pole is 1 metre. Such features nearby show partial infill by debris indicating a late Triassic age of initial formation.*

Figure 158. Entrail-like dissolution of limestone by mineralising fluids.

(**Left**) *Joint surface close to the Oxwich rockslides (SS509857) showing a succession of bulges along a ~10 mm-wide solution tube.*

(**Right**) *Mineralised solution tube on the outermost head of Worms Head (SS384876). The lower enlarged image shows that the dissolution has involved corrosion to leave a fossil tube, at* **A**, *and a cluster of calcite crystals at* **B** *grown within a bulbous part of the tube. This limonite-stained mineral lining is identical to the fracture-surface mineralisation found on the Oxwich rockslide blocks (Fig. 155 Top), interpreted there and hence here as due to hydrothermal fluids of Late Carboniferous to Early Permian age.*

Finally, in this section, an entrail-like dissolution feature is linked with the early phase of mineralisation recorded on the surface of the Oxwich rockslide boulders (Fig. 155 Top). Figure 158 shows a small entrail-like tube on a limestone joint surface, near the Oxwich rockslides, alongside what is taken to be a well-preserved analogue from the end of Worms Head.

Centimetre-scale solution tubes are characterised by successive bulges reminiscent of animal intestines. The better-preserved feature shows limestone corrosion that leaves fossils intact and pockets of well-formed calcite crystals. These are identical to the block surfaces exposed by the Oxwich rockslides (Fig. 155 Top) recording Phase 1 mineralised-fluid penetration along fractures during the late Carboniferous to early Permian folding and faulting.

So, it is inferred that the patterned rock records penetration by aggressive hydrothermal fluid, but the significance of the bulges remains unclear. It is known that during fault activity accompanied by earthquakes, hydrothermal fluids may be passively pumped along in pulses as dilated fault sections close. Alternatively, fluid pressures that build up from depth can cause fault dilation and slip, in a type of valve-release mechanism. In either case the effect is of fluid pumping along faults (Sibson 1981). It is not suggested here that each 'nodule' along these entrail-like tubes records an earthquake, but regular fluctuations of pressure and hence chemical reactivity seem likely to be involved.

Storms Notes

[1] We commonly refer loosely to bad weather with heavy rainfall and gale force winds as 'storms'. Strictly though, in terms of wind strength on the widely used Beaufort Scale, gales scale at 8 and 9 while true storms are 10 or 11. Fortunately, true storms and hurricanes (scaled at 12) are rare on Gower.

[2] Cerastoderma edule, formerly Cardium edule.

[3] There were several. Perhaps most notorious was the especially detested William (c. 1197 – 1230) known to the Welsh as *Gwilym Ddu*, Black William, hanged by order of Llewelyn the Great for some injudicious intercourse.

[4] The micrite is extremely tough and breaks into shards or splinters. On no account should these few occurrences be attacked with a hammer and chisel.

[5] This happened in all of the valleys that took glacial outwash, e.g., Ilston and Green Cwm.

[6] The deep water relatively close to the sheltered west shore was exploited during WW2 when German U-boats came into Pwll-du Bay to enable shore parties to collect fresh water unseen.

Draba aizoides, at besanded Pennard Castle.

7 APPENDICES

7.1 Acknowledgements

Part of the fun in writing this extended essay has been in the exchanges wherein so many people have been most generous with their expertise, time and patience. Most amusing has been the re-connection with like-minded explorers not met since the days of our youth. The following list of those to whom I am most sincerely grateful is alphabetical by surname, as there is no other sensible order. Contributions are mentioned here in brief although the true scale of many is considerable; numerous individuals and teams are also acknowledged in the text and notes.

I thank you all!

Gareth Acreman (Gomer): Patiently and kindly saw me through the lead up to printing the book.
Luke Ashton: Reinforced our aged cave digging team with youth and recovered data from Tooth Cave.
Dick Baynton: Sadly, a long-gone amusing friend (1944-1967), a great source, motivator and mentor to many like-minded explorers of our generation. Dick provided considerable insight.
Matthew Carroll: Freely related his experience of severe flooding of Green Cwm.
Gareth Carter (BGS): Kindly processed and provided the high-resolution coastal bathymetry images.
Roy Church (Parkmill Heritage Centre): Readily provided repeated access to his springs and borehole water for monitoring.
Dave Dunbar (Swansea University): Kindly made measurements on his mission in Tooth Cave.
Matt Eynon and Danilo Bettosi (Earth Science Partnership): Provided unpublished borehole data.
Gareth Farr (BGS): Gave tremendous support, with loaned temperature measuring equipment and help in obtaining water chemical analyses. We measured borehole temperature profiles together.
Bill Fitches: Painstakingly read and sensitively reviewed of all of the chapters, which significantly improved the scientific presentation. Bill taught me structural geology as an undergraduate at Aberystwyth, in 1970-71. Many thanks Bill.
Andy and Antonia Freem: Great support, mentorship and friendship in many ways, especially underground and in acquisition of great drone photographs (mostly above ground!). Absolutely huge contributions via underground sampling and provision of most of the cave pictures, and via the lead in digging to recover access into Llethrid without me getting stuck (again). Speleoscience guidance and reviews greatly appreciated; I learned a lot.
Clark Friend: Read, re-read and rigorously reviewed parts of the cave developments chapter.
John Harvey: Shared amusing recollections of explorations in Green Cwm.
Andy Heath (University of Liverpool): Managed image processing, especially the LiDAR shaded relief views, gave web and computation advice patiently without fail when asked, and removed eyesores.
John Heimstra (Swansea University): Positively reviewed and discussed the glaciation chapter.
Jason Hyatt (Dŵr Cymru): Provided access to Wellhead Pumping Station and Water Treatment Plant, and excellent gloves.
Brian Jorgensen: Friend who first took me rock climbing on Gower, aged 12*, and kindly provided old photographs and free use of information, including the Llethrid 1965 survey map he made with Dick Baynton. *(My mother said I was 15 in order for me to join the climbing course).
Charlie (Kokelaar): Faithful water-loving field companion (2007-2021), sorely missed.
Helen Kokelaar: Gave mostly uncomplaining assistance in mapping ice limits and surveying at Tooth Cave. Annoyingly tended to point out the overlooked obvious. Her "huge grey erratic" was a cow.
Bryan Lovell (University of Cambridge): Stimulating discussions of the uplift and subsidence of Britain.

Adrian Luckman (Swansea University): Generously provided processed LiDAR images.
Brendan Marris: Kindly gave permission to use one of his photographs of Tooth Cave.
Rhian Meara: Drew and corrected multiple versions of the South Wales Coalfield map and cross-sections.
Thomas Methuen-Campbell and Simon Bevan: Provided access to Penrice Estate land.
Neil Mitchell (University of Manchester): Freely discussed and provided unpublished marine bathymetry images from around Helwick.
Terry Moon: Shared tales of caving on Gower in the 1960s.
Ian Murphy (Dŵr Cymru): Provided immensely helpful free access to data, including all of the Wellhead records, plus advice on hydrological practices. Friend with huge positivity in shared natural interests, especially grass-roots rugby in South Wales and occasional beers.
Paddy O'Reilly: Shared tales of caving on Gower in the 1960s and collaborated on provision of 'best possible' Green Cwm cave maps. I greatly appreciated his attention to detail and his care in recording the legacy of Dick Baynton.
Julie and Adrian Parker: Land owners who provided outstanding hospitality, cooperation and real assistance in all digging endeavours on their land at Llethrid (Llethryd Barns).
Alan Price (NRW Hydrogeology): Provided water chemistry data.
Julia and Steve Robson: Owner-occupants of Green Cwm Cottage who freely provided video and photographs of Green Cwm floods and notified the author of events in progress.
Jem Rowland: Read and advised on chapters; also constructively provided access to SWCC archive photographs and contact with erstwhile adventurers.
Diana Sahy (BGS): Isotope geochemist who generously provided speleothem age analyses and good advice.
Peter Sambrook: Wellhead engineer, water diviner, and raconteur par excellence, a true friend with whom it was always delightful to re-explore Gower and to recollect funny times.
Richard Shakesby (Swansea University retired): Positively reviewed and discussed the glaciation chapter.
Gareth Smith: Reinforced our aged cave digging team with youth, and his wife Val.
Martin Smith and Dave Schofield (BGS): Authorised BGS preparation of the utilised NextMap image.
Philip Stephens: Provided considerable scholarly expertise and enlightening amusing discussions on Welsh placenames and anecdotes of Gower.
John Stevens: Provided the modern survey of Ogof Ffynnon Wyntog.
Paul Tarrant: Shared his considerable knowledge of Gower caves.
Sue and Dean Taylor: Repeatedly gave access to Butter Well / Saint Illtyd's Well in their garden.
Glanville Thomas: Helped with landscape refurbishment at the Barns Cave dig.
Paul Thomas: Gave access to the borehole drilling to 100 m (to -50 m OD) between Parkmill and Lunnon and provided later access for down-hole temperature profiling.
Phill Thomas: Kindly recovered samples from Tooth Cave.
James Utley (University of Liverpool): Provided expert high precision X-Ray Diffraction analyses of clay-grade minerals, plus scrape-sample mineral identifications.
Nicky White and Hannah Galbraith-Olive (University of Cambridge): Most helpfully provided revised figures for the book, bearing on the geophysics and evolution of the North Atlantic.
Geoff Williams: Friend and climbing partner with shared strong affinity for Gower (and the Alps); provided extremely useful guidance and feedback on the hydrogeology and helpfully reviewed other chapters. He wrote the appended stand-alone review of Wellhead data.
Jessica Winder: Gave permission to use one of her fabulous photographs of nature on Gower.
Nigel Woodcock (University of Cambridge): Long established friendly colleague who provided considerable help in interpreting the Triassic sediments located within Tooth Cave, as well as discussions on the structural evolution of Gower.
Tomasz Zalewski: Kindly provided photos and recovered samples from Tooth Cave.

7.2 References

Abesser C, Shand P, Ingram J (2005) Baseline Report Series: 22. The Carboniferous Limestone of Northern England. British Geological Survey Commissioned Report No. CR/05/076N.

Abesser C, Smedley PL (2008) Baseline groundwater chemistry of aquifers in England and Wales: the Carboniferous Limestone aquifer of the Derbyshire Dome. British Geological Survey Open Report, OR/08/028. 66 pp.

Al-Kindi S, White NJ, Sinha M, England RW, Tiley R (2003) The crustal trace of a hot convective sheet. Geology, 31, 207-210.

Bain CG, Bonn A, Stoneman R, Chapman S, and 21 others (2011) Project Report. IUCN UK Commission of Inquiry on Peatlands. Edinburgh IUCN UK Peatland Programme.

Banks D, Farr G, Inman P, Low R (2007) Groundwater quality and supply survey for the Precambrian Gwna Group, Anglesey. Environment Agency Report.

Battiau-Queney Y (1984) The pre-glacial evolution of Wales. Earth Surface Processes and Landforms. 9, 229-252.

Baynton RT (1968-1969) The hydrology of Gower. South Wales Caving Club Newsletters 58-62.

Bögli A (1980) Karst hydrology and physical speleology. Springer-Verlag, Berlin, 284 pp.

Bowen DQ, Phillips FM, McCabe AM et al. (2002) New data for the Last Glacial Maximum in Great Britain and Ireland. Quaternary Science Reviews, 21, 89–101.

British Geological Survey (1975) Carmarthen. 1:50,000 Sheet 229 Solid Geology.

British Geological Survey (1977) Ammanford. 1:50,000 Sheet 320 Solid Geology.

British Geological Survey (2002) Worms Head. 1:50,000 Provisional Series. England and Wales Sheet 246. Solid and drift geology. (Keyworth, Nottingham: British Geological Survey).

British Geological Survey (2011) Swansea. 1:50,000 Geology Series Sheet 247. Bedrock and superficial deposits. (Keyworth, Nottingham: British Geological Survey).

Brown PJ (2015) Coasts of catastrophe? The incidence and impact of aeolian sand on British medieval coastal communities. European Journal of Postclassical Archaeologies, 5, 127-148.

Bryant EA, Haslett SK (2002) Was the AD 1607 coastal flooding event in the Severn Estuary and Bristol Channel (UK) due to a tsunami? Archaeology in the Severn Estuary,13, 163-167.

Bryant EA, Haslett SK (2007) Catastrophic wave erosion, Bristol Channel, United Kingdom: impact of tsunami? Journal of Geology, 115, 253-269.

Busby J, Kingdon A, Williams J (2011) The measured shallow temperature field in Britain. Quarterly Journal of Engineering Geology and Hydrogeology, 44, 373-387.

Chambers WJ (1973) Limestone springs and individual flood events. Transactions of the Cave Research Group of Great Britain, 15, 91-97.

Chambers WJ (1983) Intensive sampling at a karst spring system: Leason, Gower, South Wales. Cave Science, 10, 188-198.

Charles J-H, Whitehouse MJ, Anderson JCØ, Shail RK, Searle MP (2017) Age and petrogenesis of the Lundy granite: Paleocene intraplate peraluminous magmatism in the Bristol Channel, UK. Journal of the Geological Society, 175, 44-59.

Clarke CD, Hughes ALC, Greenwood SL, Jordan C, Sejrup HP (2012) Pattern and timing of retreat of the last British-Irish Ice Sheet. Quaternary Science Reviews, 44, 112-146.

Cox R, Jahn KL, Watkins OG, Cox P (2018) Extraordinary boulder transport by storm waves (west of Ireland, winder 2013-2014), and criteria for analysing coastal boulder deposits. Earth-Science Reviews, 177, 623-636.

Crowther J (1989) Karst geomorphology of South Wales. In: Ford TD Limestones and Caves of Wales, Cambridge University Press, 20-39.

Daniels KE (2017) The role of force networks in granular materials. EPJ Web of Conferences 140, 01006 Powder & Grains. DOI: 10.1051/epjconf/20171401006.

Davies NS, Liu AG, Gibling MR, Miller RF (2016) Resolving MISS conceptions and misconceptions: A geological approach to sedimentary surface textures generated by microbial and abiotic processes. Earth-Science Reviews, 154, 210-246.

Davis MW, White NJ, Priestly KF, Baptie BJ, Tilmann FJ (2012) Crustal structure of the British Isles and its epeirogenic consequences. Geophysical Journal International, 190, 705-725.

Dinnis R, Bates MR, Bello SM, Buck LT et al. (2019) Archaeological collections from Long Hole (Gower, Swansea, UK) and their place in the British Palaeolithic. Cave and Karst Science, 46, 37-46.

Ede DP (1975) Limestone drainage systems. Journal of Hydrology, 27, 297-318.

Ede DP and Bull PA (1989) Swallets and caves of the Gower Peninsula. In: Ford TD Limestones and Caves of Wales, Cambridge University Press, 211-216.

Estep J, Dufek J (2012) Substrate effects from force chain dynamics in dense granular flows. Journal of Geophysical Research, 117. F01028, doi:10.1029/2011JF002125.

Evans E, Hart R, Lewis R, Locock M, Lodwick M, Owens E, Webster P (2010) Church Hill, Penmaen, Swansea. Survey and trial excavation. GGAT report 2010/033, 28pp.

Farr G, Bottrell SH (2013) The hydrogeology and hydrochemistry of the thermal waters at Taffs Well, South Wales, UK. Cave and Karst Science, 140, 5-12.

Farr G, Busby J, Wyatt L, Crooks J, Schofield DI, Holden A (2020) The temperature of British coalfields. Quarterly Journal of Engineering Geology and Hydrogeology, https://doi.org/10.1144/qjegh2020-109

Fleming K, Johnston P, Zwartz D, Yokoyama Y, Lambeck K, Chappell J (1988) Refining the eustatic sea-level curve since the Last Glacial Maximum using far- and intermediate-field sites. Earth and Planetary Science Letters, 163, 327-342.

Ford DC, Williams P (2007) Karst hydrogeology and geomorphology. John Wiley & Sons Ltd., 562 pp.

George TN (1932) The Quaternary beaches of Gower. Proceedings of the Geologists' Association, 43, 291-324.

George TN (1938) Shoreline evolution in the Swansea District. Proceeding of the Swansea Scientific and Field Naturalists' Society, 2, 23-48.

George TN (1940) The structure of Gower. Quarterly Journal of the Geological Society, 96, 131-198.

Gibbard PL, Hughes PD, Rolfe CJ (2017) New insights into the Quaternary evolution of the Bristol Channel, UK. Journal of Quaternary Science. DOI: 10.1002/jqs.2951.

Glasser NF, Davies JR, Hambrey MJ, Davies BJ, Gheorghiu DM, Balfour J, Smedley RK, Duller GAT (2018) Late Devensian deglaciation of south-west Wales from luminescence and cosmogenic isotope dating. Journal of Quaternary Science, DOI: 10.1002/jqs.3061.

Goskar KL and Trueman AE (1934) The coastal plateaux of South Wales. Geological Magazine, 71, 468-479.

Grove HPF (2008) Church Hill Parkmill: In the footsteps of the old antiquaries. Gower, 59, 7-12.

Hale M, Thompson M, Wheatley MR (1984) Laser ablation of stream-sediment pebble coatings for simultaneous multi-element analysis in geochemical exploration. Journal of Geochemical Exploration, 21, 361-371.

Halfacree SP, Williams HJ (1988) Parkmill aquifer risk assessment SW/88/27. Scientific Services Welsh Water South Western District.

Hallis LJ, Huss GR, Nagashima K, Taylor GJ, Halldórsson SA, Hilton DR, Motti MJ, Meech KJ (2015) Evidence for primordial water in Earth's deep mantle. Science, 350(6262), 795-797.

Hansen J, Sato M, Russell G, Kharecha P (2013) Climate sensitivity, sea level, and atmospheric carbon dioxide. Philosophical Transactions Royal Society A, 371, 20120294. doi:10.1098/rsta.2012.0294.

Haq BU, Hardenbol J, Vail PR (1987) Chronology of fluctuating sea level since the Triassic. Science, 235, 1156-1167.

Harvey JC, Morgan R and Webley DP (1967) Tooth Cave, Ilston, Gower. An Early Bronze Age occupation. Archaeology and Art, 22, 277-283. (*author: i.e., Ilston parish not cwm*)

Hearty PJ (1997) Boulder deposits from large waves during the last interglaciation on North Eleuthera Island, Bahamas. Quaternary Research, 48, 326-338.

Hearty PJ, Neumann AC (2001) Rapid sea level and climate change at the close of the last interglaciation (MIS5e): evidence from the Bahama Islands. Quaternary Science Reviews 20, 1881-1895.

Hearty PJ, Neumann AC, Kaufman DS (1998) Chevron ridges, and runup deposits from storms late in Oxygen-Isotope Substage 5e. Quaternary Research, 50, 309-322.

Hearty PJ, Tormey BR (2017) Sea-level change and superstorms; geologic evidence from the last interglacial (MIS 5e) in the Bahamas and Bermuda offers ominous prospects for a warming Earth. Marine Geology, 390, 347-365.

Hiemstra JF, Rijsdijk KF, Shakesby RA, McCarroll D (2009) Reinterpreting Rotherslade: implications for Last Glacial ice limits and Quaternary stratigraphy of the British Isles. Journal of Quaternary Science, 24, 399-410.

Holt H (1996) Pwlldu remembered. The story of Gower's smallest village. Published by the author, 212pp. ISBN 0 9529165 0 9.

Horsburgh K, Horritt M (2006) The Bristol Channel floods of 1607 - reconstruction and analysis. Weather, 61, 272-277.

Howells MF (2007) British Regional Geology: Wales (Keyworth, Nottingham: British Geological Survey).

Hydraulics Research Wallingford (2016) Review of aggregate dredging off the Welsh coast: Review of evidence. Report DDM7582-RT001-R05-00, 56pp.

Jansen E, Bleil U, Henrich R, Kringstad L, Slettemark B (1988) Paleoenvironmental changes in the Norwegian Sea and northeast Atlantic during the last 2.8 m.y.: Deep Sea Drilling Project/Ocean Drilling Program Sites 610, 642, 643 and 644. Paleoceanography and Paleoclimatology, 3, 563-581.

Jones SM, White N (2003) Shape and size of the starting Icelandic plume swell. Earth and Planetary Science Letters, 216, 271-282.

Jones SM, White N, Clarke BJ, Rowley E, Gallagher K (2002) Present and past influences of the Iceland Plume on sedimentation. In: Doré AG, Cartwright JA, Stoker MS, Turner JP, White N, Exhumation of the North Atlantic Margin: timing, mechanisms and implications for petroleum exploration. Geological Society, London, Special Publications, 196, 13-25.

Jouzel J, Masson-Delmotte V, Cattani O, Dreyfus G, et al. (2007) EPICA Dome C ice core 800 kyr deuterium data and temperature estimates. IGBP PAGES/World Data Center for Paleoclimatology data contribution series, 91.

Judson D (1974) Cave Surveying for Expeditions. The Geographical Journal, 140, 292-300.

Kokelaar BP, Bahia RS, Joy KH, Viroulet S, Gray JMNT (2017) Granular avalanches on the Moon: Mass wasting conditions, processes and features. Journal of Geophysical Research: Planets 122 (9), 1893-1925.

Kopp RE, Simons FJ, Mitrovica JX, Maloof AC, Oppenheimer M (2009) Probabilistic assessment of sea level during the last interglacial stage. Nature, 462, 863-867.

Lambeck K (1993) Glacial rebound of the British Isles-II. A high-resolution, high-precision model. Geophysical Journal International, 115, 960-990.

Lambeck K (1995) Late Devensian and Holocene shorelines of the British Isles and North Sea from models of glacio-hydro-isostatic rebound. Journal of the Geological Society, London, 152, 437-448.

Lees DJ, Sell S (1983) Excavation of a medieval dwelling at Pennard. Gower, 34, 44-52.

Lewis MJ, Neill SP, Elliott AJ (2014) Interannual variability of two offshore sand banks in a region of extreme tidal range. Journal of Coastal Research, 31, 265-275.

Lisiecki LE, Raymo ME (2005) A Pliocene-Pleistocene stack of 57 globally distributed benthic δ18O records. Paleoceanography, 20, doi:10.1029/2004PA001071.

Marcott SA, Shakun JD, Clark PU, Mix AC (2013) A reconstruction of regional and global temperature for the past 11,300 years. Science, 339, 1198-1201.

Mayes J, Powell J (2003) The climate of Gower, South Wales: a 40-year perspective. Weather, 58, 303-309.

Miller KG, Kominz MA, Browning JV, Wright JD, et al. (2005) The Phanerozoic record of global sea-level change. Science, 310, 1293-1298.

Miller KG, Mountain GS, Wright JD, Browning JV (2011) A 180-million-year record of sea level and ice volume variations from continental margin and deep-sea isotopic records. Oceanography, 24, 40-53.

Moreau M, Shand P, Wilton N, Brown S, Allen, D (2004) Baseline Report Series 12: The Devonian Sandstone aquifer of south Wales and Herefordshire. British Geological Survey Commissioned Report No. CR/04/185N.

Morgan R, Harvey J (1961) Tooth Cave - Gower. South Wales Caving Club Newsletter 38.

Morgan WL (1894) Discovery of a megalithic sepulchral chamber on the Penmaen burrows, Gower, Glamorganshire. Archaeologia Cambrensis, Vol. XI, No. XLI, 7pp.

Mullard J (2006) Gower. The New Naturalist Library, Collins, 445 pp.

Nash GH, van Calsteren P, Thomas L, Simms MJ (2012) A discovery of possible Upper Palaeolithic parietal art in Cathole Cave, Gower Peninsula, South Wales. Proceedings of the University of Bristol Spelaeological Society, 25, 327-336.

Oldham T (1978) The Caves of Gower, 68 pp, privately published.

O'Reilly P, Kokelaar P (202x) Tooth Cave - Gower's Longest Cave: Exploration, surveys and key features. South Wales Caving Club Newsletter, xxx (in prep.).

O'Reilly P, Kokelaar P, Moon T (2021) The continuing legacy of Dick (Richard Thomas) Baynton (1944-1967). South Wales Caving Club 75th Anniversary Newsletter, volume 1, 130-133.

Owen TR (1964) Further Thoughts on Arthur's Stone. Gower, 16, 54-55.

Pedoja K, Husson L, Regard V, Cobbold PR, et al. (2011) Relative sea-level fall since the last interglacial stage: Are coasts uplifting worldwide? Earth-Science Reviews, 108, 1-15.

Phillips FM, Bowen DQ, Elmore D (1994) Surface exposure dating of glacial features in Great Britain using cosmogenic Chlorine-36: preliminary results. Mineralogical Magazine 58A, 722-723.

Phillips MR (2008) Beach erosion and marine aggregate dredging: A question of evidence? The Geographical Journal, 174, 332-343.

Philp R (2018) Changing tides: The archaeological context of sea level change in prehistoric South Wales. Unpublished PhD thesis, Cardiff University, 429pp. http://orca.cf.ac.uk/118952/

Piper AM (1944) A graphic procedure in the geochemical interpretation of water-analyses. Eos, Transactions American Geophysical Union, 25, 914-928.

Polyak VJ, Onac BP, Fornós JJ, Hay C, et al. (2018) A highly resolved record of relative sea level in the western Mediterranean Sea during the last interglacial period. Nature Geoscience, 11, 860-864.

Powell JS (1984) The sizzling Summer of '83. Gower, 34, 62-64.

Powell JS (1985) An all-time dry. Gower, 35, 51-56.

Prestwich J (1892) The raised beaches, and 'head' or rubble-drift, of the south of England: their relation to the valley drifts and to the glacial period; and on a late post-glacial submergence. Quarterly Journal of the Geological Society, London, 190, 263-343.

Pye K, Blott SJ (2009) Coastal processes and shoreline behaviour of estuary dominated systems in Swansea Bay and Carmarthen Bay. Report to Halcrow Group Ltd. Annex A1, 54pp.

Robins NS, Davies J (2016) Hydrogeology of Wales. British Geological Survey.

Robinson PL (1957) The Mesozoic fissures of the Bristol Channel area and their vertebrate faunas. Journal of the Linnean Society (Zoology), 43, 260-282.

Roe L, Hart MB, Taylor GK, Marks A (1999) The St. Erth Formation: history of the clay workings, geological setting and stratigraphy. Geoscience in south-west England, 9, 304-309.

Rowland J (2019) The Clague Taylor Photo Collection - A detailed record of the caving activities of a remarkable trio. South Wales Caving Club Newsletter, 135, 68-73.

Rowley DB, Markwick PJ (1992) Haq et al. eustatic sea level curve: Implications for sequestered water volumes. Journal of Geology, 100, 702-715.

Schmitt T, Mitchell NC (2014) Dune-associated sand fluxes at the nearshore termination of a banner sand bank (Helwick Sands, Bristol Channel). Continental Shelf Research, 76, 64-74.

Schmitt T, Mitchell NC, Ramsay TS (2007) Use of swath bathymetry in the investigation of sand dune geometry and migration around a near shore 'banner' tidal sandbank. In: Balson PS, Collins MB (eds) Coastal and shelf sediment transport. Geological Society of London, Special Publications, 274, 53-64.

Schoonman CM, White NJ, Pritchard D (2017) Radial viscous fingering of hot asthenosphere within the Icelandic plume beneath the North Atlantic Ocean. Earth and Planetary Science Letters, 468, 51-61.

Seyler CA (1920) "Stedworlango." A Study of the Fee of Penmaen in Gower. Archaeologia Cambrensis, 134–58.

Shakesby RA, Hiemstra JF (Eds) (2015) The Quaternary of Gower: field guide. Quaternary Research Association, London, 145pp. ISBN: 0 907780 164

Shakesby RA, Hiemstra JF, Kulessa B, Luckman AJ (2018) Re-assessment of the age and depositional origin of the Paviland Moraine, Gower, south Wales, UK. Boreas, 47, 577-592. DOI 10.1111/bor.12294.

Shennan I, Bradley S, Milne G, Brooks A, Bassett S, Hamilton S (2006) Relative sea-level changes, glacial isostatic modelling and ice-sheet reconstructions from the British Isles since the Last Glacial Maximum. Journal of Quaternary Science, 21, 585-599.

Sherman A (2011) Recently discovered trackways in Swansea Bay. Studia Celtica XLV, 1-25.

Sibson RH (1981) Fluid flow accompanying faulting: field evidence and models. Earthquake prediction: an international review. 4. 593-603.

Smith R, Bowen DQ, Cope JCW, Reid A (2002) An arietitid ammonite from Gower: its palaeogeographical and geomorphological significance. Proceedings of the Geologists' Association, 113, 217-222.

Spratt RM, Lisiecki LE (2016) A Late Pleistocene sea level stack. Climate of the Past, 12, 1079-1092.

Stephens P (2021) Gower and what it means. Gower Placenames Press.

Stevens J (2020) Surveying in Ogof Ffynnon Wyntog. Chelsea Speleological Society Newsletter, 62, 90-92.

Strahan A (1907a) The geology of the South Wales Coal-field, part VIII. The country around Swansea. Memoir of the British Geological Survey of Great Britain, HMSO, London.

Strahan A (1907b) The geology of the South Wales Coal-field, part IX. West Gower and the country around Pembrey. Memoir of the British Geological Survey of Great Britain, HMSO, London.

Sutcliffe AJ (1995) Insularity of the British Isles 250 000-30 000 years ago: the mammalian, including human, evidence. Geological Society, London, Special Publications, 96, 127-140.

Sutcliffe AJ, Currant AP, Stringer CB (1987) Evidence of sea-level change from coastal caves with raised beach deposits, terrestrial faunas and dated stalagmites. Progress in Oceanography, 18, 243-271.

Tamisiea ME, Mitrovica JX (2011) The moving boundaries of sea level change: Understanding the origins of geographic variability. Oceanography, 24(2), 24-39.

Tappin DR, Chadwick RA, Jackson AA, Wingfield RTR, Smith NJP (1994) United Kingdom offshore regional report: the geology of Cardigan Bay and the Bristol Channel. (London: HMSO for the British Geological Survey.)

Taylor MC (1991) Three below Gower: the story of cave exploration by 'The Taylors'. 90pp.

Tiley R, White N, Al-Kindi S (2004) Linking Paleogene denudation and magmatic underplating beneath the British Isles. Geological Magazine, 141, 345-351.

Toft LA (1985) The twin settlements of medieval Rhossili. Gower, 35, 47-56.

Toft LA (1988) A study of coastal village abandonment in the Swansea Bay region, 1270-1540. Morgannwg, 32, 21-37.

Tucker HM (1951) Gower Gleanings. Swansea: The Gower Society, 90 pp.

Walsh PT, Atkinson K, Boulter MC, Shakesby RA (1987) The Oligocene and Miocene outliers of West Cornwall and their bearing on the geomorphological evolution of Oldland Britain. Philosophical Transactions of The Royal Society of London A, 323, 211-245.

White N, Lovell B (1997) Measuring the pulse of a plume with the sedimentary record. Nature, 387, 888-891.

Whiteside DI, Marshal JEA (2008) The age, fauna and palaeoenvironment of the Late Triassic fissure deposits of Tytherington, South Gloucestershire, UK. Geological Magazine, 145, 105-147.

Wilkinson JG (1870) Avenue and carns about Arthur's Stone in Gower. Archaeologia Cambrensis, Fourth Series, 1, 22-45.

Wilson A, Lavé J (2013) The legacy of impact conditions in morphometrics of percussion marks on fluvial bedrock surfaces. Geomorphology, 186, 174-180.

Woodcock NH, Miller AVM, Woodhouse CD (2014) Chaotic breccia zones on the Pembroke Peninsula, south Wales. Journal of Structural Geology, 69, 91-107.

Wright V, Woodcock NH, Dickson JAD (2009) Fissure fills along faults: Variscan examples from Gower, South Wales. Geological Magazine, 146, 890-902.

7.3 Springs Gazetteer

This section primarily locates sinks and springs of central Gower and it provides images of those worthy of a photograph. Some springs outside the central area and mentioned in the text are also listed.

References to the Ordnance Survey National Grid are given with prefix SS, which identifies the 100 km square.

Yellow asterisk locates Penmaen weather station SS531888. L indicates Llwyn-y-bwch sink cluster SS483916 and O indicates Oldwalls sinks cluster SS497914. 1. Leason SS483927. 2. Staffal Haegr SS491923. 3. Llanrhidian Church SS496922. 4. Butter Well / Saint Illtyd's Well SS497923. 5. West Well SS496900. 6. Holy Well SS497899. 7. Hot Well SS499903. 8. Moormills sinks SS505912. 9. East Well SS500898. 10. Decoy sinks SS519909. 11. Llethrid Swallet (sink) SS531911 and flood overflow sink into The Barns Cave SS53169112. 12. Former well at Green Cwm Cottage SS534903. 13. Willoxton sink SS539906. 14. Wellhead resurgence SS539897. 15. Kitchen Well SS539896. 16. East Wellhead SS53998971. 17. Parkmill Heritage Centre and Lunnon boreholes SS54358931 and SS54398946. 18. Lunnon East SS55228961. 19. Killy Willy sink SS55769018. 20. Trinity Well and Killy Willy Rising SS55308945 and SS55308965. 21. Sunnyside borehole SS54498892. 22. Pennard Castle (aka Saint Mary's Well) SS544885. 23. Sambrook SS53978843. 24. Notthill West and East SS53498833. 25. Tor Bay SS52738781. 26. Parc le Breos pond inlet SS52568910. 27. Nicholaston sink SS516884. 28. Nicholaston East SS516879. 29. Nicholaston West SS514879. 30. Nicholaston Woods SS510879. 31. Parsonage SS513887. 32. Perris Wood SS503890. 33. Reynoldston Post Office SS481901.

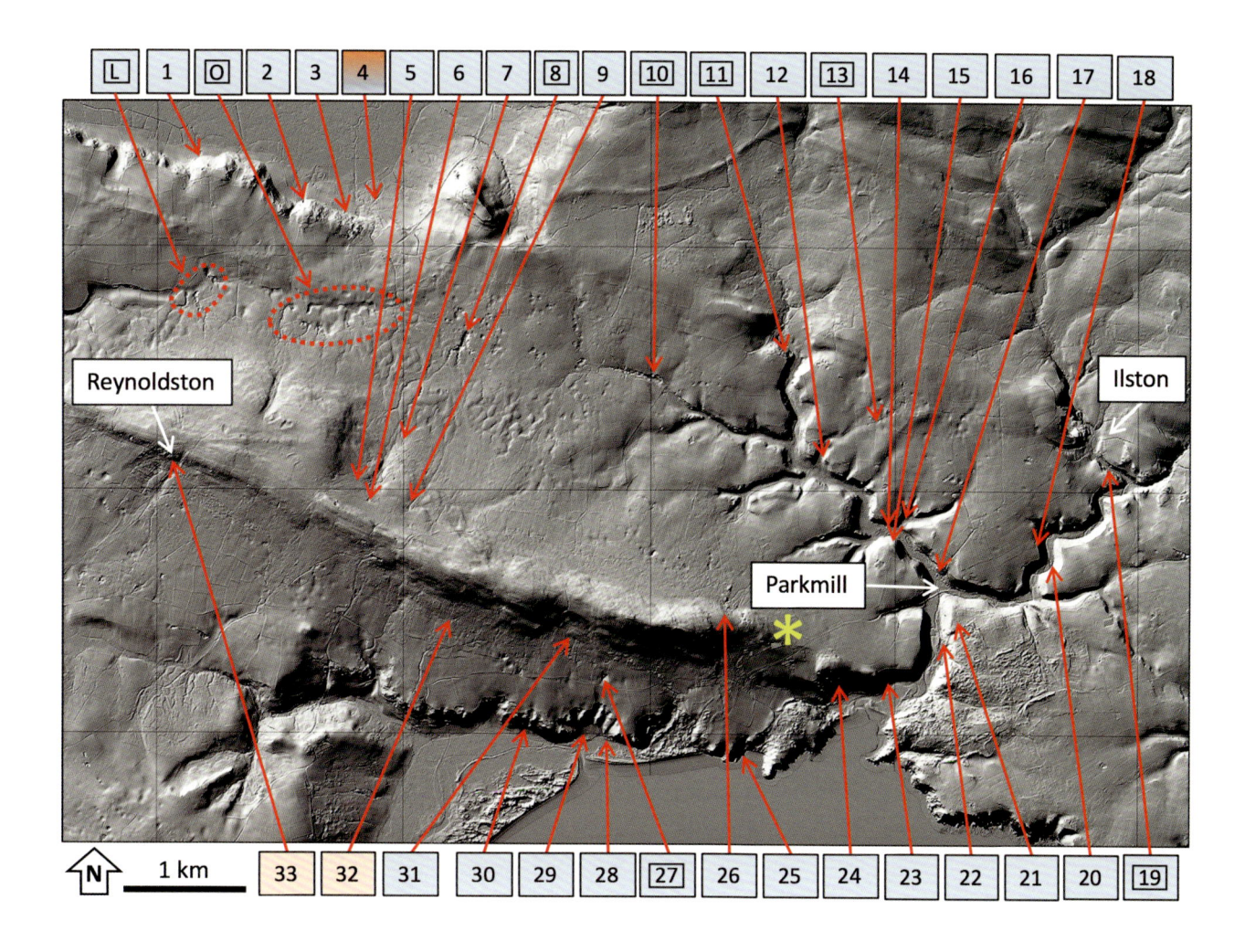

Shaded-relief LiDAR image locating the sinks and springs of central peninsular Gower that are studied in this work; some have known names, other are named here by location. **Blue boxes** *indicate a limestone setting;* **buff** *indicates conglomerate and sandstone;* **brown/blue** *indicates mudstone on limestone. Double boxes are sinks; single boxes are springs. (Image base data* contains Natural Resources Wales information © Natural Resources Wales and Database Right*).*

ALL OUR OWN WATER

[1] Leason

[2] Staffal Haegr

[3] Llanrhidian Church

[4] Butter Well/St Illtyd's Well

[6] Holy Well

[7] Hot Well

[8] Moormills Sinks

268

APPENDICES

[9] East Well
[10] Decoy Sink
[11] Llethrid Swallet
Llethrid flood overflow into Barns Cave dig
[12] Former well at Green Cwm Cottage
[15] Kitchen Well
[14] Wellhead Resurgence

ALL OUR OWN WATER

[17] Parkmill Heritage Centre

[18] Lunnon East

[19] Killy Willy Sink

[20] Trinity Well and Killy Willy Rising

[22] Pennard Castle

[25] Tor Bay

[24] Nott Hill West with tufa

[28] Nicholaston East

[29] Nicholaston West

[30] Nicholaston Woods

Non-central Gower springs mentioned in the text: Ffynnon Wyntog SS43278628, Delvid SS42439276, Lagadranta SS42599285, Raven's Well SS43029112, Caswell West SS58958757, Caswell East SS59348772, St Peter's Well SS59058836, Widegate Rising SS56728794; Devil's Spring (aka Westcliff Spring) SS5472187424, Burry Head SS45679027.

Ffynnon Wyntog at The Knave

Delvid

Raven's Well

Lagadranta

Caswell West

APPENDICES

Caswell East

St Peter's Well

Widegate Rising

Devil's Spring

Burry Head

273

7.4 Review of Wellhead data by Geoff Williams

The primary aim of this section is to investigate the relationships between Wellhead discharge, turbidity, electrical conductivity and dissolved solids, mainly using data collected in 1990, 1994-95 and during the drought of 2018. These are considered in the contexts of rainfall and water table variations within the Carboniferous limestone succession of Green Cwm.

Chemical analysis (Section 5.18) has revealed that water sampled under drought conditions in 2018 is distinct from the calcium bicarbonate water typical of Welsh Carboniferous limestones (Fig. 133) and the Wellhead samples described by Baynton (1968-1969), Halfacree and Williams (1988) and Welsh Water (1994/5) (GW Fig. 1). The drought composition has a high magnesium content (Mg/Ca ratio = 2.1), which reflects flow through the dolomitised Black Rock Limestone above which the Wellhead spring emerges. The drought sample also has slightly increased NaCl content, but its Total Dissolved Solids content (TDS), and hence its Electrical Conductivity (EC), is within the range of the other samples. Thus, these limited chemical data do not indicate a systematic change in EC in relation to discharge. However, the similarity of the samples taken at very different times reflecting the Ca-HCO$_3$ type water does suggest a degree of equilibrium with the aquifer.

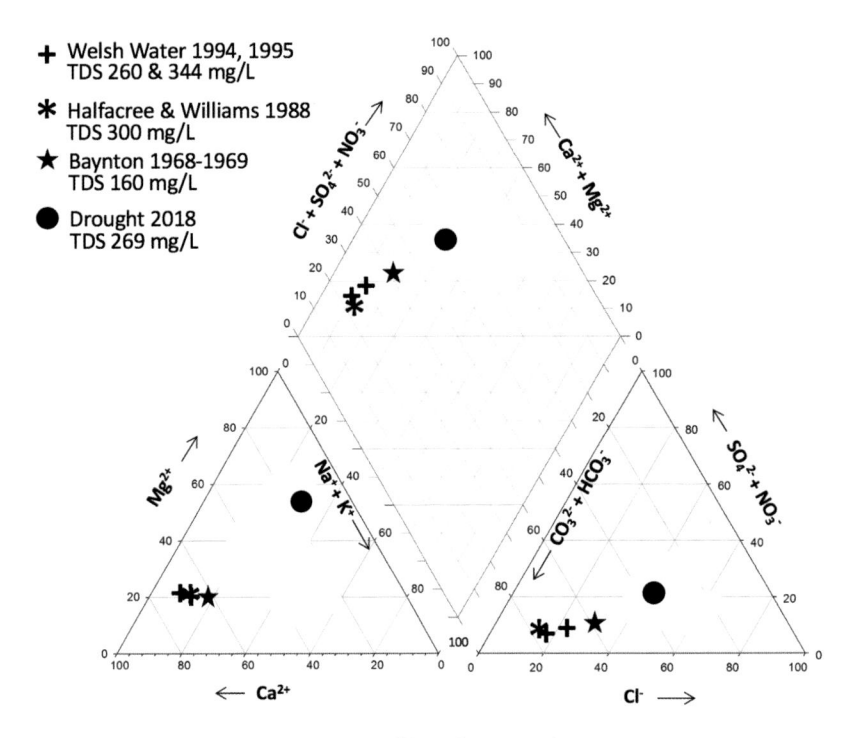

GW Figure 1. Wellhead water chemistry.

Wellhead Flow and Total Dissolved Solids data for 1990

Variations in daily discharge and Electrical Conductivity (derived from TDS) for 1990 are shown in GW Figure 2. Discharges beyond the highest recordable value of 48,000 m^3/day are omitted. Details of the measurement protocol are not recorded, but the TDS content was most likely to have been derived from the measurement of EC at some point during the day and converted to TDS using the formula TDS = 0.6 x EC. The considerable scatter between EC and discharge suggests that a single one-off daily EC measurement may not be correlated strictly with the discharge measured daily unless the flow remained fairly constant.

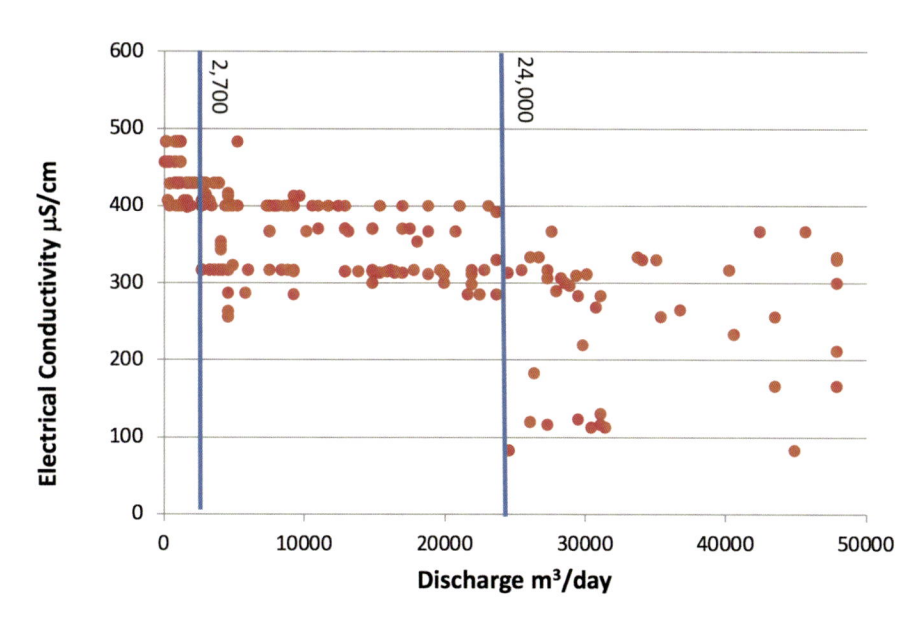

GW Figure 2. Wellhead discharge and Electrical Conductivity (EC).

The graph indicates that at low discharge the EC is high at up to 483 µS/cm but as flow increases beyond about 2,700 m³/day the EC varies between 300 - 400 µS/cm, swamping the signature of the high EC water and suggesting that the larger proportion of flow takes place through more permeable conduits at higher elevation in the limestone (e.g., as discussed previously in terms 'flood bypass'). At flows greater than 24,000 m³/day the EC becomes more scattered with occasional excursions to less than 100 µS/cm.

The annual variation of EC and discharge (GW Fig. 3) reveals that the rapid lowering of EC to below 100 µS/cm relates to increases in the discharge that in turn are dependent on rainfall.

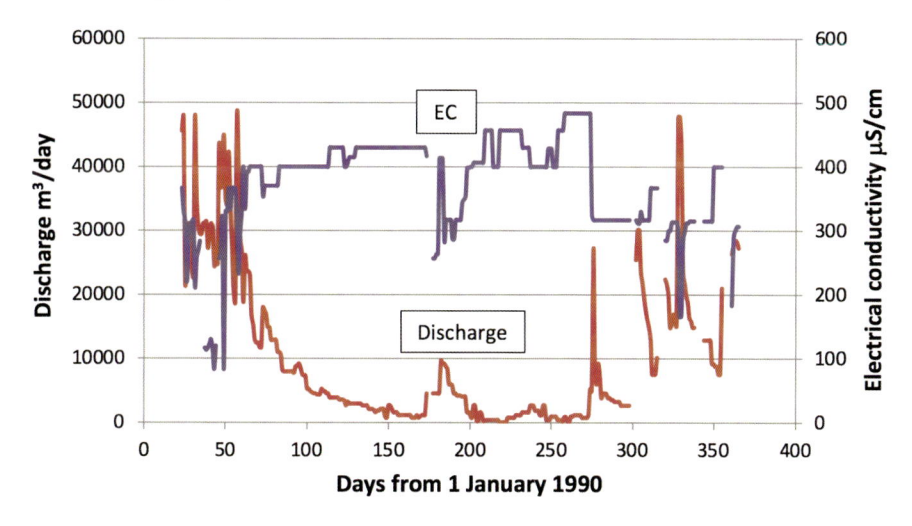

GW Figure 3. Annual variation of discharge and Electrical Conductivity (EC) at Wellhead.

Rainfall recorded daily at Penmaen is not a good predictor of flow at Wellhead, because the catchment area for Green Cwm is principally farther north on Gower and there is a lag in throughput, but the 5-day antecedent rainfall at Penmaen most likely reflects the prevailing weather, and this correlates well with the Wellhead discharge (GW Fig. 4).

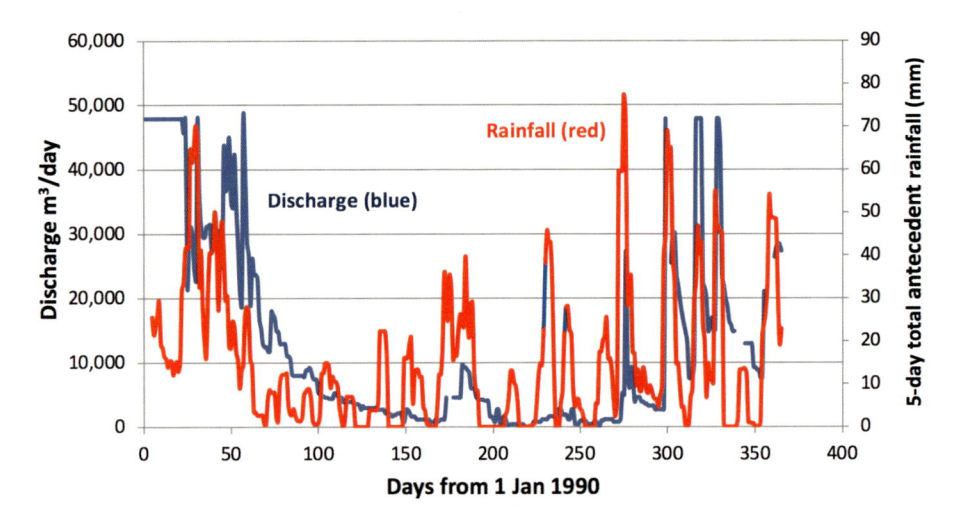

GW Figure 4. Annual variation of Wellhead discharge and 5-day antecedent rainfall at Penmaen for 1990.

Ignoring the transients in EC that follow increases in discharge (and rainfall) in GW Figure 3, there is an inverse relationship between discharge and EC during the summer months (roughly early-March to September). This can be explained by slower groundwater flow velocity allowing more time for groundwater to equilibrate with the limestone, hence the higher EC. Conversely when the Wellhead discharge is higher, up to about 24,000 m³/day, the EC appears to fall gradually as groundwater velocity increases and residence time decreases.

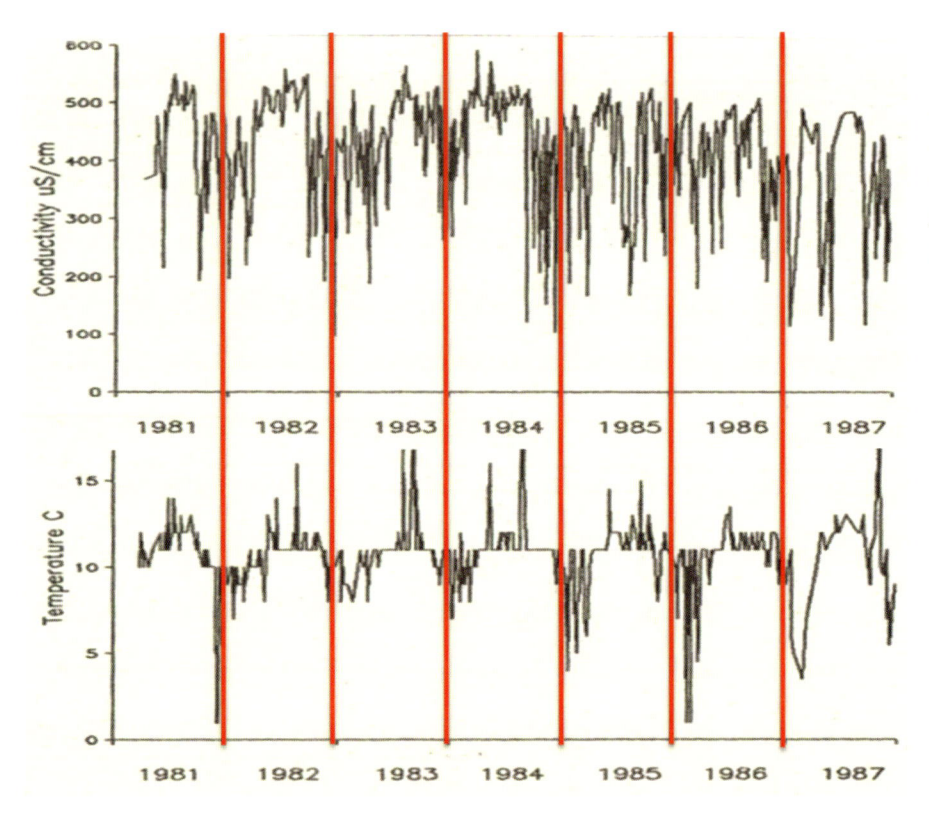

GW Figure 5. Plot of EC against groundwater temperature at Wellhead from Halfacree and Williams (1988).

Discharge rates above 24,000 m³/day occur during the winter months October to March, when rainfall is highest, and are associated with transient depressions in the EC. Similar falls in EC during the winters for 1981 to 1987 are evident in the graphs of EC against groundwater temperature reported by Halfacree and Williams (1988) (GW Fig. 5).

In karst systems where fracture permeability dominates and where porosity is low (typically <5%), large fluctuations in the water table occur rapidly in response to infiltration. In winter when the valley becomes flooded, for example near Green Cwm Cottage, surface water recharges through the cryptic sink at elevation 25 m OD (Fig. 106) at least once a year. This surface runoff will have had little contact with soil or rock and thus will have a low EC so that it dilutes the Wellhead discharge. Even without the development of surface water flow, rainfall during the winter when the water table is high would be expected to depress the EC at Wellhead in the same way giving the EC transients observed at high discharge.

Groundwater velocities from tracer tests

Tracer tests under different discharge conditions, conducted by Halfacree and Williams (1988) and discussed in Section 5.6, reveal groundwater velocities of 29 – 90 m/hour (GW Table 1).

Sink	Notional Distance to Wellhead (km)	Wellhead Discharge m³/day	Groundwater Velocity m/hr
Llethrid	1.68	29,376	61
		12,960	29
		12,960	29.5
Decoy	1.8	46,656	90
Willoxton	1	46,656	45
		46,656	37
	1.3	46,656	59
		46,656	48

GW Table 1. Groundwater flow velocity determined from the interval between adding dye at the sink and its arrival at Wellhead (after Halfacree and Williams 1988; see details in Fig. 107). The notional flow distance is the direct line whereas in reality, e.g., the cave maps (Fig. 106), the distances and velocities must be greater.

When velocity is plotted against discharge it appears that Llethrid and Decoy are comparable, whereas the apparent groundwater velocity from Willoxton is lower (GW Fig. 6). Extrapolation suggests that the groundwater flow velocity would be around 0.036 m/hr at the lowest discharge of 18 m³/day recorded at Wellhead in 1990, which is supported by cavers' reports of encountering practically static water in the Tooth and Llethrid caves. The aquifer residence time is probably a significant control on the conductivity of the water resurging at Wellhead at low discharge.

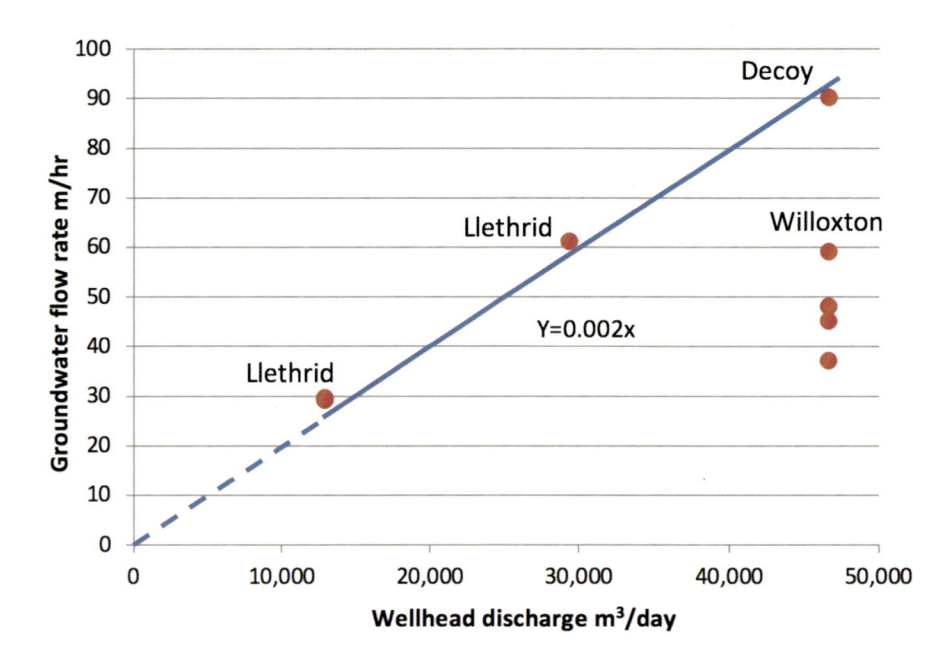

GW Figure 6. Groundwater flow velocity from tracer tests (after Halfacree and Williams 1988).

As discussed in Section 5.11, the Willoxton breakthrough curves are complex indicating multiple flow paths, but even with a range of putative flow distances between the Willoxton sink and Wellhead it appears that the flow regime is more convolute than that of Decoy and Llethrid. There are no data on groundwater levels that might help resolve the flow regimes, apart from that recorded for the well at Green Cwm Cottage, 675 m down valley to the west of Willoxton sink. Surface elevation at the well is 36 m OD and, on sinking in 1921, water was intercepted at a depth of 25m (82 ft) or 11 m OD, which, at that time, before building of the weir at Wellhead, was just 1 m above the level of the Wellhead resurgence at 10 m OD. The well subsequently dried and was abandoned to rubbish. The depth, however, is dependent upon the verbal account of one who sank the well some 40 years earlier, given to Dick Baynton (Baynton 1968-1969). Also, water was seen within 1-2 m of the surface during repair of the Parc le Breos tumulus and at a similar shallow level 200 m farther south opposite Wellhead during installation of a water supply for Scout and Guide campers.

In the absence of water level measurements, an approximation of the average hydraulic gradients from the various sinks towards Wellhead may be gathered at times of flood by assuming the hydraulic head at the sink is at ground level there (GW Table 2).

While this may be valid for the sinks near the valley bottom, it is not for Willoxton, which is at a much higher elevation on the adjacent platform. At high rainfall the Willoxton sink overflows and water recharges in several places lower down the valley towards Green Cwm Cottage. So, the Willoxton sink water is probably to a degree perched or seriously restricted, possibly by a relatively impermeable layer or infill similar to that found in the Iron Ore Series in Tooth cave (Section 4.8).

Location	Water level elevation (m OD)	Distance to Wellhead (m)	Hydraulic gradient
Wellhead	12	0	NA
Llethrid	39 (flood)	1680	0.016
Decoy	61 (flood)	2300	0.021
Cryptic Sink near Green Cwm Cottage	25 (flood)	600	0.022
Willoxton	74 (perched?)	940	0.66?

GW Table 2. Hydraulic gradients estimated assuming water table at surface during flood.

The seemingly anomalous water rise and then drying in the Green Cwm Cottage well, with the disparity in Willoxton flow regime compared with those for Llethrid and Decoy, emphasises the complex nature of flow in the limestone. Rather than assuming a regional water table to be present, the limestone should perhaps be viewed as containing disparate fracture networks that may or may not be interconnected depending on groundwater levels, so understanding how the water moves will always be uncertain. While it might be assumed that the aquifer is unconfined this does not necessarily mean that water will recharge uniformly over its surface.

However, from the current analysis we can construct a conceptual model for flow at Wellhead related to discharge (GW Table 3).

Wellhead discharge	Water table (WT) m OD	Groundwater velocity m/hr	Flow regime
Low Up to 2,900 m³/day	Low WT ≥ 12	0 - 6	Saturated thickness of the aquifer limited to below 12 m OD in a zone near the contact with the underlying dolomitised Black Rock Limestone. High EC is probably due to long residence times.
Intermediate Up to 24,000 m³/day	Intermediate WT ≫ 12	6 - 50	Water dominantly flows through higher more permeable conduits not dolomitised. EC decreases but is variable as residence times and infiltration rates fluctuate.
High More than 24,000 m³/day	At or near surface WT ≤ 25 near Green Cwm Cottage	50-90 +	The water table approaches ground surface but occasionally the flow capacity of the aquifer is exceeded when low EC rain water recharges through sinkholes giving transients of low EC at Wellhead.

GW Table 3. Conceptual model for flow at Wellhead.

Wellhead Turbidity and Electrical Conductivity 1994-95

Simultaneous measurements of EC and turbidity were recorded during 1994-95 (GW Fig. 7). Highest EC of 524 µS/cm occurred during the summer months while EC decreased to a minimum of 305 µS/cm during the winter when flow was greater. These accord with the variation of EC with discharge seen in GW Figures 2 and 3, but low EC transients are not picked up at this sampling frequency.

Turbidity is measured by nephelometry and is expressed as Formazin Turbidity Units (FTU). The observation that turbidity increases with flow (Section 5.10) implies a negative correlation between turbidity and electrical conductivity, as can be seen in GW Figure 8. However, most of the data points are below a FTU value of 5 and do not suggest that the linear correlation extends to higher FTU values.

Attempts to calculate possible mixing between water of differing EC values (EC of 524 µS/cm to represent the low flow component, and 300 µS/cm to represent the higher flow character) have been considered to link EC and turbidity with discharge rate. However, the large dilution effect of the low EC water at flows above 2,900 m^3/day masks the low flow EC signature and restricts its use to discharges below this value, which thus has little practical use.

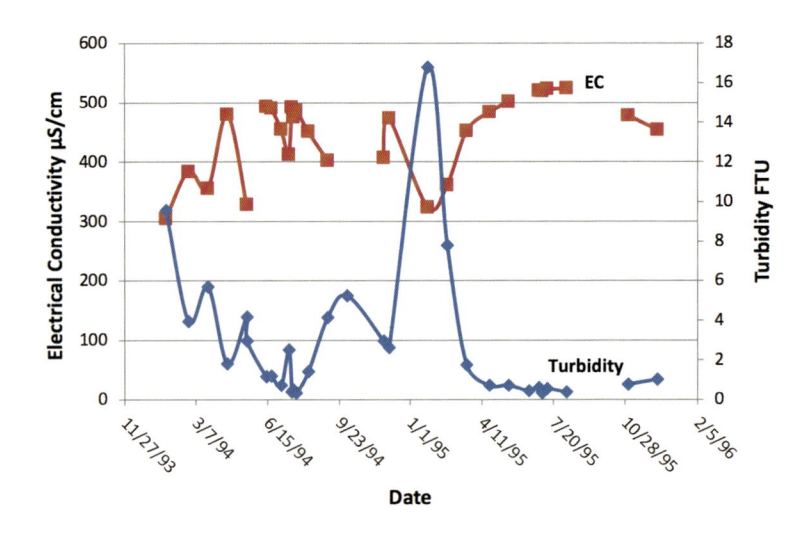

GW Figure 7. Variation in Electrical Conductivity (EC) and Turbidity with time.

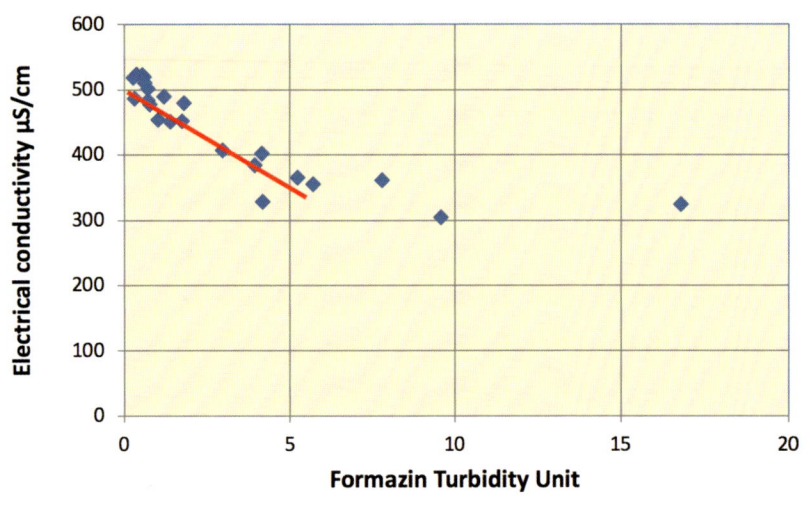

GW Figure 8. Correlation between Electrical Conductivity and Turbidity.

Conclusion

A conceptual model for groundwater flow is presented using the data collected while Wellhead treatment works was in operation. When Wellhead discharge is low groundwater velocities are minimal, up to 6 m/hr, and dissolution of the dolomite present in the Black Rock Limestone group produces a Mg-rich bicarbonate water. As discharge increases, the water table rises and the groundwater velocity increases so the major component of flow takes place through the higher conduits in the limestone, which contain little or no dolomite. The low-flow dolomite signature is quickly lost and the water attains a Ca-bicarbonate character. At discharges higher than 24,000 m^3/day the water table approaches surface level and conduit flow capacity is exceeded. Ultimately, in winter when rainfall and Wellhead discharge are highest, water flows along the surface and recharges near Wellhead in cryptic sinks resulting in low EC transients.

Flow to Wellhead takes place via the three sinks Llethrid, Decoy and Willoxton. The latter seemingly entails some physical restriction with a complex connection with Wellhead.

There is an inverse linear relationship between electrical conductivity and turbidity at low values of turbidity but there are insufficient data to extend the correlation above values of 5 FTU.

Wellhead.

7.5 Placenames index

This index lists sections, figures and pages that have significant information bearing on that place. Maps that may be useful are on pages: vi, 5, 79, 125 and 172.

Arthur's Stone (Maen Ceti) 3.8
Bennett's Pill iv
Blue Anchor 3.14; Fig. 67
Broughton Bay 2.6; p. 112
Burry 3.10; p. 104
Caswell 6.4
Cefn Bryn Fig. 5; 2.4; p. 69; 3.8; 3.9; 5.13-5.17
Clyne Common Fig. 23
Crofty 3.14
Deep Slade 2.6; 2.14; Fig. 68
Dunvant Fig. 71
Fairy Hill 3.14
Fall Bay Fig. 6; 2.4; p. 66; Fig. 138
Great Tor 2.11; Figs 26 & 27; 4.10; 4.11
Green Cwm Fig. 51; 4.4-4.9; Fig. 97; p 165; 5.6; 5.10
Hardings Down 2.4
Heatherslade (Southgate) 2.15
Hills Tor Fig. 142; Fig. 145
Horton 2.6; 2.10; 3.10
Hunts 2.14; Fig. 68
Ilston (Gelli-hir) 3.14
Ilston Cwm Fig. 97
Landimore iv
Langland Fig. 148
Limeslade 3.4
Little Tor 2.6; Fig. 20
Llangennith Fig. 55
Llanmadoc Hill 2.4; Fig. 55; 3.12
Llanrhidian 3.9; 5.19
Llethrid Fig. 42; 4.4; 4.12
Loughor Estuary iv
Mewslade Bay Fig. 68; Fig. 138; 6.5
Mumbles Head 3.6
Nicholaston Fig. 20; 6.2
North Hill Tor Fig.70
Oldwalls 3.14; Fig. 137

Overton Mere Fig. 29
Oxwich Bay 6.8
Oxwich Point 2.6; 2.10; 2.14
Parkmill Fig. 36; 4.10; 5.1; 5.2; 5.5; 5.6
Paviland Fig. 6; Fig. 30; 3.10
Penclawdd 3.15; Fig. 67
Pengwern Common 3.14; Fig. 83; Fig. 108
Penmaen 6.2
Pennard 2.6; Fig. 36; 6.2
Penrice Fig. 63
Port-Eynon 2.4
Port-Eynon Point Fig. 6; Fig. 29; Fig. 41; Fig. 149
Pwll-du 3.6; 6.7
Pwll-du Head Fig. 15; Fig. 23; Fig. 48
Ram Grove 4.3
Reynoldston 3.11
Rhossili 6.2
Rhossili Bay p. 69; Fig. 6; Fig. 138
Rhossili Down 2.4; Fig. 60
Rotherslade 3.4
Scurlage 3.13
Slade 2.10
Southgate 2.14; Figs 33 & 35; Fig. 69; 4.10
Tears Point Fig. 6; 2.14; Fig. 34
The Knave 4.3
The Sands Fig. 25
Three Cliffs 2.6; Fig. 36; 4.10
Three Crosses 3.14
Welsh Moor Common 3.14
Whiteford Sands Fig. 140; 6.3; Fig. 143
Worms Head Fig. 6; 2.4; 2.6; Figs 41, 138 & 149